Fire Retardant Building Products and Coatings 1970

Dr. M.W. Ranney

Thirty-Five Dollars

NOYES DATA CORPORATION
Park Ridge, New Jersey, U.S.A.

Noyes Data, S.A. Zug, Switzerland London, England

Library of Congress Catalog Card Number: 79-117447
SBN: 8155-0312-1
Printed in the United States

FOREWORD

The detailed, descriptive information in this book is based on U.S. Patents relating to the production of fire retardant building products and coatings.

This book serves a double purpose in that it supplies detailed technical information and can be used as a guide to the U.S. Patent literature in this field. By indicating only information that is significant, this book then becomes an advanced review of processes for fire retardant building products and coatings.

The U.S. Patent literature is the largest and most comprehensive collection of technical information in the world. There is more practical, commercial timely process information assembled here than is available from any other source. The technical information obtained from the patent literature is extremely reliable and comprehensive; sufficient information must be included to avoid rejection for "insufficient disclosure".

The patent literature covers a substantial amount of information not available in the journal literature. The patent literature is a prime source of basic commercially utilizable information. This information is overlooked by those who rely primarily on the periodical journal literature. It is realized that there is a lag between a patent application on a new process development and the granting of a patent, but it is felt that this may roughly parallel or even anticipate the lag in putting that development into commercial practice.

These publications are bound in paper in order to close the time gap between "manuscript" and "completed book". Industrial technology is progressing so rapidly that hard cover books do not always reflect the latest developments in a particular field, due to the longer time required to produce a hard cover book.

The Table of Contents is organized in such a way as to serve as a subject index. Other indexes by company, inventor, and patent number help in providing easily obtainable information.

CONTENTS AND SUBJECT INDEX

INTRODUCTION

The National Fire Protection Association statistics published in 1967, indicate that in the United States a fire starts every two seconds; a home is destroyed by fire every two minutes; and every 44 minutes someone dies from burns received in these fires. Annual property damage due to fire is estimated to be 1.5 to 1.8 million dollars. Increasingly, the construction industry is faced with the necessity of complying with more stringent local and national fire protection standards.

The processing of wood to impart fire retardency has been studied in considerable depth over the years. Because of economic considerations, much of the commercial effort has centered on the use of inorganic salts such as mono- and diammonium phosphates, zinc chloride and borax-boric mixtures. The method of processing generally involves the vacuum treatment of wood in a suitable vacuum chamber, immersion of the wood in the treating solution while still under reduced pressure, and finally, pressurizing of the vessel with the lumber submerged in the salt solution. In addition to controlling cost, care must be exercised to avoid excessive changes in the physical properties, hygroscoposity, paintability and other desirable properties of the treated wood. A number of phosphorus based organic systems have given very promising results, using impregnation methods.

Approaches to imparting fire retardancy to fiberboard products generally involve treatment of the pulp slurry with the inorganic salts with considerable effort being made to increase the degree of deposition of the fire-retardant salt during normal slurry processing. In situ halogenation of cellulosic materials, perhaps followed by phosphorylation also provides built-in retardancy. Similar efforts have been made to develop fire retardant acoustical ceiling tile and panels.

The demand for fire-retardant coatings is increasing rapidly due to the large scale use of wood and plywood wall finishes. The rate of flame spread for these products is already relatively high, and may be even further increased by the use of decorative, but highly flamnable coatings. The most common type of fire-retardant coating comprises: (a) phosphorus compounds such as ammonium phosphates and phosphorus-containing amides capable of producing phosphoric acid during thermal degradation. (b) Nonresinous materials, referred to as carbonific agents, commonly carbohydrates or polyfunctional alcohols such as pentaerythrital which yield large volumes of nonflamnable gas and carbon in the presence of phosphoric acid. (c) Gas-producing components, such as urea or dicyandiamide, which release large quantities of gas under the influence of the phosphoric acid produced at elevated temperatures. To improve the generally poor water resistance of these coatings, formulations incorporating scrub resistant latex binders have been developed.

Fire-retardant asphaltic products such as roofing, coatings, shingles, backing for thermal insulation and for general use on utility poles, etc., have been produced using inorganic fillers, phosphorus and chlorine containing additives and general intumescent formulations.

This report summarizes the patent literature through 1969 as it relates to some 145 different processes for imparting fire retardancy to wood, fiberboard, asphaltic materials, coatings, and other common building products.

WOOD IMPREGNATION

PHOSPHORUS COMPOUNDS

Dicyandiamide and Phosphoric Acid

In a process described by I.S. Goldstein and W.A. Dreher (U.S. Patent 2,917,408; December 15, 1959; assigned to Koppers Company, Inc.) wood is impregnated with a solution of dicyandiamide and phosphoric acid, and dried and cured by heating it to a temperature of 70° to 100°C. It is possible to impart nonleachable fire retardance to wood by using a treating solution containing 17.5 to 25% by weight of the combined dicyandiamide and phosphoric acid, 20% by weight being preferred. For interior construction, nonleachable fire retardant properties are generally not required. In such cases, an impregnating solution of 10 to 17.5% by weight of dicyandiamide and phosphoric acid gives suitable results. The mol ratio dicyandiamide to phosphoric acid ranges from 1:1 to 3:1; below 1:1, the treating solution tends to be too strongly acid and will thus impair the treated product. It cannot be higher than 3:1 because of the limited solubility of the dicyandiamide in water. To offset the relatively low water solubility of dicyandiamide, impregnating solutions having a 2:1 to 3:1 mol ratio are prepared by decreasing the amount of phosphoric acid in the formulation. The 1:1 to 2:1 formulations give better results than do the 3:1 formulations, since this reduction of phosphoric acid tends to diminish the effectiveness of the impregnator as a fire retardant.

Dicyandiamide and phosphoric acid react in a heated aqueous solution to form guanylurea phosphate. According to the process, the reactants of the impregnating solution may be prereacted by combining them at temperatures between 80° to 100°C. The prereacted solutions offer a slight advantage in that they are of lower acidity and tend to insure no degradation of the wood during treatment. Both the prereacted solutions and the unreacted solutions react further with the cellulose of the wood during the curing step. This interreaction with the wood helps to provide a good retention of the impregnator within the wood, even after leaching. Furthermore, it tends to reduce the degree to which the wood can swell, thus enhancing the dimensional stability of the treated wood. It is necessary, to add the dicyandiamide and phosphoric acid as independent components to the impregnating solution, rather than merely to dissolve preformed guanylurea phosphate in water. Guanylurea phosphate is less soluble and, therefore, less reactive than the reactants of the prereacted aqueous solution of dicyandiamide and phosphoric acid. If preformed guanylurea phosphate is added to water and wood is impregnated, a low retention of the impregnator is obtained. This low retention is reflected in an unsatisfactory fire tube weight loss of 57% after leaching. The process is illustrated by the following examples.

Example: Fire Retardance and Impact Strength — The wood specimens were treated under a variety of conditions, using the process, to evaluate their fire retardance and impact strength. To determine fire retardance, the specimens used for the impregnations were of clear sap southern pine, 3/8" x 1 5/8" x 22" long. Ten of these specimens were provided for each treatment, five of which were used to determine fire retardance without leaching and five to obtain a comparative measurement after leaching with tap water. After impregnation, drying and curing, the specimens which were not to be leached were placed in an air-conditioned room to equilibrate at 23°C. and 50% relative humidity. The specimens which were to be leached were separated by untreated sticks and submerged in a tank of tap water. Tap water was allowed to flow through the tank for seven days. At the end of the seventh day, the specimens were placed in a forced convection oven to dry for 48 hours at 50°C.

The leached specimens were then placed in an air-conditioned room to equilibrate. When the specimens had reached equilibrium, they were ripped lengthwise and assembled into the required 3/8" x 3/4" x 40" fire tube specimens. They were then tested in the fire tube apparatus in accordance with the fire tube test, ASTM Standard E69-50. This test measures the loss in weight of the sample on exposure to a standard flame for a standard period of time. Low weight losses are indicative of good fire retardant properties. Weight losses of less than 35% meet railroad specifications. Weight losses of above 50% are considered poor in respect of their fire retardance. The weight loss of untreated wood is 80 to 90%.

To determine the impact strength of the wood, 10, 15, or 20 southern yellow pine impact test specimens (1/2" x 1/2" x 5") were provided for each treatment. After impregnation, drying, and curing, the impact specimens were open stacked in the air-conditioned room to equilibrate at 23°C. and 50% relative humidity. When equilibrium had been attained, the specimens were tested on a standard impact tester known as a "Tinius Olsen plastics impact tester" using the "Charpy head" on a four inch span. Alternate specimens were broken in the radial and tangential directions. The average of 10, 15 or 20 values is reported in the table below.

In the table there is given a description of the treatment used, the composition of the impregnation formulation, the cure conditions, leaching conditions, the weight increase caused by treatment, the weight increase resistant to leaching, the weight percentage loss in the fire tube test, the time of afterflame in seconds, and the impact strength in inch-pounds. Impregnation formulations were prepared by dissolving the indicated amount of organic base in water and adding the required amount of phosphoric acid. Impregnation was accomplished using a full cell cycle comprising a 15 minute vacuum of 25 to 26 inches Hg followed by a two hour pressure period of 160 psi at 50° to 60°C. After the impregnation step, the wood specimens were placed in a drying oven, dried and cured under the conditions indicated in the table.

Description of Treatment	Formulation	Cure Conditions, °C.	Leaching	Weight Increase caused by Treatment, Percent of original dry weight	Weight Increase Resistant to Leaching, Percent of original dry weight	Percent Weight Loss in Fire Tube Test	After-Flame, seconds	Impact Strength, In.-lbs.	
								Specimen	Control
23% dicyandiamide: Phosphoric acid. Prereacted, 50° cure	1,440 g. H_3PO_4 (85%) 1,040 g. dicyandiamide 7,490 g. H_2O Heated to exotherm	48 hrs. at 50°	No	24		14	0		
			Yes	26	3.5	62	60		
			No	39				52	67
23% dicyandiamide: Phosphoric acid. Prereacted, 70° cure	1,440 g. H_3PO_4 (85%) 1,040 g. dicyandiamide 7,490 g. H_2O Heated to exotherm	24 hrs. at 70° C.	No	29		14	0		
			Yes	26	7.9	41	26		
			No	36				65	67
23% dicyandiamide: Phosphoric acid. Prereacted, 85° cure	3 lb. 7 oz. dicyandiamide 4 lb. 12 oz. H_3PO_4 (85%) 24 lb. 13 oz. H_2O Heated to exotherm	24 hrs. at 85° C.	Yes	36	8.8	29	0		
			No	32				73	67
23% dicyandiamide: Phosphoric acid. Prereacted, 90° cure	1,440 g. H_3PO_4 (85%) 1,040 g. dicyandiamide 7,490 g. H_2O Heated to exotherm	24 hrs. at 90° C.	No	28		16	0		
			Yes	32	8.4	32	0		
			No	37				70	67
23% dicyandiamide: Phosphoric acid. Prereacted, 100° cure	1,440 g. H_3PO_4 (85%) 1,040 g. dicyandiamide 7,490 g. H_2O Heated to exotherm	24 hrs. at 100° C.	No	31		14	0		
			Yes	33	8.1	23	2		
			No	44				63	67
23% dicyandiamide: Phosphoric acid. Prereacted, 130° cure	3 lb. 7 oz. dicyandiamide 4 lb. 12 oz. H_3PO_4 (85%) 24 lb. 13 oz. H_2O Heated to exotherm	20 hrs. at 100° C. 2 hrs. at 130° C.	No	32		14	0		
			Yes	31	15	22	0		
			No	37				54	67
23% dicyandiamide: Phosphoric acid. Prereacted, 150° cure	1,440 g. H_3PO_4 (85%) 1,040 g. dicyandiamide 7,490 g. H_2O Heated to exotherm	20 hrs. at 100° C. 2 hrs. at 150° C.	No	25		13	0		
			Yes	26	5.6	26	0		
			No	33				41	67
20% dicyandiamide: Phosphoric acid. 90° cure	9.3 parts dicyandiamide 12.6 parts H_3PO_4, (85%) 78.1 parts H_2O	20 hrs. at 70° C. 4 hrs. at 90° C.	No	32		16.5	0		
			Yes	31	8.5	28.0	0		
			No	32				62	57
25% guanylurea phosphate. 90° cure	25 parts guanylurea phosphate, 75 parts H_2O (insoluble crystals formed).	24 hrs. at 90° C.	No	20		20	0		
			Yes	21	7	57	53		
			No	24				67	67
19% urea-phosphoric acid. Phosphate. 140° cure	3,450 g. H_2O 372 g. guanylurea phosphate 450 g. urea.	20 hrs. at 100° C. 2 hrs. at 140° C.	No	15		23	0		
			Yes	16	4.6	56	70		
			No	19				63	68
16% urea-phosphoric acid, 3:1. 100° cure	14,860 g. H_2O 1,242 g. H_3PO_4 (85%) 1,890 g. urea	22 hrs. at 100° C.	No	21		20	0		
			Yes	19		62	58		
19% urea-guanylurea, 3:1.	6,900 g. H_2O 690 g. H_3PO_4 (85%) 1,080 g. urea	20 hrs. at 100° C. 2 hrs. at 130° C.	No	13		20	0		
			Yes	15	4.4	42	62		
			No	18				46	61

In studying the data given in the table, it will be noted that wood treated with the prereacted solutions of dicyandiamide and phosphoric acid, when cured within the prescribed range of 70° to 100°C., had nonleachable flame retardant properties and virtually undiminished impact strength. It will be observed that the specimens cured at 50°C. showed poor fire retardance after leaching and diminished impact strength. Those cured at 150°C. showed excellent fire retardance after leaching, but their impact strength was considerably impaired. In the test the impregnation formulation was not prereacted, excellent fire retardance was also obtained, and there was likewise no loss in impact strength. The purified guanylurea phosphate, when preformed and then mixed with water to form an impregnating solution, provided a wood product having an unsatisfactory flame retardance after leaching.

In contrast with the method of the process, the urea-phosphoric acid system failed to give a nonleachable flame retardant product of undiminished impact strength. In the case of specimens which were cured at a temperature of at least 140°C. for one hour or more, suitable nonleachable flame retardance was obtained, but the impact strength of the product was markedly diminished. When a specimen was given a final cure of 140°C. for only 1/2 hour, the flame retardance of the product after leaching was unsatisfactory and the impact strength was also reduced. Using a 100°C. curing temperature with an impregnation formulation comprising a mixture of urea and phosphoric acid gave a product having poor flame retardance after leaching, and somewhat diminished impact strength. Formulations comprising phosphoric acid in combination with other bases such as acetamide, cyanoacetamide, semicarbazide hydrochloride, and guanidine carbonate likewise gave an unsatisfactory product using a final curing temperature of 90°C.

In a continuation of the above work, I.S. Goldstein and W.A. Dreher (U.S. Patent 3,159,503; December 1, 1964; assigned to Koppers Company, Inc.) have found that a small, controlled amount of formaldehyde can be used to modify the impregnating solutions described in U.S. Patent 2,917,408 in such a way that blooming is controlled and the water-solubility of dicyandiamide is improved without adversely affecting the other beneficial properties imparted to the wood as a result of the impregnation.

The amount of formaldehyde used in the impregnator ranges between 0.05 and 0.15 mol per mol of dicyandiamide. If the amount of formaldehyde is less than 0.05 mol the solubility of the dicyandiamide is not enhanced; thus it is not possible to prepare the more concentrated solutions of active ingredients and the blooming which tends to occur as a result of the impregnation treatment is not eliminated. If the amount of formaldehyde is greater than 0.15 mol, the treated wood becomes considerably more hygroscopic than the untreated wood, and the resulting product is subject to considerable warping and checking.

There has been considerable interest shown in the use of the impregnating solutions of U.S. Patent 2,917,408 as impregnators for wood fiber products, particularly particle board, hardboard and insulating board. The most frequently mentioned criticism of such use, the limited solubility of the formulation resulting in inability to get enough fire retardant chemical in fibers that have been saturated previously with water, has been overcome by the use of formaldehyde. The following example illustrates the process.

Example: An experiment was designed to measure the fire retardant efficiency of a formulation of the process containing a 0.15 mol addition of formaldehyde in comparison with a 1:1 molar combination of dicyandiamide and phosphoric acid prepared according to U.S. Patent 2,917,408. Untreated specimens were used as a control. The two formulations were mixed to contain 20% solids, using the proportions listed in the table below. Both solutions were prereacted by heating to 90°C. for about 5 minutes, after which they were allowed to cool to the 50°C. treating temperature. Five southern pine fire tube specimens 3/8 x 3/4 x 40 inches long were cut for use with each series of tests. The wood specimens were placed in an experimental treating cylinder designed for pressure impregnation. The 20% treating solution was poured over the wood specimens, which were prevented from floating by the weight of steel restraining bars. A 15 minute vacuum of 26 inches of mercury was applied to the wood. Then a pressure of 150 psi of air pressure was applied for a period of four hours at a solution temperature of 50°C. This procedure was repeated to prepare specimens for each of the two formulations being tested.

The treated wood specimens in their wet condition were next placed in a small experimental dry kiln where they were heated for 24 hours at 65°C. dry bulb temperature and 55°C. wet bulb temperature. They were then heated for four hours at 93°C. dry bulb with no wet bulb control (ambient humidity prevailing). All treated specimens and a set of matched untreated control specimens were next equilibrated in a room maintained at 23°C. and 50% relative humidity. After moisture equilibrium had been attained, the specimens were individually tested in a so-called "fire tube" apparatus, according to ASTM Standard E69-50. Averaged results for these tests are presented below.

TABLE 1: RESULTS OF ASTM FIRE TUBE TESTS (STANDARD E69-50) ON SOUTHERN PINE SPECIMENS PRESSURE IMPREGNATED WITH TWO FIRE RETARDANT SOLUTIONS

Formulation of Treating Solution	Fire Tube[1] Weight Loss, Percent	After-Flame,[1] Seconds	After-glow,[1] Seconds
20% 0.15:1:1 Molar Ratios of Formaldehyde:Dicyandiamide:Phosphoric Acid:			
9.1 Parts Dicyandiamide; 12.3 Parts 85% Phosphoric Acid; 1.2 Parts 40% Formaldehyde Solution; 77.4 Parts Water	17.2	0	0
20% 1:1 Molar Ratios of Dicyandiamide:Phosphoric Acid:			
9.2 Parts Dicyandiamid; 12.7 Parts 85% Phosphoric Acid; 78.1 Parts Water	16.5	0	0
Untreated Pine Controls	84.8	32	124

[1] Values reported are averages of five specimens.

The values reported above indicate that the addition of the 0.15 mol of formaldehyde does not result in any significant reduction in the fire retardance imparted to wood. Both sets of test results indicate an excellent fire retardant effect on the wood.

Bis-(2-Bromoethyl)-2-Bromoethane Phosphonate

D.L. Kenaga and A.J. Erbel (U.S. Patent 2,725,311; November 29, 1955; assigned to The Dow Chemical Company) describe a treating composition containing a bis-(haloalkyl)haloalkane phosphonate. The phosphonates which may be used in the process are those in which the alkyl and alkane groups contain 2 or 3 carbon atoms and in which the halogens are chlorine or bromine. As preferred compounds may be mentioned bis-(2-bromoethyl)-2-bromoethane phosphonate and bis-(2-chloroethyl)-2-chloroethane phosphonate.

These compounds are easily prepared by adding phosphorous trihalide to an alkylene oxide preferably in a solvent, and then, raising the temperature of the reaction mixture to 150°C. or higher to cause a rearrangement of the phosphite to a phosphonate. The use of a solvent is not necessary, but better control of the reaction is possible when one is used. As suitable solvents may be mentioned o-dichlorobenzene or aromatic mineral spirits. The yields of the phosphonates when prepared in this manner are almost quantitative so that in many cases when the above solvents are used, the resulting solution may be employed directly for treating the cellulosic articles. The following examples illustrate the process.

Example 1: Blocks of ponderosa pine sapwood 1/2 inch by 1/2 inch by 3 inches were treated with a 25% solution of bis-(2-bromoethyl)-2-bromoethane phosphonate in aromatic mineral spirits ("Penola 100"). The impregnation was accomplished using a modified full cell process which used atmospheric pressure instead of the superatmospheric pressure normally used. The impregnation was carried on until 6 pounds of phosphonate was retained per cubic foot of the wood. The wood blocks were then allowed to season in the air for one month to remove substantially all of the solvent. They were then tested according to the procedure described in ASTM test E160-50. In that test 24 blocks are piled in a wire crib and a calibrated Meker burner placed under the crib for 3 minutes. The times required for the flames to cease, and for glowing to cease after the flame is removed are noted. The blocks are then weighed and their loss in weight recorded. When the blocks treated in this example were subjected to the above described crib test, flaming ceased after 2 minutes; there was no afterglow, and the blocks had lost 40% of their weight. By way of contrast, when blocks were treated similarly using tris-(2-chloroethyl)-phosphate they were found to have lost 66% of their weight. The blocks continued to burn for 4.5 minutes and there was no afterglow.

Example 2: Twelve inch squares of 1/20 inch birch face veneer were cut and treated with the compounds of this process and with other similar compounds dissolved in aromatic mineral spirits. The impregnation of the panels was accomplished by soaking in the solution for 2 minutes then allowing them to season in air for 24 hours to remove the aromatic mineral spirits. The panels were then subjected to an inclined plane flame test. In that test, a 12 inch square panel of wood is placed on a wire rack so as to be at 45° to the horizontal. A small cup holding one milliliter of alcohol is placed 1/4" below the veneer. The alcohol is ignited and the flame allowed to impinge on the veneer. The effectiveness of the compounds as fire retardants for the veneer was rated according to an arbitrary scale which included the time for the flame to penetrate through the veneer, the amount of area charred, and the time for the flames to extinguish themselves. According to this rating 0 means no combustion, 1 means the flames were self extinguishing, 2 means the flames continued but with evident retardancy, and 3 means flaming never ceased. The results of the above test on the compounds used are listed below.

Compound	Concentration (% by Wt. of Wood)	Retarding Characteristics
Bis-(2-bromoethyl)-2-bromoethane phosphonate	2.2	0
Bis-(2-chloroethyl)-2-chloroethane phosphonate	2.2	0
Bis-(3-chloropropyl)-3-chloropropane phosphonate	2.1	2
Diethyl 2-bromoethane phosphonate	5.1	3
Triphenyl phosphate	2.9	3
Tetrabromobutane	4.1	3
N"(2,4,6-tribromophenyl)-N,N,N',N'-tetramethyl phosphoramide	2.5	3

In related work, A.J. Erbel and D.L. Kenaga (U.S. Patent 2,803,562; August 20, 1957; assigned to The Dow Chemical Company) describe the treatment of wood with a water, fire, and weather resistant composition containing a bis(2-haloalkyl)alkenyl phosphonate. The bis(2-haloalkyl)alkenyl phosphonates which may be used in the process are those in which the haloalkyl and alkenyl groups contain the same number of from 2 to 3 carbon atoms and in which the halogen is chlorine or bromine. As typical examples of such compounds may be mentioned bis(2-chloroethyl)vinyl phosphonate, bis(2-bromoethyl)vinyl phosphonate, bis(2-bromopropyl)propenyl phosphonate, and bis(2-chloropropyl)propenyl phosphonate. The following example illustrates the process.

Example: Southern pine sapwood stakes 3/4 inch by 3/4 inch by 18 inches were treated with bis(2-chloroethyl)vinyl phosphonate, triphenyl phosphate, and tricresyl phosphate. The compounds were dissolved at 10% concentration in a wood preservative consisting of 45% of a heavy petroleum oil equivalent to a #5 fuel oil and 55% of a medium residue coal tar creosote. The impregnation was carried out by a full cell pressure process to a retention of 15 to 25 pounds of the solution per cubic foot of wood. After treatment, the stakes were placed upright in a cup. One milliliter of alcohol was placed in the cup and ignited. The losses in weight of the stakes are listed below.

Fire Retardant	Weight Loss, %
Bis(2-chloroethyl)vinyl phosphonate	0.8
Triphenyl phosphate	54.7
Tricresyl phosphate	58.1

As can be seen, the fire retardant compounds of this process give far greater fire retardancy when combined with common wood preservatives than do either triphenyl phosphate or tricresyl phosphate.

Phosphoric Acid Esters

S.S. Sakornbut (U.S. Patent 2,769,730; November 6, 1956; assigned to Monsanto Chemical Company) has found that volatile hydrocarbon solutions of chlorinated phenol, highly chlorinated hydrocarbons and esters of phosphoric acid are especially effective as water repellent and fire retardant preservative compositions for porous cellulosic products if the products have been previously treated with ammoniacal zinc solutions. These zinc solutions are prepared by dissolving zinc oxide, zinc phosphate, zinc borate or other zinc compound in aqueous ammonia, which converts the zinc into a soluble complex, thereby enabling an effective penetration of the zinc compound into the porous wood, paper or fiberboard article. A subsequent drying operation precipitates the zinc in the interstices of the porous article. The product subject to this pretreatment is then retreated by any of the conventional methods, brushing, spraying, dipping, or vacuum and/or pressure treatment, with the volatile hydrocarbon solution of chlorinated phenol, highly chlorinated hydrocarbons and esters of phosphoric acid. By this method products usually subject to destruction by fire, and damage by weathering, and by fungal and insect attack are rendered highly resistant to these agencies. The following examples illustrate the process.

Example 1: A 20% aqueous ammonium hydroxide solution was used to dissolve 10% by weight of zinc phosphate. This solution was then applied to a Ponderosa pine wafer by immersion. After drying, the sample contained about 4% by weight of the zinc compound. The sample was then retreated in the same manner with a solution comprising 75% by weight of mineral spirits, 5% of pentachlorophenol, 14.5% of a commercially available chlorinated paraffin hydrocarbon (Chlorowax 70), 5% cresyl diphenyl phosphate and 0.5% of a microcrystalline wax. The sample was subjected to the Standard Swellograph Method of the National Wood Products Manufacturers Association (NWMA-M-2-51) and was found to have a water repellency of 73%. When tested by the horizontal flame spread method, the average flame-out period was 0 seconds, indicating substantial fireproofness.

Example 2: To demonstrate the effect of the preimpregnation with the water solution of soluble zinc compounds, samples were treated without the zinc solution application. When tested as described in the preceding example, it was found to have a water repellency of 68.3% and a flame-out time of 15 seconds.

S.S. Sakornbut (U.S. Patent 2,926,096; February 23, 1960; assigned to Monsanto Chemical Company) has also found that chlorinated phenols, and particularly pentachlorophenol, can be made in a concentrated form which is capable of being diluted with volatile solvents to form useful impregnating compositions, if all or a substantial portion of the solvent is an ester of phosphoric acid. The esters of phosphoric acid may be aliphatic or aromatic, including tributyl phosphate, triamyl phosphate, triphenyl phosphate, tricresyl phosphate, trioctyl phosphate, and other near homologues of the triesters of phosphoric acid. Suitable mixed esters which may be used are:

Dibutyl phenyl phosphate
Butyl diphenyl phosphate
Amyl diphenyl phosphate
Diamyl cresyl phosphate
Cresyl diphenyl phosphate
Phenyl dicresyl phosphate
Butyl dicresyl phosphate

In the preparation of useful impregnating chlorinated phenol solutions by the process, there is a critical relationship between the chlorinated phenol and phosphate esters. In general, the impregnating solutions will have a relatively high concentration of the chlorinated phenols, for example, from 3 to 65% by weight and 1/2 to 5 parts of the phosphate ester for each part of the chlorinated phenol. An important modification of the process is the concentrate which may be diluted with conventional petroleum solvents to form an effective impregnating solution, which may consist of only the chlorinated phenol and the phosphate ester.

The optimum proportions of chlorinated phenol and phosphoric acid esters will vary with the esterifying group in the phosphoric acid ester. The lower aliphatic phosphates, such as tributyl phosphate, will tolerate a larger proportion of the chlorinated phenol, whereas with the aromatic esters and particularly those containing alkyl substituted aromatic groups, more of the phosphoric acid ester will be required in the preparation of a concentrate of optimum properties. The preferred minimum ratio of ester to chlorinated phenol may be determined by the following empirical formula: n/20 where n is the total number of carbons in the phosphoric acid ester. The maximum ratio of ester to chlorinated phenol for obtaining optimum performance may be represented by the following empirical formula: n/10 where n is the number of carbon atoms in the phosphoric acid ester. The solutions are prepared merely by mixing the phosphoric acid esters with the pentachlorophenol with agitation until a homogeneous solution is obtained. The following examples illustrate the process.

Example 1: A pentachlorophenol concentrate for dilution with 10 volumes of mineral spirits to a fire retardant paintable and nonblooming 5% pentachlorophenol solution was prepared by dissolving 37.8 parts of pentachlorophenol, 41.4 parts of tributyl phosphate and 20.8 parts of an aromatic petroleum oil (AMSCO-SOLV HT). This concentrate, having a solution point of -11°C., was satisfactory with respect to resistance to cold weather. The concentrate was found to be capable of dilution with mineral spirits to form clear diluted solutions, which property was not lost by standing in moist atmospheres. Wood samples treated with the solution were dried for 24 hours to permit the evaporation of the mineral spirits. No evidence of blooming was observed, and the wood was found to be fire resistant and free from afterglow when heated in a bunsen flame. The treated wood was readily paintable.

The advantages of the use in accordance with this example are evident from a comparison with an impregnating solution which utilizes diacetone alcohol, well known for its high solvency for pentachlorophenol in place of the phosphate. The solution was susceptible to contamination by water and upon dilution with the petroleum oil a cloudy solution resulted. Wood treated with the solution will show serious blooming upon drying and is not fire retardant.

Example 2: The procedure described in the preceding experiment was repeated using 33.8% pentachlorophenol, 44.1% dibutyl phenyl phosphate and 22.1% of chlorinated biphenyl (sold under the trademark Aroclor 5460). The solution had a solution point of -6°C. and was an effective impregnant for wood treatment. Wood treated in this manner contained an effective proportion of pentachlorophenol which did not bloom or otherwise affect the surface, and is fire retardant.

Di- and Monoammonium Phosphates

A process described by W.P. Green, Jr. and W.N. Meek, Jr. (U.S. Patent 2,766,139; October 9, 1956; assigned to Masonite Corporation) is directed to impregnating hardboard products with a mixture of diammonium phosphate and monoammonium phosphate. It has been found that lignocellulose hardboard sheets, having densities of about 1 or greater, may be first immersed in a hot concentrated aqueous solution of the phosphate mixture, and then immersed immediately in a cooler concentrated aqueous solution of the phosphate mixture whereby the sheets retain, upon drying and/or baking, from 14 to 20% of the phosphate salts based on the dry weight of the hardboard sheets. The treated sheets are characterized by greatly improved resistance to fire.

The proportions of the two ammonium phosphates to be used are critical since they tend to precipitate from the concentrated solutions employed in the method unless the limits established are closely adhered to. Broadly, the range of proportions suitable is from 2:1 to 1:2 parts by weight of diammonium phosphate to monoammonium phosphate. Within this operative range, it has been found that optimum results are obtained by employing proportions of 1:1 to 1.25:1 parts by weight of diammonium phosphate to monoammonium phosphate. The hot bath of concentrated aqueous phosphate mixture is preferably maintained at about 80% concentration although the amount of salts in solution will obviously vary somewhat as hardboard sheets are immersed. The boiling point of this concentrated solution is about 240°F. and it is preferable to maintain the bath at this temperature although the desired results are obtainable over a bath temperature range of about 230° to 250°F.

The cooler second bath is preferably maintained at about 60% concentration of phosphate salts although the amount of salts in solution may vary by about 5% above or below the optimum concentration. It has been determined that the second bath should be maintained at about 120°F. plus or minus 20°. In carrying out the process, the selected hardboard sheet is first immersed in the hot bath from about 3 minutes for the most dense sheet to about 1 minute in the case of a sheet of relatively low specific gravity such as about 1 or slightly less. The sheet is immediately immersed in the cooler bath for about 2 to 0.5 minutes depending on the board density.

Upon removal of the sheet from the second bath, the solution remaining on the sheet surfaces may be removed as by washing with water and the impregnated sheet may be dried and/or baked. Uptake of phosphate salts lies within the range of about 14 to 20% based on the dry weight of the hardboard sheet. Where the fire resistant sheet is to be coated with an air drying agent, it is only necessary to dry the sheet to remove water. However, where the impregnated sheet is to be finished with a baked enamel or similarly heat treated agent, it is preferable to bake the sheet to prevent gassing during curing of the enamel or other coating. It has been found that the baking cycle must be carried out over a progressively increasing temperature range of 140° to 275°F. extending over a period of 12 to 16 hours. Subsequent to this baking treatment the sheet may be subjected to additional heat treatment without adversely affecting the coating material and without appreciable loss of fire resistance.

It is believed that immersion of a hardboard in the hot bath serves to drive out some of the air and moisture present in the sheet. In this respect, it is important that the hardboard to be impregnated should have a minimum moisture content of 3.5% and preferably about 5%. Some impregnation of the hardboard occurs during this hot bath immersion. However, when the hot sheet is immersed in the second and relatively cool bath, there is apparently a vacuum effect on the voids present in the sheet. The cooler solution is thus enabled to penetrate the sheet very rapidly and thoroughly. In any event, much greater uptake of salts is obtained than by any single impregnation method. The result is the provision of hardboard containing sufficient ammonium phosphate salts to render the treated product fire retardant. The following example illustrates the process.

Example: Several pieces of 49" square lignocellulose hardboard were prepared. The square hardboard sheets had specific gravities of about 1.12 and were smooth on both surfaces. In sequence, the boards were immersed in two aqueous solutions containing 1:1.25 proportions by weight of diammonium phosphate and monoammonium phosphate under the following conditions:

Solution 1 — approximately 80% concentration of phosphate mixture. Temperature 241° ± 1°F. Boards immersed for 3 minutes.

Solution 2 — approximately 60 ± 1.5% concentration of phosphate mixture. Temperature 122° ± 9°F. Boards immersed for 2 minutes.

Escape of the evolved gases beneath the boards was permitted by use of a tilted hold-down for the boards. After being removed from the second solution, the boards were washed with water, allowed to drain, and dried employing a graduated temperature 16 hour baking cycle at 140° to 275°F. The impregnated board products had an average specific gravity of 1.28 and contained an average of 18.3% of ammonium phosphate by weight. The treated boards were tested for flame resistance according to Federal Specification SS-A-118b of August 4, 1954, paragraphs 4.3.3, 4.3.3.1, 4.3.3.3 and 4.3.3.4. All results represent average figures for three tests:

SS-A-118b Federal Fire Test
(40 minute exposure)

Weight loss	34.0%
Initial flame	3 min. 28 sec.
Cracked through	5 min. 20 sec.
Pieces falling	8 min. 45 sec.
Char diameter	27 inches.
After flame time	None.
Afterglow time	None.
After smoke	2.25 min. Light amount.
Rated	Fire retardant.

In the above test, the sample is rated as fire retardant where the charred area does not extend to the edges of a 30 square inch area exposed to the flame employed.

Tetrakis(Hydroxymethyl)Phosphonium Chloride

W.A. Reeves and J.D. Guthrie (U.S. Patent 2,927,050; March 1, 1960) describe the use of cross-linked polymers containing reoccurring structural units that are composed of phosphorus atoms, which are members of pentavalent phosphorus radicals of the group trimethylene phosphine oxide and tetramethylene phosphonium chloride radicals, and which are linked to trivalent nitrogen atoms by connecting structures of the group $-CH_2OCH_2-$ and $-CH_2$.

The polymers are produced by reacting phosphorus compounds of the group tris(hydroxymethyl)phosphine oxide and tetrakis(hydroxymethyl)phosphonium chloride and mixtures with monomeric organic compounds containing at least two members of the group, hydrogen and $-CH_2OH$, attached to trivalent nitrogen atoms. U.S. Patent 2,809,941 describes the polymers, their production and certain of their uses. The compounds, tetrakis(hydroxymethyl)phosphonium chloride and tris(hydroxymethyl)phosphine oxide are referred to by their initials THPC and THPO respectively. The term "phosphorus compounds" is employed to refer to THPC or THPO or mixtures of THPC and THPO.

In general the process comprises: impregnating the surface and at least a portion of the volume of a piece of wood with an incompletely polymerized THPC-resin and subjecting the impregnated piece of wood to conditions under which the polymerization of the resin is completed in situ. Preferred THPC-resins are made by condensing the phosphorus compounds with nitrogen compounds of the group urea, melamine, guanidine, and alkenoyl amides, such as acrylamide, and water-soluble methylolated and related alkylated methylolated derivatives. A catalyst is not needed when THPC or mixtures of THPC and THPO in which THPC is the major phosphorus-containing component is reacted with amine compounds, but when THPO is used alone, or used as the major phosphorus-containing component, an acidic or basic catalyst may advantageously be used. Preferred acid reacting catalysts are acetic acid, hydrochloric acid and the acid catalysts conventionally used in the production of amine resins; preferred basic reacting catalysts are alkali metal carbonates and bicarbonates, triethanolamine, and diethanolamine.

Solutions of incompletely polymerized THPC-resins can be caused to penetrate into the cell wall structure of the wood. The polymers with which the treated wood is impregnated can be insolubilized in situ by heating. The resin impregnated wood is nonhydroscopic, resists decay, termites, and acids, and is very hard, and is flameproof and glowproof. The following examples illustrate the process.

Example 1: Aqueous solutions of THPC dissolved in water and neutralized to pH of about 6.8 with triethanolamine were added to aqueous solutions containing trimethylolmelamine and urea. These final solutions were applied to various types of wood by various methods and cured by heating. These woods were all flameproof and glowproof and remained flameproof and glowproof after they had been submerged in hot running water for several hours and dried in an oven. The solutions were applied to these woods by one of the following methods: (1) by submerging or soaking the wood in the solution for about 16 hours, (2) by brushing the solution on the wood, and (3) by reducing the pressure for about 12 hours in a vessel which contained the wood submerged in the solution then releasing the vacuum and allowing the wood to remain in the solution for an additional two hours.

In order to determine if the woods were flameproof and glowproof, they were held in the flame of a Bunsen burner for two minutes and if the wood did not support a flame nor glow when removed from the Bunsen burner, it was considered flameproof and glowproof. The conditions of the experiments and the results of the tests are given in the table below. The flameproofing resins penetrated the pine and fir woods to a considerable depth when applied by the vacuum method. Samples of the wood taken from about 5 mm. beneath the surface would not burn when held in an open flame.

Sample No.	Type of wood treated	Solution concentration			Method of applying solution to wood	Method of cure	Flamming and glowing properties of treated wood (one min. over Bunsen burner)
		THPO, percent	Urea, percent	Trimethylolmelamine, percent			
1	Pine	19	9	11.5	Vacuum	Heat 30 min., 120° C.	Flameproof and glowproof.
2	do	19	9	11.5	do	do	Do.
3	Cypress	19	9	11.5	do	do	Do.
4	do	10.5	5	6.4	do	do	Do.
5	Pine	10.5	5	6.4	do	do	Do.
6	Oak	10.5	5	6.4	do	do	Do.
7	Fir plywood	10.5	5	6.4	do	do	Do.
8	Cypress	10.5	5	6.4	Soak	do	Do.
9	Pine	10.5	5	6.4	do	do	Do.
10	Oak	10.5	5	6.4	do	do	Do.
11	Fir plywood	10.5	5	6.4	do	do	Do.
12	Pine	10.5	5	6.4	Brushed	do	Do.

Example 2: An aqueous solution of THPC (35 parts THPC and 10 parts of water) was made to pH of about 5 with a concentrated aqueous solution of sodium carbonate and then added to an aqueous solution of trimethylolmelamine and urea (20 parts of trimethylolmelamine, 12 parts of urea, and 23 parts of water). This viscous solution was brushed on a piece of pine wood. After 4 days at room temperature the solution had formed a thin clear insoluble resin over the wood. The treated wood was flameproof and glowproof.

Water-Soluble Organic Phosphorus Esters and Polyesters

A process described by S.C. Temin (U.S. Patent 3,179,522; April 20, 1965; assigned to Koppers Company, Inc.) involves a flame retardant cellulosic composition comprising cellulosic material impregnated with water-soluble pentavalent phosphorus diesters having the formula:

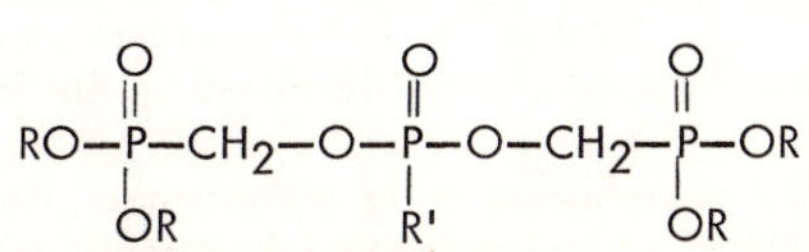

where R is a member selected from the group consisting of hydrogen and lower alkyl and R' is a member selected from the group consisting of lower alkyl, halo lower alkyl, cycloalkyl of 5 to 6 carbon atoms, phenyl, halophenyl, alkaryl of up to 8 carbon atoms and aralkyl of up to 8 carbon atoms and from 1 to 20% of the diester is retained by the cellulosic material. The diesters are made as described in U.S. Patent 3,065,183, by reacting at a temperature of 25° to 150°C. a phosphonic acid of the formula as shown on the following page.

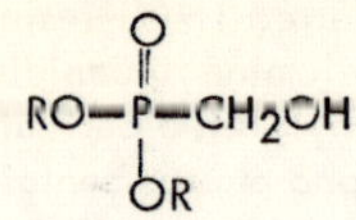

Here R is as defined above. This is reacted with a compound of the formula:

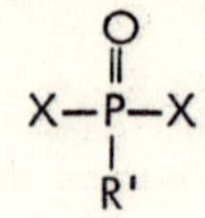

where R' is as defined above and X is halogen. Another flame retardant cellulosic composition has been found which comprises cellulosic material impregnated with a water-soluble poly(phosphonate-phosphinate) of the formula:

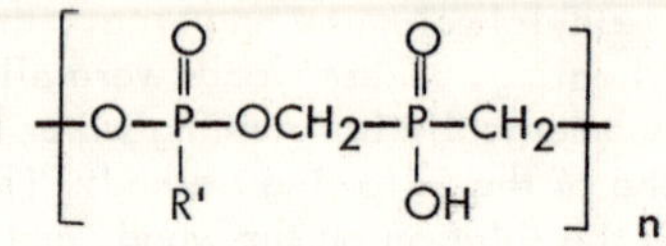

where R' is a member selected from the group consisting of cycloalkyl of 5 to 6 carbon atoms, lower alkyl, halo lower alkyl, phenyl, halophenyl, alkaryl of up to 8 carbon atoms and aralkyl of up to 8 carbon atoms and n is an integer having a value of at least three and from 1 to 20% of the polymer is retained by the cellulosic material. The method of making this polymer comprises reacting at a temperature of 25° to 150°C. a phosphonic acid of the formula:

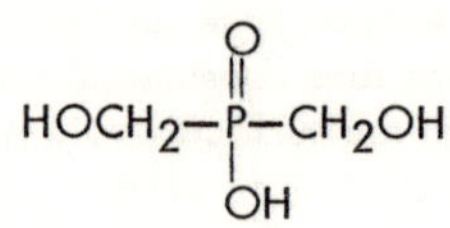

with a phosphonic dihalide of the formula:

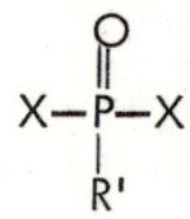

where R' is as defined above and X is halogen.

The phosphonate diesters are usually characterized by solubility in water and methanol and insolubility in benzene, gasoline, carbon tetrachloride and other oil solvents. The diesters are further characterized by high viscosity at room temperature with no tendency to crystallize. For preparing the polymers, bis(hydroxymethyl)phosphinic acid is used in conjunction with the phosphonic dihalide as previously defined. Homologues in which the hydroxymethyl groups (i.e., $-CH_2OH$) are replaced by higher hydroxy alkyl groups (i.e., $-CHROH$) have not produced the desired water-soluble polymers.

Polymerization to form the phosphonate-phosphinate occurs under the same conditions of temperature and pressure as given above in the preparation of the diesters. The chain length of the polyphosphonate-phosphinate is determined by the stoichiometry or molar ratios of the reactants. Higher molecular weight products are obtained by avoiding excess of either reactant and carefully purifying these same reactants. Further, higher molecular weights are formed by the use of lower subatmospheric pressures during the reaction. The degree of polymerization represented by the symbol "n," has a minimum value of 3 and a maximum value of about 100. By controlling the value of "n", a polymer having a molecular weight of over 550 to about 25,000 may be obtained. The polymer thus produced is generally characterized by the properties of solubility in water and polar solvents, insolubility in hydrocarbons, high viscosity at room temperature and no tendency to crystallize at room temperature. The process is illustrated by the following examples.

Example 1: Bis(phosphonomethyl)chloromethyl phosphonate — To a three-neck flask, fitted with a mechanical stirrer, dropping funnel and outlet tube, was added 112 g. of hydroxymethylphosphonic acid. The flask was heated by means of an oil bath maintained at 110°C. and, under continuous stirring, 84 g. of chloromethylphosphonic dichloride were gradually added from the dropping funnel. During the reaction, hydrogen chloride gas was evolved. After two hours, the outlet tube was connected to a vacuum source and the melt was heated to about 110°C. and under a reduced pressure (20 mm. of mercury) for an additional two hours. The yield of bis(phosphonomethyl)chloromethylphosphonate obtained was 159 g. The product was a viscous liquid, $n_D^{22.5}$, 1.5050. Elemental analyses were as shown on the next page.

Element	Found (percent)	Calculated for $C_3H_{10}O_9P_3Cl$ (percent)
C	11.3	11.3
H	3.3	3.1
P	29.6	29.2

Example 2: Bis(phosphonomethyl)phenylphosphonate — In an apparatus similar to that described in Example 1, 81 g. of hydroxymethylphosphonic acid was heated and stirred at 115°C. and at a pressure of 30 mm., while 73.5 g. of phenylphosphonic dichlorides were added over a period of two hours. The product, 128 g. was a colorless syrup which barely flowed at room temperature. Elemental analyses were as follows:

Element	Found (percent)	Calculated for $C_8H_{13}P_3O_9$ (percent)
C	28.3	27.8
H	3.9	3.8
P	27.7	26.9

The infrared spectrum was consistent with that of the triphosphonate compound.

Example 3: Poly[phosphinico bis(methyl)]phenylphosphonate — In an apparatus similar to that described in Example 1, 50.4 g. of bis(hydroxymethyl)phosphinic acid was heated to a temperature of 100°C. To the acid was added, with stirring, over a period of one hour, 78.0 g. of phenylphosphonic dichloride. The temperature was then raised to 150°C. and the pressure reduced to 20 mm. After 3 hours, a slightly amber, viscous liquid was obtained which on cooling became a stiff semisolid material. The yield was 95 g. Infrared analysis corresponded to a polymer with the structure:

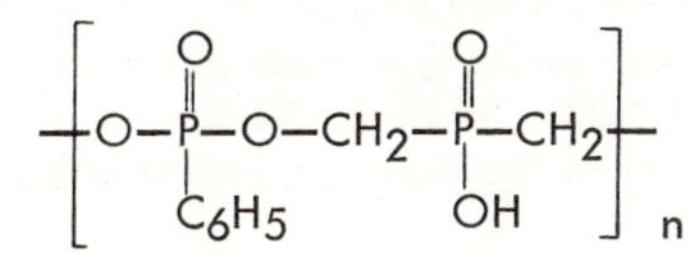

The fluid had n_D^{22}, 1.530. It analyzed 24.6% phosphorus compared with a calculated value of 25.0%.

Example 4: The effect on fire resistance imparted to wood — A 15% aqueous solution of bis(phosphonomethyl)chloromethylphosphonate was applied to Southern yellow pine. The percent retention of the triphosphonate, based on the oven dried weight of the wood, was 20%. Using the conventional fire tube apparatus (ASTM E69-50) the percent weight loss on burning was found to be 22%. In a similar test with poly[phosphinico bis(methyl)]chloromethylphosphonate the weight retention was again 20% and the percent weight loss (ASTM E69-50) was only 18%. A value of only 18% weight loss approaches a fire retardant classification of "excellent."

Example 5: The effect on fire resistance imparted to paper — From the reaction of formaldehyde with dimethyl hydrogen phosphite was obtained dimethyl hydroxymethylphosphonate, $HO—CH_2—P(O)(OCH_3)_2$. To 24.8 g. of this compound was added 11.5 g. of p-chlorophenylphosphonic dichloride over a period of 30 minutes. The stirred reaction mixture was maintained at 65° to 70°C. while gaseous hydrogen chloride was evolved. After about 150 minutes, the pressure in the reaction vessel was reduced to 10 mm. This pressure and a temperature of 85° to 90°C. was maintained for two hours. The clear viscous product remaining gave an elemental analysis conforming to the calculated values for $C_{12}H_{20}O_9P_3Cl$. A piece of filter paper impregnated with 10% (of its original weight) of this product no longer supported combustion after removal from the flame of a Bunsen burner.

Linear Tertiary Phosphine Oxide Polymers

A.Y. Garner (U.S. Patent 3,127,357; March 31, 1964; assigned to Monsanto Chemical Company) describes linear tertiary phosphine oxide polymers and methods for their preparation. The following examples illustrate the process.

Example 1: Twenty grams of a viscous linear interpolymer having a number average molecular weight of about 600 and comprised of substantially equimolar proportions of phenyl phosphine and diallyl ether are dissolved in 150 ml. of acetone together with 60 ml. of a 3% by weight aqueous potassium permanganate solution. The resulting solution is stirred at about room temperature for 15 hours; after which the remaining permanganate is inactivated by adding, dropwise, a 1 normal aqueous sodium bisulfite solution until the characteristic permanganate color disappears. Inorganic salts are removed by filtration, followed by vacuum distillation of the filtrate to remove water and acetone. A colorless, viscous polymer, soluble in ethanol and benzene, is obtained. Analysis and infrared spectroscopy establish the polymer as being comprised of a plurality of structural units corresponding to the following formulae, each moiety being present in substantially equimolar proportions.

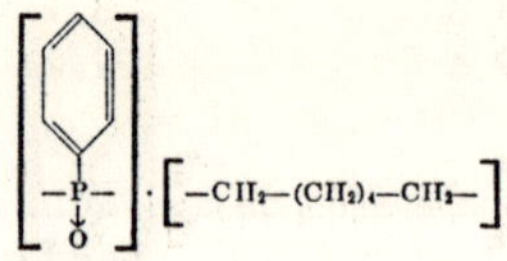

Example 2: Twenty grams of a solid linear interpolymer having a number average molecular weight of about 180,000 and comprised of substantially equimolar proportions of phenylphosphine and diallyl ether are dissolved in 150 ml. of glacial acetic acid together with 1 gram of lead tetraacetate. The resulting solution is heated at reflux for about 10 hours; then is cooled and poured into an excess of hexane to precipitate a tough, flexible polymer. This polymer is soluble in ethanol and benzene. Upon testing for flammability by holding a lump of the polymer in the flame of a Meeker burner until ignited, the polymer is found to be self-extinguishing. Analysis and infrared spectroscopy establish the polymer as being comprised of a plurality of structural units corresponding to the following formulae, each moiety being present in substantially equimolar proportions.

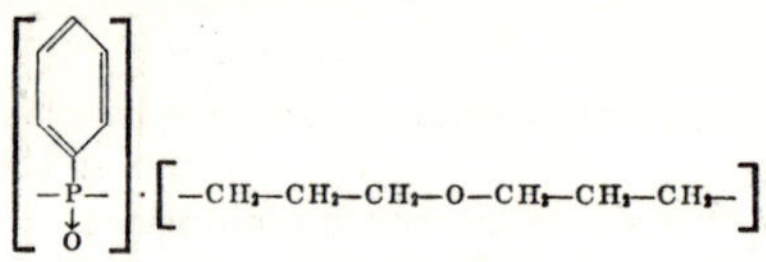

Example 3: A rectangular chip of ponderosa pine measuring about 1/4" x 1" x 3" is immersed for 4 hours in 100 ml. of a 75% by weight solution of the linear tertiary phosphine oxide polymer prepared in Example 1 dissolved in benzene. The pine chip is then heated in an air oven at 80°C. for about 3 hours to remove all adsorbed solvent. Testing for flammability by holding the pine chip vertically in the flame of a Meeker burner until ignited, the chip is found to be self-extinguishing.

Tris(1-Aziridinyl)Phosphine Oxide and Thiourea

R.B. Le Blanc (U.S. Patent 3,376,160; April 2, 1968; assigned to The Dow Chemical Company) describes the treatment of wood with a mixture containing tris(1-aziridinyl)phosphine oxide and at least one thiourea compound. The use of tris(1-aziridinyl)phosphine oxide (APO) in combination with urea for the production of flame resistant cellulosics is described for example, in U.S. Patent 2,859,134. A process for fixing pigments on fibrous materials using a mixture of urea precondensation products and polyfunctional ethylenimine compounds is described in U.S. Patent 2,961,349. One of the disadvantages of such processes, however, is that comparatively large amounts of expensive APO are required in order to obtain good flame retardant properties.

It has now been found that the amounts of APO required to achieve satisfactory flame retardance for various cellulosic materials can be reduced considerably by using mixtures of thiourea (and/or substituted thiourea compounds) along with APO as the major components of the treating composition. The following example illustrates the process.

Example: A bath composition consisting of the following ingredients was prepared by simply mixing the components.

Ingredient	Grams
80% APO in ethyl alcohol	216
Thiourea	117
Water	1150
Total	1483

Red cedar shingles (approximate dimensions: 7" x 16" x 1/4") were dipped in the freshly prepared bath for 30 minutes, drained and weighed to determine wet pickup. The average wet pickup (6 samples) was 20%. The drained specimens were cured for 30 minutes at 300°F. in an oven. The treated specimens charred when placed in a luminous open flame (Bunsen burner), but would not burn when the flame was removed.

Tris[2,2,2-Tris(Chloromethyl)Ethyl] Phosphate

S. Carpenter and E.R. Witt (U.S. Patent 3,403,049; September 24, 1968, assigned to Celanese Corporation) describe a flame-resisting material comprising flammable organic material with a coating of

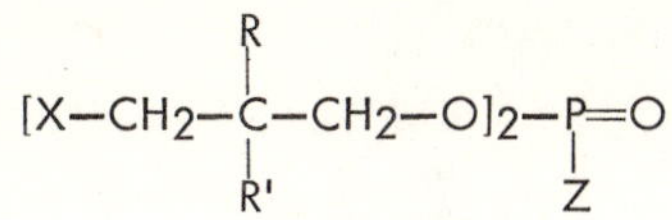

or

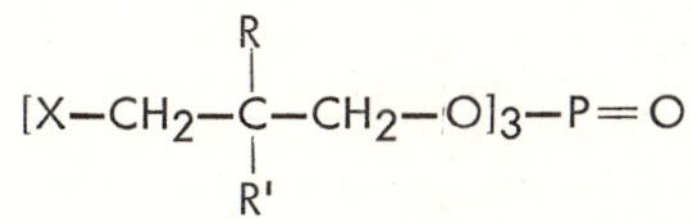

where X is a halogen selected from the group consisting of chlorine and bromine, Z is selected from the group consisting of alkyl and aryl radicals; R and R' are each selected from the group consisting of saturated and unsaturated aliphatic hydrocarbon radicals having from 1 to 18 carbon atoms, halogenated saturated aliphatic hydrocarbon radicals having from 1 to 18 carbon atoms, aromatic hydrocarbon radicals having up to 10 carbon atoms, and halogenated aromatic hydrocarbon radicals having up to 10 carbon atoms. The following examples illustrate the process.

Example 1: This example illustrates the preparation of tris[2,2,2-tris(chloromethyl)ethyl] phosphate, the formula for which is

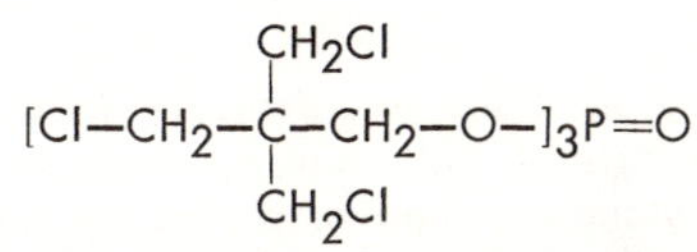

A mixture of phosphorus oxychloride, 76.3 g. (0.5 mol), 3,3-bis(chloromethyl)oxetane, 233 g. (1.5 mols), and dry (anhydrous) benzene (500 ml.) is refluxed for 1 hour. The reaction mass is then stripped of light ends to a pot temperature of 100°C. at 1 mm. Hg. There remains in the pot 144 g. of a pale yellow, viscous oil comprising tris[2,2,2-tris(chloromethyl)ethyl] phosphate. Analysis — Theoretical: Cl, 50.3%; P, 5.0%. Found: Cl, 50.3%; P, 5.8%.

Example 2: Tris[2,2,2-tris(bromomethyl)ethyl] phosphate is prepared as described under Example 1 with the exception that instead of the reactants employed in that example there are used 0.5 mol of phosphorus oxybromide and 1.5 mols of 3,3-bis(bromomethyl)oxetane.

Example 3: This example illustrates the preparation of a phosphate of the kind described under Example 1 using a catalyst for the reaction and in the absence of an inert, anhydrous, liquid reaction medium. Phosphorus oxychloride, 76.3 g. (0.50 mol), 3,3-bis(chloromethyl)oxetane, 233 g. (1.5 mols), and a catalyst for the reaction, specifically titanium tetrachloride, 0.61 g., are heated together at 80°C. for 1 hour and then heated to 140°C. At this higher temperature the reaction became highly exothermic and the temperature of the reaction mass rises to 300°C. From the reaction mass 146 g. of crude, off-white crystals comprising tris [2,2,2-tris(chloromethyl)ethyl] phosphate is obtained by crystallization from an acetone-heptane solvent. The crude crystals are recrystallized from heptane to give 97g. of white crystals having an MP of 106.4 to 108.6°C. and which show the following upon analysis: Theoretical for tris[2,2,2-tris(chloromethyl)ethyl] phosphate, percent: Cl, 50.3; P, 5.0; C, 29.2; H, 3.9. Found, percent: Cl, 49.8; P, 5.7; C, 33.8; H, 4.6.

A portion (60 g.) of the recrystallized material is dissolved in benzene (200 ml.), washed first with dilute potassium permangnate solution with no reduction of the permanganate, and then with 2 weight percent of sodium hydroxide solution. An equal volume of heptane is added, and 40 g. of white crystals are obtained that analyze as follows: Theoretical for tris[2,2,2-tris(chloromethyl)ethyl] phosphate, percent: Cl, 50.3; P, 5.0; C, 29.2; H, 3.9. Found percent: Cl, 50.5; P, 5.1; C, 30.3; H, 3.9.

Example 4: A piece of filter paper, more particularly cellulosic filter paper, is coated and impregnated by immersing it in a small amount of the molten phosphate of Example 3 as finally purified. The treated filter paper is self-extinguishing when removed from a flame. Similar results are obtained from other articles of manufacture comprising a flame-resisting material comprising a normally flammable, organic material having at least on its external surfaces a flame-resisting coating which includes essentially tris[2,2,2-tris(chloromethyl)ethyl] phosphate or similar compounds of the process.

Fluorooxyphosphoranes

R.A. Braun (U.S. Patent 3,368,000; February 6, 1968; assigned to E.I. du Pont de Nemours and Company) describes fluorooxyphosphoranes having the following structures:

(1)

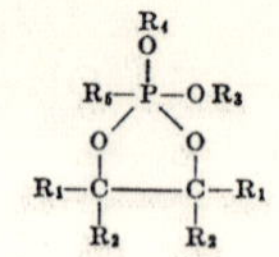

(2)

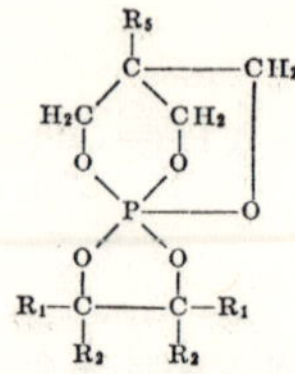

where R_1 and R_2 can be perfluoroalkyl having 1 through 5 carbon atoms, R_3 and R_4 can be alkyl radicals of 1 through 12 carbon atoms; alkyl radicals of 1 through 12 carbon atoms substituted with chlorine, bromine or cyano; cycloalkyl radicals of 5 through 7 carbon atoms; aryl radicals; or aryl radicals substituted with chlorine, bromine, cyano or alkyl of 1 through 5 carbon atoms; and R_5 can be an alkyl radical of 1 through 12 carbon atoms; an alkyl radical of 1 through 12 carbon atoms substituted with chlorine, bromine, or cyano; an O-alkyl radical of 1 through 5 carbon atoms; a cycloalkyl radical of 5 through 7 carbon atoms; an aryl radical; or an aryl radical substituted with chlorine, bromine, cyano or alkyl of 1 through 5 carbon atoms.

The compounds of the process are useful as fire and flame retardants for wood, natural fibers, paper and other combustible substrates. Such materials can be made more resistant to burning by applying a 1 to 10%, by weight, solution of a compound of the process in petroleum ether or acetone and then air-drying the material. Preferred for the superior fire and flame retarding characteristics they confer are the compounds of structures (1) and (2) where R_1 and R_2 are $-CF_3$. Highly preferred for this use are the compounds of structure (1) where R_1 and R_2 are $-CF_3$ and R_3 and R_4 are bridged. The following example illustrates the process.

Example: Phenyl ethylene phosphite (55.2 grams, 0.3 mol) and hexafluoroacetone (100 grams, 0.6 mol) were heated in a bomb at 150°C. for 6 hours. The resulting liquid product was fractionated to give 105.8 grams of a colorless liquid having a boiling point of 102° to 105°C. at 0.3 mm. of pressure. On analysis the product gave the following data: Calculated for $C_{14}H_9F_{12}PO_5$: C, 32.57; H, 1.76; F, 44.17; P, 6.00. Found: C, 32.93; H, 2.01; F, 43.87; P, 6.30. The product had the structure:

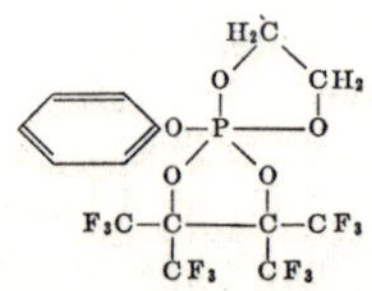

A 5%, by weight, solution of this compound in benzene was prepared. A piece of 1/16-inch hardwood veneer was dipped into this solution and air dried. The wood so treated did not burn, even when exposed to an open flame.

Immersion in Ammonia Solution Prior to Phosphate Treatment

J.J. Franck and B. Sandor (U.S. Patent 2,994,620; August 1, 1961; assigned to Ignicel Societe Anonyme, France) describes the fireproofing of cellulosic materials by forming the waterproofing salts in situ in such a manner that for a given period of soaking or other treatment, the quantity of salt retained in the treated cellulosic or lignocellulosic material, is considerably increased. A simplified flow diagram of the process for fireproofing insulating boards is shown on the following page.

The fireproofing method consists in incorporating in the material at least one fireproofing salt such as an ammonium hydrogen phosphate, i.e., mono- or diammonium phosphates or compounds capable of forming such a salt while simultaneously causing ammonium phosphates to be formed and retained in situ in sufficient quantity without altering the mechanical and surface properties of the treated material. Prior to the introduction of the waterproofing compounds, it may be advantageous to cause the channels and cells in the material to expand, such an expansion being obtained

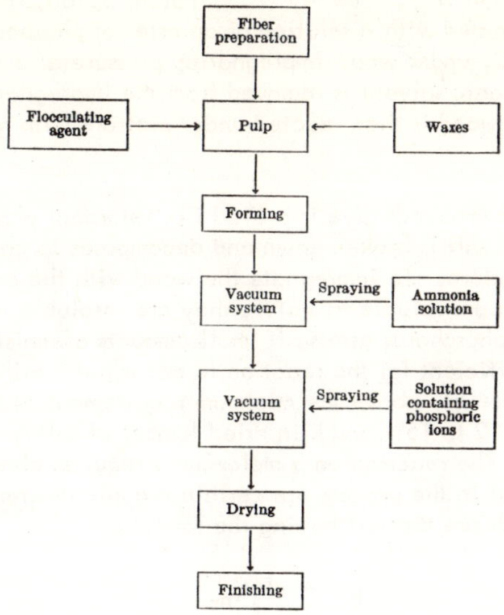

without alteration of the fibers by soaking the material in a solution of ammonia having a concentration comprised between 3 and 10% according to the nature of the material. The immersion time of the material in the solution as well as the concentration, is determined in such a manner as to avoid an alteration of the fibers while expanding the channels, then the cells of the mass, to cause a penetration of ammonia, which makes possible a rapid penetration of a sufficient quantity of PO_4 ions which, by combination with the ammonia from the expanding solution will form in situ the waterproofing ammonium phosphates. When the material is first expanded the second bath furnishing PO_4 ions preferably consists of an aqueous solution of ammonium hydrogen phosphate $PO_4(NH_4)_mH_n$, where m and n are 1 to 2 and m + n = 3, having a concentration comprised between 10 and 30% and a pH comprised between 3 and 7 preferably between 3.9 and 6.2 according to the nature of the treated material.

The concentrations of the fireproofing salts should be such that they lead to the presence in situ of quantities of H_3PO_4 comprised between 5 and 15% by weight of the dry material which is to be fireproofed. The concentration of the waterproofing solutions or emulsions should be such that the quantity of dry waterproofing agent, fixed on the treated material is comprised between 0.5 and 5% by weight of dry material. The following examples illustrate the process.

Example 1: Deal boards of 0.3 x 2.5 x 0.01 m. were immersed for 24 hours in 5% ammonia solution. After a quick draining, the boards were immersed in 20% ammonium phosphate solution having a pH of 3.95. After drying, the boards containing 10% by weight, of ammonium phosphates were subjected to inflammability tests and proved to be fireproof.

Example 2: Oak floor boards of 0.4 x 0.7 x 0.005 m. were immersed for 12 hours at room temperature in a 5% ammonia solution. After having been quickly drained, the boards were immersed during 12 hours in a 22% aqueous bath of ammonium phosphates having a pH value of 3.9, at a temperature of 65°C. After drying, the boards containing 9.5% by weight of fireproofing salt were submitted to the standard tests and proved to be fireproof.

Example 3: Beech plies of 2.1 mm. thick were immersed for 5 minutes in a 5% ammonia solution, then, after draining were immersed for the same time in an ammonium phosphate solution having a specific gravity of 1.13 and a pH value of 3.9, the said solution being maintained at a temperature of 35°C. After drying in air at room temperature, the treated plies had absorbed 11.1% by weight of fireproofing salts and proved to be noninflammable.

Example 4: 8 kg. of a 50% aqueous emulsion of an inorganic microcrystalline wax were added to 10 $m.^3$ of lignocellulosic fiber paste containing 20 g./l. of dry material and the pH of which was brought between 8.5 and 9 by careful addition of an expanding ammonia solution and the wax was precipitated on the fibers by adding a strong mineral acid until the pH value of the mixture reached 5.5. The paste was then formed into panels and the panels were placed on the band of a suction filter and sprinkled with a solution of ammonium phosphate containing 55 kg./$m.^3$ of phosphoric acid and having a pH value comprised between 5.5 and 6. When the volume of filtered phosphate solution reached the panels, the sprinkling was stopped and the panels drained. The phosphate solution was recovered and used for a further impregnation after adding H_3PO_4 and adjusting the pH at above indicated values. Samples taken from the panels showed acceptable absorption of fireproofing salts. The panels proved to be fireproof.

In Situ Reaction of Ammonia and Phosphorus Esters

In a process described by I.S. Goldstein and W.J. Oberley (U.S. Patent 3,160,515; December 8, 1964; assigned to Koppers Company, Inc.), wood is impregnated with a solution of an ester of phosphoric or phosphorous acid in an organic solvent having a boiling point below 80°C. under wood-impregnating pressure at a temperature below the boiling point of the solvent under pressure, and the organic solvent is removed from the impregnated wood. The organophosphorus ester within the impregnated solvent-free wood is then reacted under pressure with gaseous ammonia and the unreacted ammonia is removed.

The ammonium salts of organophosphorus compounds give improved fire retardant properties. A possible theory is that when the wood is burned, the ammoniated salt is broken down and decomposes to gaseous ammonia which dilutes the combustible gases to a noncombustible mixture. To impregnate the wood with the prepared ammoniated salts presents a problem since although these compounds are soluble in water, they are insoluble in organic solvents. Gaseous ammonia will react in the wood with organophosphorus esters, if small amounts of moisture are present in the wood. The moisture present in the wood usually is sufficient for the reaction to occur but a minimum of about 4% is required. The moisture content of dried lumber is considerably above the minimum requirement as shown by air dried lumber which has a minimum moisture content ranging from 12 to 15% and kiln dried lumber of soft and hard woods which has a minimum moisture content ranging from 6 to 12%. The reaction on a molar basis requires about one mol water for each mol of organophosphorus ester. Compounds useful in the process are certain organic diesters of phosphoric and phosphorous acid and trimethyl phosphite. These compounds are diesters having the formula:

$$(RO)_2\overset{[\overset{\Vert}{O}]_{n-1}}{P}X$$

where R is member selected from the group consisting of lower alkyl and chlorinated lower alkyl, X is a hydroxy radical and n is an integer having a value of from one to two; and trimethyl phosphite. Particularly useful compounds are: dimethyl phosphite, diethyl phosphite, diethyl phosphate, bis(beta-chloroethyl) phosphate, bis(1,3-dichloropropyl) phosphite, bis(1,3-dichloropropyl) phosphate, and trimethyl phosphite. The amount of ammoniated flame retardant chemical retained by the treated wood, in order to give satisfactory flame retardance, should be within a range of about 5 to 20% by weight of the dry wood. Less than 5% retention does not offer sufficient protection, whereas more than 20% retention is found unnecessary.

Impregnation of wood is conveniently accomplished by a full cell cycle. The object of the full cell process is to fill the structure of the wood with the treating solution. This is accomplished by subjecting the treating reactor containing the wood to be treated to a vacuum of about 22 to 26 inches of mercury for a period of time ranging from about 15 minutes to an hour. The reactor is then filled with treating solution and subjected to a pressure of about 125 to 200 psi at 25° to 100°C. for about one and a half to three hours. The larger and more dense pieces of wood require either a longer vacuum period, a longer pressure period or perhaps both. For most woods, a 15 minute vacuum of 25 to 26 inches of mercury followed by a two hour pressure process of 160 psi at ambient temperature is adequate. The flame retardance of wood may be conveniently measured by the percentage weight loss in the fireproof test. A standard test is the ASTM method E69-50. Using this test, a fire tube weight loss of approximately 40% is indicative of moderate fire retardance and less than 30% is considered good. In addition, the rate and extent of flame spread may be observed and estimated. The process is illustrated by the following examples.

Example 1: Twenty pieces of southern yellow pine, having the dimensions 3/8" x 3/4" x 40", were impregnated at ambient temperatures with a 15% solution of dimethyl phosphite in benzene. Using the full cell method, the samples were placed in a treating cylinder under an initial vacuum of 27 inches Hg for 15 minutes followed by a pressure of 150 psi applied for one and one-fourth hours. The solvent was then removed from the samples by evaporation under reduced pressure. It was determined that the dimethyl phosphite retained by the dry wood was approximately 15%. The samples were divided into two groups of ten pieces each. One group was placed in a cylinder and subjected to reduced pressure for 15 minutes. Dry ammonia gas was, thereafter, forced into the cylinder under a pressure of 12 psi for a period of 30 minutes, after which the excess ammonia was removed by diffusion at atmospheric pressure.

During the ammonia treatment, ammonium methyl phosphite was formed in situ within the impregnated samples. Both groups of samples were tested in a fire tube apparatus according to the ASTM E69-50 test method. A comparison between the nonammoniated and ammoniated samples showed that (1) dimethyl phosphite impregnated samples gave rapid and complete flame spread, and had a weight loss of 45%, and (2) dimethyl phosphite impregnated samples treated with ammonia gave no flame spread and had a weight loss of 20%.

Example 2: The procedure of Example 1 was repeated using other compounds instead of dimethyl phosphite. The dried samples retained approximately 15% by weight of the solute. A table is given on the following page which shows the results of the tests.

Wood Impregnation

Test	Compound	% Wt. Loss	Flame Spread
A	Bis(beta-chloroethyl)phosphite	30	Slight
	Bis(beta-chloroethyl)phosphite + NH_3	20	None
B	Diethyl phosphite	55	Fast and complete
	Diethyl phosphite + NH_3	28	Slight
C	Diethyl phosphate	45	Fast and complete
	Diethyl phosphate + NH_3	26	Slight
D	Triethyl phosphite	72	Fast and complete
	Triethyl phosphite + NH_3	75	Fast and complete
E	Tris(dichloropropyl)phosphate	73	Moderately fast and complete
	Tris(dichloropropyl)phosphate + NH_3	75	Moderately fast and complete

Phosphorus Oxychloride with Vinylchloride-Vinyl Acetate Binder

C.F. Perizzolo (U.S. Patent 3,371,058; February 27, 1968; assigned to Stauffer Chemical Company) has found that wood treated with phosphorus oxychloride imparts to the treated wood certain desirable and durable flame retarding properties. It has also been found that these effective and durable flame retarding properties of phosphorus oxychloride in wood may be extended favorably by the incorporation of a suitable binding resin. Such a supplemental treatment protects the primary flame retarding agent, phosphorus oxychloride, against excessive leaching by water. Excellent durability results may be obtained from the use of phosphorus oxychloride alone. The following examples illustrate the process.

Example 1: This example illustrates the impregnation of a selected representative wood specie with a loading of phosphorus oxychloride sufficient to be effective both before and after exposure to severe weathering conditions. Samples of red cedar, 4 inches x 6 inches x 5/16 inches were accurately weighed and used as the test specimen. The pieces of wood were placed in a pressure vessel. The impregnation technique used was a modified full cell procedure. The chamber was evacuated to a vacuum of about 15 to 20 inches for approximately 30 minutes. A solution of 4% phosphorus oxychloride in methylene chloride as solvent was introduced into the pressure vessel sufficient to fill the pressure chamber containing the wood specimen. Hydrostatic pressure of approximately 90 pounds per square inch was then applied to the liquid in the pressure vessel until the wood samples no longer absorbed the liquid. The pressure was then released and the unabsorbed solution was removed from the chamber. The entire process was carried out at room temperature. The wood specimens were dried to constant weight at 65°C. in an oven. The load of phosphorus oxychloride absorbed was equivalent to approximately 7.43 pounds per 100 pounds of wood, after drying. Analysis indicated 1.50% phosphorus present in the wood.

Prior to testing the durability of the system in wood, the wood samples were exposed to a gas/air flame 1 1/4 inches by 6 1/2 inches of approximately 1300°F. An external air supply equivalent to a 12 mile per hour wind was superimposed on the test flame and wood samples. The sample pieces of wood were arranged in a 3-member deck simulating a roof structure. Wood samples were exposed for four minutes and for three cycles comprising 1 minute flame exposure and 2 minutes rest. Once the samples passed these flame exposure tests satisfactorily they were subjected to the following durability tests.

Various sets of wood samples thus treated were subjected to accelerated durability tests before exposure to flame. The rain/heat test consisted of alternate exposure over a 12 week period to 3.5 days of "rain" at 12.5 inches/hour and 3.5 days of heat at 65°C. The severe heat test consisted of 90 days exposure to a temperature of 80°C. The weatherometer test consisted of a 600-hour exposure to alternating cycles of 18 minutes of spray and 102 minutes of arc. The temperature during the arcing cycle was 145°F. The wood samples subjected to the durability tests were then exposed to the above described intermittent gas/air flame test.

Observations were made such as the spread of the flame beyond area of flame impingement, afterglow on removal of the flame, structural strength of the specimen, and the like. There was essentially no flame spread, no afterglow or at best several small short lived glow areas and the specimen did not disintegrate or become substantially weakened after the flame removal. The wood samples treated with $POCl_3$ did exhibit flame retardant characteristics in the testing procedure.

Example 2: This example illustrates the use of a typical binder resin in conjunction with a lower loading requirement of phosphorus oxychloride sufficient to be durable and present excellent flame retardant characteristics. In the same manner as Example 1, samples of red cedar, 4 in. x 6 in. x 5/16 in., were accurately weighed and used as representative test specimen. Using the modified full cell procedure the wood samples were impregnated with a loading of phosphorus oxychloride and vinyl chloride-vinyl acetate copolymer as the binder resin. The impregnant solution was 2% phosphorus oxychloride, 5% copolymer resin (13% vinyl acetate-87% vinyl chloride) and 93% methylene chloride. The loading obtained was equivalent to 6.7 pounds of phosphorus oxychloride and copolymer resin per 100 pounds of wood. This is equivalent to 1.13 pounds of phosphorus oxychloride, 0.23% phosphorus. Similarly, wood samples were impregnated with a solution consisting of 4% phosphorus oxychloride, 10% copolymer resin (defined above), and 86% methylene chloride. The loading obtained was equivalent to 8 pounds of phosphorus oxychloride and copolymer resin per 100 pounds of wood. This is equivalent to 1.58 pounds of phosphorus oxychloride, 0.32% phosphorus. Both sets of samples were dried to constant weight.

The wood samples thus treated were arranged in the usual 3-membered deck simulating a roof structure and subjected initially to the gas/air test flame for four minutes. No adverse burning or other degradative effects were observed. Sample sets were now subjected to the previously described accelerated weathering conditions. After nine weeks in the rain-heat test the samples of lower loading, i.e., 1.12 pounds of phosphorus oxychloride per 100 pounds of wood, satisfactorily resisted the three cycle 1-minute-2-minute flame-rest exposure test. The sample set with 1.58 pounds of phosphorus oxychloride per 100 pounds of wood satisfactorily passed the three cycle 1-minute-2-minute flame-rest exposure test after twelve weeks of subjection to the rain/heat test.

Pinene-Phosphorus Pentasulfide Reaction Products

C.M. Shaw (U.S. Patents 2,991,182; July 4, 1961 and 2,990,427; June 27, 1961; both assigned to California Research Corporation) has found that flammable articles are effectively flameproofed by impregnation with a composition comprising a major proportion of liquid vehicle, more particularly a preservative oil, and at least 5% by weight of a member of the groups consisting of (1) reaction product of phosphorus pentasulfide and pinene, (2) zinc dialkyl dithiophosphate containing from 3 to 8 carbon atoms in each of the alkyl groups and zinc dialkylphenyl dithiophosphate containing from 14 to 24 carbon atoms in each of the alkylphenyl groups, and (3) combination of members of each of group (1) and group (2) the combination consisting of from 10 to 90% by weight of reaction product and from 10 to 90% by weight of zinc dithiophosphate.

The reaction product of phosphorus sulfide and pinene is the material resulting from the condensation of a terpene such as present in turpentine with a phosphorus pentasulfide. In the preparation of these condensation products, the ratio of turpentine to phosphorus pentasulfide is usually within the range of about 1:1 to about 3:1. The condensation is exothermic and proceeds spontaneously after heating slightly to initiate the reaction. Usually, the turpentine or other pinene or terpene-containing material is heated to about 200°F. and the phosphorus pentasulfide is added slowly with stirring. Cooling may be employed if necessary to avoid undesirably high temperatures in excess of 300°F. In the process, addition agents such as aliphatic alcohols, for example, lauryl alcohol may be combined with the turpentine and phosphorus pentasulfide.

The pinene-phosphorus sulfide reaction products and the metal dithiophosphates of the flameproofing compositions, as mentioned above, are generally known as lubricating oil additives. They are described in considerable detail in various texts on lubricating oil additives. A brief description of such additives is to be found at pages 167 to 178 of "Motor Oils and Engine Lubrication" by Georgi, published in 1950 by Reinhold Publishing Corporation, New York. The following compositions are illustrative of the process.

Composition	Carrier or Wood Preservative	Percent	Compound (1)	Percent	Compound (2)	Percent
1	Diesel Fuel Oil Pentachlorophenol	90.0 1.5	Zinc ditetradecylphenyl dithiophosphate.	4.25	Turpentine P_2S_5 Reaction Product.	4.25
2	Creosote	90.0	Zinc butyl hexyl dithiophosphate.	1.0	Pinene-P_2S_5 Reaction Product.	9.0
3	Grade 1 Naphthenic Fuel Oil.	85.0	Zinc didodecyl dithiophosphate.	10.0	do	5.0
4	Diesel Fuel Pentachlorophenol	90.0 5.0	Zinc butyl amyl dithiophosphate.	2.5	do	2.5
5	Creosote	80.0	Zinc dodecylphenyl hexyl dithiophosphate.	18.0	Turpentine-P_2S_5 Reaction Product.	2.0
6	No. 2 Furnace Oil	95.0	Zinc methyl ethyl dithiophosphate.	2.0	Pinene-P_2S_5 Reaction Product.	3.0
7	Water	85.0	Zinc ditetradecylphenyl dithiophosphate.	5.0	Turpentine-P_2S_5 Reaction Product.	5.0
8	Creosote	90.0	Zinc butyl amyl dithiophosphate.	10.0		
9	do	90.0			Pinene-P_2S_5 Reaction Product.	10.0
10	Boiled Linseed Oil	90.0	Zinc butyl amyl dithiophosphate.	5.0	do	5.0

The effectiveness of the flameproofing compositions is illustrated by the following examples.

Example 1: A flameproofing composition is prepared consisting of 90% by weight of creosote and 10% by weight of a typical concentrate of a combination of thiophosphorus compounds as described above. In this instance, the concentrate consists of 42.5% by weight of zinc butyl hexyl dithiophosphate, 42.5% by weight of phosphorus pentasulfide and alpha-pinene reaction product and 15% by weight of a medium boiling liquid aromatic petroleum hydrocarbon SO_2 extract. The flameproofing composition thus prepared is found to give excellent fire retarding properties to wood. The standard crib burning tests show the effect of the flameproofing compositions on a field scale.

In these tests, a comparison is made by the use of four different treatments. The treatments are: (1) straight creosote, (2) an aqueous solution containing about 8% of water soluble salts consisting of 10% diammonium phosphate, 60% ammonium sulfate, 20% sodium tetraborate anhydrous and 10% boric acid, (3) a commercially available flameproofing composition consisting of 88% creosote and 12% tricresylarylphosphate (8.6% phosphorus), and (4) a composition according to the process consisting of 10% by weight of the concentrate of Example 1, and 90% by weight creosote. In the tests, the time for the fire to go out is measured, as well as the time for any collapse of the timbers. The weight of the remainders of the pieces after scraping off char is also measured. The results of the tests are given in the following table.

Treatment Composition	Creosote Or Solution, lbs./cu. ft.	Percent Weight Lost In Burning	Time (min.) For Fire To Go Out Or To Collapse of Crib
Set No. 1—4¼" x 7" x 40" pieces			
(1)	15.4	100	Collapsed 104'.
(2)	15.8	38	Out in 45'.*
(3)	15.5	42	Do.*
(4)	14.9	34	Out in 33'.
Set No. 2—4" x 4" x 24" pieces			
(1)	20.4	100	Collapsed.
(3)	23.0	27	Out in 34'.
(4)	22.7	28	Out in 29'.
Set No. 3—2" x 4" x 24" pieces			
(1)	25.2		Collapsed 35'.
(3)	25.4		Collapsed 48'.
(4)	25.2		Collapsed 66'.

*Very small flames remained at end of a few places.

It is readily seen from the above test results that the thiophosphate flameproofing compositions are more effective than the commercial compositions. The need for flameproofing compositions is clearly shown in the test results. Wood treated with wood preservative oil alone such as the creosote is completely consumed in the test. Employing the flameproofing compositions of this process, it is seen that a substantial amount of the flameproofer is retained by the treated wood, there is less weight lost in the wood and the flames burn themselves out and the fire is self-extinguished in a shorter period of time.

Example 2: In this example, the widely accepted panel fire test is employed to evaluate flameproofing compositions. In the tests, analogous flameproofing compositions employing the two different thiophosphorus compounds separately and in combination are carried out to show the effectiveness of the compositions. Douglas fir panels are made up of three pieces of 2 in. x 10 in. x 14 ft. long wood and two pieces of 2 in. x 6 in. x 14 ft. long. The panels are treated under the usual temperature and pressure conditions with 10% by weight of each flameproofing agent in a 50:50 mixture of creosote and residual fuel oil (topped naphthenic crude). The panels are allowed to weather for a period of at least 60 days. Following this, the panels are installed as part of the flooring in a mock-up railroad trestle section. The burning of the panels is accomplished by the ignition of tumbleweed stacked tightly under the trestle section. During the burning, temperatures between 1500° and 1600°F. are reached.

The following table shows the results of the tests. In Composition No. 1, the flameproofing agent added to the 50:50 creosote and residual fuel is 85% by weight alpha-pinene and phosphorus pentasulfide reaction product in combination with 15% by weight of medium boiling petroleum aromatic hydrocarbon extract. The agent of Composition No. 2 is 85% by weight zinc butyl hexyl dithiophosphate and 15% extract. Composition No. 3, illustrative of the preferred flameproofing compositions of the process, uses a typical concentrate consisting of 42.5% by weight alpha-pinene and phosphorus pentasulfide reaction product, 42.5% by weight zinc butyl hexyl dithiophosphate and 15% by weight aromatic hydrocarbon extract.

	No. 1	No. 2	No. 3
Retention, Pounds per cu. ft.	20.69	20.31	22.30
Wood Burning, Minutes	20	12	11
Weight Before Burning, lb	450½	436½	442½
Weight After Burning, lb	435	425	434
Weight Lost, Pounds	15½	11½	8½
Wind, m.p.h	3	3	3
Depth of Char, inches	⅛	⅛	⅛
No. of Panels Treated	2	2	2

ZINC CHLORIDE AND BORATES

Zinc Chloride, Boric Acid, Ammonium Sulfate

A process described by R.H. Bescher (U.S. Patent 2,545,222; March 13, 1951; assigned to Koppers Company, Inc.) provides wood which is preserved against destruction by organisms, is fire resistant and is waterproof and weather-resistant. The preferred fireproofing treating solution is composed of approximately 80% chromated zinc chloride, 10% boric acid and 10% ammonium sulfate. All of these compounds are readily soluble in water, the impregnating solution being made up of from 10 to 16% by weight of the compounds dissolved in 90 to 84% water. The impregnating solution is forced into the wood to be treated by inserting the wood in a treating chamber, filling the impregnating chamber around the wood with the treating solution, heating the solution to a temperature of 150° to 200°F., and placing the solution in the cylinder under a pressure of 100 to 250 lbs. per square inch. The wood is held under heat and pressure for a sufficient period of time so that the outer layers of the wood will contain three lbs. of solid salts per cubic foot of wood.

The impregnating solution defined above is quite resistant to weather and leaching but tests have shown that the wood will lose a considerable portion of the treating compounds when leached over an extended period by being immersed in water, either standing in the water or being treated by running water, or being treated in a plurality of successive portions of water. To avoid leaching and weathering of the preserved and fireproofed wood a fireproof coating is placed over the wood which is waterproof and weatherproof. The preferred coating for the impregnated wood is made up of 80% of coal tar sold on the market under the trade name "RT-12" and 20% of trichlorobenzene or 20% dichloronaphthalene. When these chlorinated hydrocarbons are dissolved in the coal tar, the coal tar is fire resistant and will not burn or give an afterglow.

Basic Zinc Chloride

E.E. Pershall (U.S. Patent 2,637,691; May 5, 1953; assigned to T.J. Moss Tie Company) has found that wood products can be made which are fire resistant (initially as well as after leaching), insect and fungi resistant, free from objectionable odor, paintable, and safe to handle by employing an ammoniated basic zinc chloride. The following examples illustrate the process.

Example 1: An aqueous solution of approximately 80% zinc chloride by weight was heated and zinc oxide was added in a proportion of approximately 5% of the weight of the zinc chloride present in the solution. The zinc oxide-zinc chloride-water solution was then cooled and a basic zinc chloride having a composition of about $1ZnO{:}1ZnCl_2{:}XH_2O$ separated from the solution as a precipitate. This precipitated salt was filtered from the solution and washed with a small amount of water, and then dissolved in aqueous ammonia of approximately 25% concentration, by weight. A treating solution was made by diluting the above ammoniated basic zinc chloride solution with water until the content thereof was approximately 16 to 18% of ammoniated salt in 12 to 15% aqueous ammonia solution. The composition of this soluble ammoniated basic zinc chloride salt on a dry basis (when dried at approximately 75°C.) was approximately $0.75ZnO{:}1ZnCl_2{:}2NH_3$.

Example 2: A sample wood stick (cut to the standard dimensions described and approved by the U.S. Department of Agriculture Forest Service, Forest Products Laboratories in Madison, Wisconsin, in the 1931 article "The Fireproofing of Wood," published in 1931 Proceedings of the Fire Protection Association by T.R. Truax, Senior Wood Technologist) was impregnated with the ammoniated basic zinc chloride solution of Example 1 by any conventional treating apparatus until the retention was approximately 7.7 pounds of ammoniated basic zinc chloride salt per cubic foot of the wood sample. The treated wood was highly fire resistant, paintable, free from objectionable odor, possessed of desirable mechanical and chemical properties and retained at a high degree of fire resistance after exposure to leaching. Such treated wood sample had initially a fire tube loss (as determined by the method described and approved by the Forest Products Laboratories in the 1931 article cited above) of approximately 17%, and after leaching the sample totally submerged in briskly running water (e.g., water is completely changed at least every 2 hours) for 15 days, approximately 20%. After 30 days leaching the fire tube loss was only approximately 22%.

Basic Zinc Salts Formed in Situ

H.A. Depew, P.D. Quinn and G.R. Waitkins (U.S. Patent 2,868,673; January 13, 1959; assigned to American Zinc, Lead & Smelting Company) describe a process by which a basic zinc salt is formed and deposited in place within the pores of porous material, and in and on the surface of fibers of fibrous materials.

The term "basic zinc salt" is used to describe a coprecipitate of zinc hydroxide with a normally water-soluble zinc salt. The formula for such a salt may be represented as $ZnAn \cdot nZn(OH)_2$, in which An is the anion of a water-soluble zinc salt and n is a number, the size of which indicates the basicity of the material. The basicity (n) is preferably in the neighborhood of 2 to 3, although it may, for certain applications, vary from one-quarter to six or seven or more. The pH of the basic zinc salt is less than eight, and in the preferred range of basicities, is about 6. The method is a two-step process, where the pores of wood are first impregnated with one reactant, and subsequently impregnated with a second reactant, in such a way as to cause, upon the introduction of the second reactant, an initial formation of zinc hydroxide and the subsequent coprecipitation of zinc hydroxide and a water-soluble zinc salt, as a water-insoluble basic zinc salt, in situ.

In treating either wood or fabric, the material to be treated may first be impregnated with a solution of water-soluble zinc salt. The solution may be either an aqueous solution or a solution of water-soluble zinc salt in an organic solvent such as alcohol. The excess solution is then preferably removed, and the material is subsequently impregnated with a basic material in an amount sufficient to react with the water-soluble zinc salt to produce zinc hydroxide, under such conditions as to permit the zinc hydroxide then to combine with the residual water-soluble zinc salt, to form the basic zinc salt.

The precipitated basic zinc salt in the case of fibrous material, is deposited in and on the surface of the fibers in such a way as to strongly resist dislodgment by washing or weathering. The basic zinc salt in the porous material such as wood, is deposited on the walls of the pores, in such a way that the deposit is extremely difficult to dislodge. Of the water-soluble zinc salts, zinc chloride is preferred. However, the sulfate or nitrate, bromide, iodide or fluoride may be substituted in whole or in part for the chloride. The following examples illustrate the process.

Example 1: Fire tube test sticks are placed in a pressure tank and subjected to an air pressure of 50 psig for fifteen minutes. Sufficient of a 33% water solution of zinc chloride (one part of zinc chloride and two parts of water, by weight) is introduced into the pressure tank while the pressure is maintained unchanged, to cover the sticks. Air pressure of approximately 125 psig is then applied to the solution within the tank for one hour. The pressure is then released and the unabsorbed solution withdrawn from the tank. A vacuum of about 22 inches gauge is applied for 5 to 10 minutes, and any additional free solution is withdrawn from the tank.

An aqueous solution of sodium hydroxide and ammonia, containing 4.7% NaOH, 7.8% NH_3 and the balance water, is then introduced into the pressure tank while the vacuum is maintained. An air pressure of approximately 125 psig is then applied to the basic solution for one hour. The pressure is then released and the unabsorbed solution removed from the cylinder. A vacuum of about 22 inches gauge is applied to remove from the pores any solution which is likely to cause bleeding. The entire process is carried out at room temperature. The results are indicated in the following table. The retention of zinc was measured as the difference between the amount of solution introduced into the pressure tank and the amount of unretained solution withdrawn. The actual retention was less, because the residual drawn-off $NaOH-NH_3$ solution contained 1% of zinc.

Retention, Lbs./Cu.Ft. as ZnO	Percent Fire Tube Losses After 15 Days' Leaching: 3 Minutes	Percent Fire Tube Losses After 15 Days' Leaching: ASTM Final
3.9	25	34

The ammonium chloride is preferably kept to a minimum and this has been accomplished, in part, by replacing part of the ammonia with NaOH. In large amounts, ammonium chloride can interfere with the paintability of the treated wood, cause warping of the wood, and dissolve zinc compounds during leaching. These objections are less applicable to sodium chloride, which is produced when sodium hydroxide is used. However, the use of sodium hydroxide increases the pH of the treating solution. A mixture of sodium and ammonium hydroxide is, therefore, preferred over a solution of only sodium hydroxide. The use of a strong organic base such as the ethanol amines or their equivalents in place of the sodium hydroxide is more expensive, but technically superior. A zinc sulfate solution can be used in place of the zinc chloride solution in the process of Example 1. To obtain a solution of $ZnSO_4 \cdot H_2O$ of the same concentration of zinc as the zinc chloride solution, it is necessary to raise the temperature to 50° to 60°C. to hold the zinc sulfate in solution. Otherwise, the process may be the same. When zinc sulfate is used in place of zinc chloride, the resultant wood has 10 to 15% greater fire loss, but the resultant wood is of better color and holds paint even better than the zinc chloride treated wood. Mixtures of half zinc chloride and half zinc sulfate give intermediate results.

Example 2: Test sticks are impregnated with ammoniacal basic zinc chloride to an equivalent ZnO retention of 1.5 lbs. ZnO per cubic foot of wood by the empty cell procedure as in Example 1, and then after partial drying, with a 33% zinc chloride solution by the full cell procedure as in Example 1. The sticks are leached 15 days and the excess of zinc chloride is removed. Part remains undissolved and combines with the basic zinc chloride. The table on the following page shows the amounts of zinc, calculated as ZnO, and of chlorine, retained by the wood, and also the fire tube loss, after 15 days' leaching. The mechanism of the treatment of this example is thought to be as follows: the zinc chloride reacts with the residual ammonia to form zinc hydroxide, which reacts with more zinc chloride to form basic zinc chloride. Residual zinc oxide, hydroxide, and very basic zinc chloride react with added zinc chloride to form basic zinc chloride with a basicity, for example, of one or two to one.

ZnO Lbs./Cu. Ft. Added As—		Total	Cl Total Lbs./Cu. Ft.	Percent Fire Tube Loss	
Ammoniacal Basic Zinc Chloride	Zinc Chloride			3 Min.	ASTM Final
1.8	1.0	2.8	0.4*	22	37

These results are excellent for only 2.8 lbs. ZnO per cubic foot.

In a continuation of this work, H.A. Depew and G.R. Waitkins (U.S. Patent 2,838,424; June 10, 1958; assigned to American Zinc, Lead & Smelting Company) describe a process in which the wood is impregnated with a single impregnant solution and in which the basic zinc salt is formed and deposited from the solution within the pores of the wood. In the process, wood is impregnated with a solution of a zinc salt and a potentially basic material, reaction between the zinc salt and the potentially basic material in the solution being restrained during the impregnation, the potentially basic material is subsequently converted to a basic material within the pores of the wood, and a basic zinc salt it precipitated from the solution in situ in the wood. The resulting deposit is uniformly distributed and is characterized by high density and hardness, tenacious adherence, and a pH of around 6.

When urea is the potentially basic material, wood may be impregnated with a solution of zinc chloride and urea at room temperature. Then the impregnated wood is heated to a temperature at which the urea hydrolyzed, either with the water in the impregnating solution, if an aqueous solution is used, or with moisture in the wood if the impregnating solution is not aqueous. A reaction then takes place within the pores of the wood which may be represented by the following equations:

$$NH_2{-}CO{-}NH_2 + H_2O \xrightarrow{\text{Heat}} 2NH_3 + CO_2$$

$$2NH_3 + ZnCl_2 + 2H_2O \longrightarrow Zn(OH)_2 + 2NH_4Cl$$

$$nZn(OH)_2 + ZnCl_2 \longrightarrow ZnCl_2{\cdot}nZn(OH)_2$$

When the potentially basic material is zinc nitrite, the wood is impregnated with a cold solution of unstable zinc nitrite and zinc chloride. The impregnated wood is heated, and basic zinc chloride is precipitated in the pores of the wood. The reactions may be represented as follows:

$$ZnNO_2 + H_2O \longrightarrow Zn(OH)_2 + NO$$

$$nZn(OH)_2 + ZnCl_2 \longrightarrow ZnCl_2{\cdot}nZn(OH)_2$$

The following are illustrative examples of the process.

Example 1: A water solution of zinc chloride and urea, analyzing 8.5% zinc, 9.4% chlorine, 15.5% urea and the remainder water, by weight, is prepared. The molecular ratio of zinc chloride to urea is 1 to 2. Fire tube test sticks are placed in a pressure tank. A vacuum of about 22 inches gauge is applied for five to ten minutes. Sufficient of the zinc chloride-urea solution is introduced to ensure that the wood is covered, while the vacuum is maintained. An air pressure of approximately 125 psig is then applied to the solution for one hour. The pressure is then released and the unabsorbed solution removed from the cylinder. A vacuum of about 22 inches gauge is then applied to remove from the pores any solution which is likely to cause bleeding. The impregnation is carried out at room temperature. The test sticks are then divided into two groups, A and B. The sticks of group A are heated at 80°C. at atmospheric pressure for seven days. The sticks of group B are heated to the same temperature for 19 days. The sticks in both groups are then leached in running water for 15 days. The leach water used has a pH range from 9.2 to 10.5. The results are indicated in the following table.

Group	Initial retention ZnO lbs./cu. ft. wood	Days leaching	Percent fire tube loss	
			3 min.	ASTM final
A	4.8	15	21	26
B	4.9	15	24	34

Example 2: A water solution of zinc chloride and urea analyzing 7.1% zinc, 8.1% chlorine, 9.5% urea and the remainder water (a mol ratio of zinc chloride to urea of 1 to 1 1/2) is prepared. Three groups of test sticks, A, B and C,

are impregnated with the solution by the full cell procedure as in Example 1. The sticks are dried for 4 days at about 35°C. and then heated at 80°C. for seven days. The sticks of group A are not leached at all, the sticks of group B are leached for 15 days, and the sticks of group C are leached for 30 days. The results are shown in the following table

Group	Initial retention lbs. ZnO/cu. ft. wood	Percent fire tube loss	
		3 Min.	ASTM final
A	6.2	10	12
B	6.2	23	27
C	6.0	25	31

Tests run under the same conditions as the test of Example 2 except that the test sticks are air dried from 0 to 14 days before heating at 80°C., indicate that preliminary drying is not a significant factor. It is to be noted from Examples 1 and 2 that the use of more than a 1 to 1 1/2 mol ratio of zinc chloride to urea, and long-continued heating of a zinc chloride-urea impregnated wood did not improve the results obtained. On the contrary, the results obtained with the use of more urea and longer heating are less desirable.

The following table illustrates the effect of long-continued heating upon wood treated with solutions containing two different molecular ratios of zinc chloride and urea. Test sticks are impregnated as in Example 1. The amounts of retained zinc and chlorine were then determined. Separate groups of sticks were heated for various periods as indicated in the following table. All of the sticks were leached for 14 days and the amounts of retained zinc and chlorine redetermined.

Molecular ratio: 1 $ZnCl_2$:1 urea

Percent of initial chemicals left in wood	Heating at 80° C.						Heating at 100° C.			
	0 days	7 days	14 days	21 days	28 days	35 days	0 days	7 days	14 days	21 days
Zn	0	42	52	59	61	61	0	71	69	68
Cl	0	14	19	24	21	15	0	6	6	6

Molecular ratio: 1 $ZnCl_2$:2 urea

Percent of initial chemicals left in wood	Heating at 80° C.				Heating at 100° C.			
	0 days	7 days	14 days	21 days	0 days	7 days	14 days	21 days
Zn	0	53	63	67	0	75	78	78
Cl	0	15	12	2	0	10	0	0

In a related process, described by P.D. Quinn (U.S. Patent 3,463,665; August 26, 1969; assigned to American Zinc Company) wood or fabric is impregnated with a zinc salt the anion of which is a complex fluoride, such, for example, as zinc silicofluoride, zinc phosphofluoride or zinc borofluoride, of which the preferred salt is zinc silicofluoride. In the preferred form, the wood or fabric is impregnated with a mixture of zinc sulfate and zinc silicofluoride, and the impregnants are insolubilized within the pores or interstices of the impregnated material by one of the methods described in U.S. Patents 2,838,424 and 2,868,673 above. In any case the concentration of zinc in the treating solution should be in the range of 5 to 10% by weight.

The zinc complex fluoride salts are water soluble, as is the zinc sulfate. For example, as much as 70 grams of zinc silicofluoride can be dissolved in 100 grams of warm water, so that the treating solution is easily made up. The treating composition is effective, in the case of wood shakes and shingles, up to a ratio of zinc sulfate to zinc silicofluoride of about 5:1. The preferred range is from 1:1 to 4:1 molar ratio of zinc sulfate to zinc silicofluoride. The use of the zinc sulfate not only reduces the cost of treatment, but, more importantly minimizes the adverse effect of locally excessive amounts of ammonia (in the form of gaseous ammonia, ammonium hydroxide or the decomposition product of urea) used in insolubilizing the salts. With a 2:1 molar ratio of zinc sulfate to zinc silicofluoride, a zinc content of the impregnated shingle of the two pounds per cubic foot of wood before leaching will produce leached shingles which will pass the Class C Burning Brand Test. The following example illustrates the process.

Example: An impregnating solution was made up as follows —

	Kilograms
(A) $ZnSiF_6 \cdot 6H_2O$	188
H_2O	318
(B) $ZnSO_4 \cdot H_2O$	108.2
H_2O	398

The zinc silicofluoride is dissolved in the water to form the solution (A). The zinc sulfate is dissolved in the water to form the solution (B). Each of the solutions contains 39 kilograms of zinc. The two solutions are then mixed to form a solution with a 1:1 molar ratio of zinc silicofluoride and zinc sulfate, and a 7.7% concentration, by weight, of zinc in the solution. Bundles of red cedar shingles were placed in a treating cylinder. The treating cylinder was filled with the impregnating solution through a pressure-measuring tank connected to the treating cylinder. Air pressure was then applied, through the pressure-measuring tank to the solution in the treating cylinder, at approximately 100 pounds per square inch gauge, to force the solution into the pores of the shingles. The amount of solution introduced to the treating cylinder is accurately measured through the measuring tank.

Approximately an hour after the pressure stabilized indicating that the shingles were impregnated, the solution remaining was blown back to a storage tank. A vacuum of 25" of mercury was applied for one half hour to "wipe" excess solution from the shingles. The shingles were permitted to stand overnight in the treating cylinder, at atmospheric pressure, to permit excess solution to drain out. The total amount of unabsorbed solution was measured, the difference between the amount introduced and the amount removed from the cylinder being a measure of the amount absorbed by the shingles.

In this particular experiment, the shingles were then removed from the treating cylinder, weighed, and permitted to dry, partially, in the air for one hour. They were then replaced in the cylinder. A vacuum of 25" of mercury was then applied for one-half hour, after which gaseous ammonia in the amount of 2 mols of ammonia to 1 mol of zinc, was introduced. In terms of weight, 0.78 kilograms of ammonia was introduced for each kilogram of zinc absorbed in the shingles. The shingles were permitted to remain in the test cylinder, with the ammonia, overnight, which ensured complete reaction of the ammonia with the zinc salts, so that little offensive ammoniacal odor was given off by the shingles. The shingles were then removed and permitted to air dry. They contained 2.02 pounds of zinc per cubic foot of wood. The shingles, in bundles arranged with every shingle spaced to permit the circulation of water, were then leached in running tap water for two weeks, dried, nailed on test roofs, and subjected to the Class C Burning Brand Test. This test is the same as the "fire test of roof covering" ASTM E-108. The leached shingles of this example pass the test.

For red cedar shingles in general, about 8 to 9% zinc in the treating solution is preferred. Six and a half percent has been found satisfactory for the treatment of high pickup shingles. A higher percentage of zinc than nine is quite operative but needlessly expensive. The partial drying step between the impregnation with the solution and the insolubilization of the impregnant with the ammonia, in the two-step process as described, has advantages in saving of material. When anhydrous ammonia is absorbed in water, the absorption is accompanied by a large expansion in volume. If free impregnating solution is present in the pores, this expansion tends to cause bleeding. By evaporating much of the water, all but a trace of the bleeding can be eliminated. For further examples, one may merely substitute the use of the impregnate solution of this Example, varied, if desired, between a 0 to 1 molar ratio of zinc sulfate to zinc silicofluoride and 5 to 1 ratio, or the zinc borofluoride or phosphofluoride substituted in the various examples given in U.S. Patents 2,838,424 and 2,868,673.

Ammonium Fluoroborates

D. Bauer (U.S. Patent 3,247,059; April 19, 1966; assigned to Farbenfabriken Bayer AG, Germany) describes the ammonium fluoroborates for wood protection. In comparison with common wood protection agents, the ammonium fluoroborates are characterized by an exceptionally high solubility in water. At 20°C., about 900 parts (referred to the solid starting materials) dissolve in 100 parts of water. This has the result that wood or wood-containing materials, such as wood shavings which are to be worked up to chip boards, take up so much wood protection agent by a single treatment with the wood protection agent, for example, by a single dipping or by painting or spraying, that they become difficultly inflammable. As a result of the fungicidal and insecticidal properties of the ammonium fluoroborates, the wood is also substantially protected against the attack by biological destructive agencies, such as fungi and insects.

As a rule, a good protection of the wood is obtained when about 50 to 250 grams, and especially about 100 to 200 grams, of the wood protective agent (referred to the solid content of the agent) is applied to square meter of wood or with an addition of about 3 to 10% by weight of wood protective agent, referred to absolutely dry wood, in the case of wood materials. The following examples illustrate the process.

Example 1: 570 grams of ammonium bifluoride are mixed with 1260 grams of boric acid, with the addition of 255 grams of a 25% ammonia solution, to give a solution which contains 925 parts by weight, referred to the solid starting materials, in 100 parts by weight of water. Rough sawn pine planks of 20 millimeters thickness are impregnated with the solution by painting once with an amount of about 170 grams/m.^{2}. After drying, the treated wood is difficulty inflammable and also protected against wood destroying fungi, such as *Poria vaporaris*, and against wood destroying insects, such as the larvae of the death watch beetle.

Example 2: To a mixture of 6 mols of boric acid and 3 mols of ethanol amine there are dropped under cooling 6 mols of hydrofluoric acid as a 40% aqueous solution. There is obtained a highly concentrated aqueous solution of the wood protection agent which is ready to be used for impregnating wood. For the production of laminated wooden plates, for instance, chip boards, the fluoroborate solution is mixed with an adhesive, for instance, an urea-formaldehyde resin, and applied to the chip board by known methods.

Trimethyl Borate

In a process described by G.A. Martin (U.S. Patent 3,342,629; September 19, 1967; assigned to Callery Chemical Company), wood which is partially moist, preferably near its fiber saturation point, is impregnated with a tri(lower) alkyl borate. Trimethyl borate is preferred because of its ready availability and low cost. Also satisfactory are the triethyl, tripropyl, and tributyl borates. The wood is treated by permitting the alkyl borate to soak into the wood to a suitable depth, preferably under pressure so that the penetration time is substantially lessened. The trialkyl borates are volatile liquids which react with water to form boric acid and the corresponding alcohol. As the trialkyl borate penetrates the wood during impregnation a substantial amount of the trialkyl borate reacts with the water in the wood to form boric acid. Other boron-containing compounds, which have not been fully characterized, are also formed in the wood.

There is also evidence that the boric acid itself reacts with wood fibers or constituents over a period of time. By these reactions the boron is transformed into a nonvolatile boron constituent in the wood imparting to it not only a fire retardancy but also an insecticidal and fungicidal property. Furthermore in treating the wood the reaction with the water by the trialkyl borate serves as a seasoning or dehydrating agent. Wood so treated exhibits a striking degree of dimensional stability under varying conditions of temperature and humidity. It is believed that the trialkyl borate reaction products effect a blocking of the hygroscopic centers inside the wood which are responsible for the attraction of water under conditions of high humidity.

Heavily treated wood should contain at least about 0.7 pound of boron per cubic foot of wood in the form of nonvolatile boron reaction products which is equivalent to 4 pounds of boric acid or 6 1/2 pounds of trimethyl borate reactant per cubic foot of wood. Superior results are obtained if about one pound of boron reaction product, measured as boron, is deposited per cubic foot of treated wood. Wood so treated exhibits a striking degree of fire retardancy. The process is illustrated by the following examples.

Example 1: A four foot length of green loblolly pine being 4 inches square in cross-section and containing 92% water was heated and dried under vacuum to a moisture content of 29.8% water prior to treatment. This wood was placed in a pressure chamber and after pulling a high vacuum to eliminate the air a mixture of trimethyl borate-methanol azeotrope was introduced and left to stand for 24 hours at a pressure of 25 psig. The treating solution was then removed and by calculation 4,730 milliliters of trimethyl borate was consumed by the treatment. A high vacuum was placed on the chamber and the vapors removed were condensed in a water cooled condenser and a cold trap for 70 hours. This treatment removes unreacted trimethyl borate and the methyl alcohol from the wood. The condensed material was measured and analyzed for boron content. Calculations showed that 1.15 pounds of boron per cubic foot of wood in the form of nonvolatile boron reaction products was deposited in the wood.

The wood was sanded on all sides to remove the outside 1/16 inch and the chips from a 1/2 inch hole bored through the center of the wood 14 inches from one end were analyzed for boric acid. It was ascertained that the boric acid content was 1 pound per cubic foot of wood (15% of the total boron) with the remaining boron material in the form of nonvolatile boron reaction products other than boric acid. The wood was cut into longitudinal boards and representative pieces were submitted to a combustion test in a fire tube tester in accordance with procedures adopted by the Forest Products Laboratory of the U.S. Department of Agriculture.

These pieces demonstrated excellent flame retardant properties including the strip found in the center of the treated piece of wood. Another representative piece was tested after standing in the open air for 18 days. Three other pieces were put in distilled water and left to stand for 5 days. Two were taken out and dried for two days and burned and the third was taken out and dried for 30 days and submitted to a flame test. In all cases these boards demonstrated excellent flame retardant properties. Analysis of the wood which was submitted to the distilled water leaching treatment showed that approximately 1/3 of the boron had been leached out yet there was no appreciable decrease in the fire retardant properties. The tests also indicated that standing time improved flameproofing.

Example 2: A piece of loblolly pine similar in size to that of Example 1 was dried to about 30% moisture under closely controlled conditions and then was treated with a solution of trimethyl borate azeotrope at 25 psig and 60°C. for 20 hours. A total of 4,800 milliliters of treating solution was absorbed over this period of time. Of this total 2,040 milliliters were absorbed in the first hour, 3,900 in the first 2 hours, 4,540 in the first 4 hours, and only 50 milliliters were absorbed in the last 10 hours. This provides a basis for establishing minimum treating time. After treatment this wood was dried at 65°C. and 26 inches of mercury for 24 hours to remove volatiles.

Chips from a 1/2 inch hole bored into the center of this wood were analyzed and found to contain 0.92 pound of boron per cubic foot of wood. This wood was cut into longitudinal sticks and was submitted to flame tests over a period of time. The wood that was burned the day after final drying possessed excellent flameproofing. The remaining wood was burned 27 days later and possessed even a better flameproofing than the wood burned the day after treatment. It appears that time improves the flame-retarding properties of the treated wood.

Silicate-Borate Treatment Followed by Carbon Dioxide Reaction

E.R. Du Fresne and D.L. Campbell (U.S. Patent 3,306,765; February 28, 1967; assigned to General Dynamics Corporation) has found that wood can be made fire resistant by impregnating it with a solution containing a soluble silicate and then treating the impregnated wood with sufficient carbon dioxide to form a silicon dioxide polymer. Treatment of the impregnated wood produces a reaction between the carbon dioxide and the silicate which results in the formation of an insoluble fire resistant substance which cannot be washed from the wood. It is believed that the carbon dioxide treatment forms a polymer of silicon dioxide of sufficient size and length that it cannot be easily washed from the interstices of the wood.

A solution of a crystalline silicate is preferred to a colloidal silicate , such as a water glass solution, which has a higher viscosity that makes impregnation difficult and impractical for certain types of wood. If an alkali metal silicate is used, the ratio of the silicon dioxide to alkali metal oxide in the silicate solution may vary somewhat, but it is preferably less than about 1.6 to 1. Also, aqueous solutions of crystalline alkali metal silicates are preferred. Examples of suitable solutions include those prepared from compounds such as sodium metasilicate (Na_2SiO_3) and sodium orthosilicate (Na_4SiO_4). A sodium metasilicate solution is most preferable because it is less subject to loss as a result of weathering after treatment.

Although impregnation with an alkali silicate solution, and subsequent treatment with carbon dioxide, imparts fire resistant properties to wood, preferably the impregnating solution contains a soluble borate along with the silicate. The combined silicate-borate solution produces a wood product having even more satisfactory fire resistance. It is believed that the borate combines with the silicate so that when the silicate is polymerized, the borate is at least partly bound up in the silicate polymer. The addition of the borate increases the resistance of the final product to glowing combustion, that is, when a product treated with the combined solution is removed from a source of flame, no glowing embers remain. Preferably, the borate is present as sodium tetraborate ($Na_2B_4O_7$), however, other soluble alkali metal tetraborates can be used, and other soluble borates, such as the metaborates, may also be used.

Impregnation so that the wood contains from about 1 pound to about 15 pounds of residual silicate-borate per cubic foot is considered satisfactory. Impregnation in the range of about 1 pound to about 4 pounds of silicate-borate per cubic foot of wood is preferable. Although the fire resistant properties of the wood naturally continue to increase with the further addition of impregnant, it is felt that impregnation to about 4 pounds per cubic foot provides maximum protection in relation to impregnating cost.

After impregnation, the wood is treated with sufficient gaseous carbon dioxide to react with the solution dispersed in the wood. Treatment of the wood with carbon dioxide may also be carried out by various suitable methods, but mere exposure to ordinary air is insufficient because the percentage of CO_2 is too small. Exposure of the impregnated wood in a substantially carbon dioxide atmosphere at atmospheric pressure for a period of about 8 hours is sufficient to cause the desired polymerization of silicate. Alternately, a few minutes exposure of the impregnated wood to carbon dioxide at a pressure of about 300 to about 800 psi produces sufficient polymerization. The following examples illustrate the process.

Example 1: A suitable aqueous solution of sodium silicate is prepared in a one molar concentration using sodium metasilicate. Pieces of red cedar having dimensions of 2" x 12" x 1/2" are immersed in the solution so that they are below the surface and completely covered by the solution. The temperature of the solution is raised to about 100°C. and held at this point. After about 5 hours under these conditions, the samples are removed from the solution. Two of the samples are set aside and two other samples are exposed to a carbon dioxide atmosphere for about 8 hours. One sample which was treated with CO_2 and one untreated sample are tested to determine the amount of silicate that is in the samples as a result of the impregnation step. The results are the same, about 5 pounds per cubic foot.

Both remaining samples are then washed in running water at a temperature at about 45°F. for 24 hours. At the end of this period, the two samples are removed and suitably dried. The dried samples are tested to determine the effect of the washing upon the samples of silicate which has been respectively lost from each sample. Testing shows that the carbon dioxide-treated sample lose only about 16% of its silicate, whereas the untreated sample loses about 27% of its silicate. The dried, treated sample is then exposed to a Bunsen flame, at a temperature of about 1500°C., for about 30 seconds. The sample chars but does not ignite. When removed from the flame, the sample does not support a flame, and a slight amount of glowing combustion immediately present dies out in a matter of seconds. A similar red cedar sample that has not been impregnated with silicate bursts into flame after a few seconds over the Bunsen burner. The treated sample is adjudged to have satisfactory fire resistance. The structural properties of the sample are checked against those of a similar piece of red cedar from which the sample was cut. No noticeable change in structural properties of the wood is detected.

Example 2: An impregnating solution is prepared from equal parts of a one molar solution of sodium metasilicate and a 0.5 molar solution of sodium tetraborate. Similar pieces of red cedar as used in Example 1 are impregnated by first subjecting them to vacuum conditions, a pressure of about 1 inch of mercury for about 2 minutes. The vacuum chamber is then filled with sufficient solution to completely cover the wood samples and the vacuum is broken. After about 10 minutes impregnation at atmospheric pressure, the samples are removed from the solution and drained. Two of the samples are set aside and two other samples exposed to carbon dioxide for about 5 minutes under a pressure of about 500 psi. The samples are tested, washed, and dried as set forth in Example 1. Final testing of the washed samples shows that an untreated sample loses about 27% by weight of the silicate with which it has been impregnated and about 31% of the borate. The samples which were treated with carbon dioxide only lose about 16% of the silicate and about 26% of the borate.

A dried, treated sample is then exposed to a Bunsen flame, at a temperature of about 1500°C., for about 30 seconds. The sample chars but does not ignite. When it is removed from the flame, no further combustion is apparent. The sample does not support a flame nor is there any glowing combustion. The treated sample is adjudged to have satisfactory fire resistance. The structural properties of the sample are checked and show no decrease in structural strength. The treated sample material is considered to be well suited for use in building construction.

Borate and Chlorinated Phenol Emulsions

S.M. Draganov (U.S. Patent 3,378,381; April 16, 1968; assigned to United States Borax & Chemical Corporation) describes stable emulsion compositions and a single-step method for preserving and fireproofing wood. The emulsion comprises a water phase containing from 5 to 40% by weight of an inorganic oxygen-bearing borate, and from 0.5 to 2% by weight of an alkaline lignin sulfonate as an emulsifying agent. The oil phase contains from 0.1 to 10% by weight of a chlorinated phenol, and from 0.5 to 25% by weight of a hydrocarbon wax. The weight ratio of the oil:water phases is in the range of from 1:1 to 2.5:1.

The emulsions of the process are of the oil-in-water type where the oil phase is dispersed in the water in a finely divided state. The oil phase must have an average particle diameter of not more than about 7 microns and, preferably, in the range of from 1 to about 3 microns. It has been found that only with such an oil-in-water emulsion is it possible to obtain a stable emulsion for impregnating timber in which there is an absolute minimum of "bleeding" when the timber is exposed to the weather. Additionally, with the oil-in-water emulsions it is possible to introduce into the wood relatively greater quantities of the inorganic borates giving the wood greater protection against fire. Also, it is possible to obtain a high degree of protection against rot and insects with lesser amounts of the chlorinated phenols.

In the impregnation of wood with the emulsions of this process, the wood and emulsion are placed in an autoclave or like pressure vessel and heated to a temperature of from about ambient to about 120°C. at an elevated pressure of from atmospheric pressure to about 200 psig. At a preferred pressure of from 125 to 175 psig and temperature in the range of 50° to 100°C., substantially all timbers can be impregnated with desirable amounts of ingredients in from 2 to 12 hours. The following examples illustrate the process.

Example 1: A saturated brine was prepared by dissolving 5.4 pounds of boric acid and 5.4 pounds of borax (sodium tetraborate decahydrate) in 3 gallons of water. The solution was placed in a colloid mill and 0.7 pound of "Marasperse C" (a calcium lignin sulfonate) was added. The mill was turned on and an oil solution consisting of 16 pounds of Chevron 110 (a petrolatum wax) and 4 pounds of pentachlorophenol dissolved in a mixture of 3.5 gallons of diesel oil and 2.0 gallons of kerosene, was slowly added at a slower rate. The emulsion was recycled through the mill until all of the oil solution was added. The stable, oil-in-water emulsion had an oil:water phase weight ratio of 1.75:1 in which the average particle diameter of the oil phase was about 2 microns. The emulsion and a block of seasoned red oak were placed in an autoclave. The autoclave was kept under a pressure of 150 psig at a temperature of 90°C. for a period of twelve hours. The block was removed and found to have an emulsion retention of 29 lbs./ft.^{3}. Deposited in the wood were:

	lbs./ft.^{3}
Pentachlorophenol	0.92
Wax	3.7
Borate	4.4

Example 2: A saturated brine was prepared as described in Example 1. An oil solution containing 0.6 pound of pentachlorophenol and 0.6 pound of paraffin wax dissolved in 5 gallons of light mineral spirits (sp. gr.= 0.784) was added at a slower rate than the aqueous phase as described above. The resultant stable oil-in-water emulsion had an oil:water phase ratio of 2:1 and oil phase particle diameter of 1 to 4 microns. The emulsion and a block of seasoned southern pine were placed in an autoclave and heated at 82°C. for 2 hours under a pressure of 175 psig. Deposited in the wood were:

	lbs./ft.^{3}
Borate	3.8
Pentachlorophenol	0.25
Wax	0.25

FIBERBOARD

IMPREGNATION, PULP TREATMENTS

Ammonium Sulfamate, Sodium Tetraborate Impregnation

H.C. O'Brien, Jr. (U.S. Patent 2,648,615; August 11, 1953) describes a two step treatment which comprises impregnating the board with a mixture of substantially neutral flameproofing substances and then coating or covering the surface of the board with a vapor impervious and flame-resisting sealing preparation. As an example of a satisfactory impregnant, one may use an aqueous solution of ammonium sulfamate and sodium tetraborate and a wetting agent and treat the wallboard so as to incorporate about one-eighth pound of each of these substances per square foot of one-half inch cellulosic wallboard. Impregnation was effected by soaking the board in a hot aqueous solution (125° to 150°F.) containing approximately 7 parts of water and 1.175 parts each of ammonium sulfamate and sodium tetraborate and a suitable wetting agent which reduced the soaking time for complete impregnation. The board was allowed to remain in the solution until saturated. It was then heated to evaporate the water. The total weight of the resulting dried board was increased about 25% of dry solids.

After the impregnated board was dried the surface was painted with a coating preparation having flameproofing properties as well as providing a vapor seal. The material which was used for the coating was an organic sol of pyrobitumen. This material is resistant to flame and may have incorporated within it a pigment, such as finely divided metallic aluminum, to produce a more attractive surface as well as to increase the resistance of the coating to flame. Wetting agents manufactured and sold by Carbide and Carbon Company under the names "Tergitol F G", which is a 25% aqueous solution of the sodium salt of 2-ethyl hexene 1-sulfonic acid, and "Tergitol #7", which is a 25% solution of the sodium salt of 3-9 diethyl di-decanol 6 sulfate, are satisfactory for the process.

The coating material, an organic sol of pyrobitumen, sold under the trade name of "Roskote Resins", by Royston Laboratories, Inc., contains a major proportion of ultra divided solid highly polymerized and nonreactive asphaltene particles which are fairly saturated hydrocarbon materials primarily napthanic in nature. The major portion of these colloidal particles are under 0.01 micron in diameter. This material is relatively free from inorganic contamination and from reactive organic constituents such as oil and low molecular weight fractions. Freedom from inorganic contaminants is especially important for the presence of such materials causes the coating film to be subject to attack by acids or alkalies.

Insulating board or wallboard when treated according to the process will char but will not burn with a flame when subjected to intense heat such as the flame of a blowtorch. It is not affected by the presence of moisture. There is no deterioration of the flameproofing properties of the impregnating materials or the coating material over long periods of time.

Boric Acid Impregnation Followed by Clay Coating

In a process described by L.R. Dunn and L.W. Eckert (U.S. Patent 2,875,044; February 24, 1959; assigned to Armstrong Cork Company) a cellulosic fiberboard product can be produced which is resistant to blooming at high humidities and which is adequately resistant to flaming to meet United States Department of Commerce commercial standard specification CS-42-49, class F. This is accomplished by treating the surface of the cellulosic fiberboard, preferably while still in a wet condition prior to drying, with a solution of boric acid so that the same penetrates into the body of the board below the surface. The product is then heated to react the boric acid with the cellulosic fibers, preferably to a temperature of 240°F. The dry product is then coated with a composition including clay or similar finely divided inorganic particles and a flame resistant binder. The felted fibrous body or board may be formed essentially from wood fibers. A wood fiberboard prepared as described in U.S. Patent 2,642,359 forms a suitable base or body.

The fiberboard may be formed on a conventional fiberboard making machine, such as an Oliver forming machine. In this type of machine, a slurry of fibers suspended in water is deposited upon a rotating screen, and a continuous felted fibrous sheet is formed from which a portion of the water is extracted. The formed sheet is then delivered to a press, such as a Downingtown press, and additional water is removed from the sheet, and the sheet is consolidated to the desired thickness in the press.

The boric acid treatment may be effected by applying a solution of boric acid to the formed sheet as delivered from the Downington press in a wet condition, prior to drying. The coating may be applied by a roll coater, operating in advance of a pool of the boric acid solution which has been applied to the surface of the wet sheet. In order to assist in the control of the penetration of the boric acid treating material into the wet board, a thickening agent such as sodium alginate or the like may be incorporated into the boric acid solution. In the preparation of a suitable treating material, the following ingredients are combined, all parts being given by weight:

Boric acid	8.00
Sodium alginate (Keltex sold by Kelco Co.)	0.25
Water	91.75

This provides an aqueous acid solution, the effective or active component of which consists essentially of boric acid. In the preparation of the treating solution, the water is heated to 140° F., and the powdered sodium alginate is stirred in for 5 to 15 minutes, until a uniform distribution of it in the water has been obtained. Thereafter, the powdered boric acid is added, and the mixture is stirred for 5 minutes. The solution is maintained at a temperature of 140° F. in order to insure that the boric acid will remain in solution. If the quantity of boric acid is substantially greater than 8%, then the temperature of the solution should be maintained at a higher level in order to insure complete solution of the boric acid in the water.

The boric acid solution in heated condition is delivered to the surface of the fiberboard sheet, and the speed of movement of the sheet through the machine, the viscosity of the solution, the size of the pool applied behind the roll, and other variable factors affect the quantity of boric acid solution deposited onto the fiberboard and penetrating into the body. Generally, a quantity equivalent to 4 to 20 grams of boric acid per square foot of fiberboard surface (on a dry weight basis) is adequate for the purposes of the process. A quantity equivalent to 10 to 15 grams per square foot is preferred. The application of the heated boric acid solution to the wet surface of the fiberboard (the board as delivered from the Downingtown press may have a surface temperature of 100° F.) results in immediate interspersion of the boric acid treating solution and the water contained in the upper portion of the board; a slight penetration of the boric acid solution into the body of the board occurs, the depth of penetration being controlled by the temperature of the solution, its viscosity, and other factors previously mentioned.

The surface treated board is then delivered to a conventional drier, and water is removed from the board and the boric acid reacts with the fibers in the area of penetration of the boric acid solution. In a typical drier installation, the first and middle zones may be heated to 450° to 350° F. (from entrance to exit of each zone), and the exit zone may be similarly heated to 370° to 260° F. The treated surface of the board should be at some stage in the process elevated to a temperature of at least 220° F. to insure an adequate reaction between the boric acid and the cellulosic fibers. After the board has been dried, it is delivered to a work zone where a clay or equivalent top coat is applied, preferably in two stages; first, a thin prime coat and following that a finish coat. A suitable top coat may be formulated as follows, all parts being given by weight:

Water	1248.0
Trisodium pyrophosphate	3.0
Clay	1000.0
Casein	75.0
Ammonia	6.0
Water	16.0
Pine oil	10.0

In the preparation of the coating composition, the trisodium pyrophosphate is dissolved in water and the clay is added. Mixing is continued until the clay is dispersed in the water. The trisodium pyrophosphate serves as a dispersing agent. The casein or other binder is added to the dispersion of clay and water, and mixing is continued. Following this, the ammonia, dissolved in the 16 parts of water, is added and mixing continued for a period of two hours. The pine oil which serves as a wetting and leveling agent for the paint may be added at any convenient time during the process, and preferably during the mixing in of the ammonia treated casein solution. Preferred practice is to apply the primer portion of the top coating by means of spray guns, followed by the application of heat and pressure to the surface by the use of heated ironing rolls. This tends to level the surface of the board by pressing down any projecting fibers which may extend above the surface. The heated rolls also elevate the temperature of the surface of the board, removing moisture from the prime coating. The ironing rolls may be heated by gas flames and have a surface temperature of 900° F. They serve to elevate the temperature of the surface of the board to 140° F. in a period of one second

The finish portion of the top coat may be of the same composition as the prime coat, and it may be applied by spray guns and then oven dried. Satisfactory results will be achieved if 4 g., dry weight, of the coating are applied per square foot of board surface in the prime coat and 10 g., dry weight, per square foot in the finish portion. The total quantity of clay or equivalent inorganic coating material applied over the boric acid treated layer will vary with the type of fiber in the board, the smoothness of the board texture, the size of the clay or other inorganic particles, and other variable factors. Preferably with a wood fiberboard such as discussed above a total of 12 to 25 g. per square foot on a dry basis will be used. For practical purposes, about 30 g. per square foot is a reasonable upper limit, and as little as 4 g. will give some improvement with a smooth finished fiberboard.

It has been found that the combination of the boric acid treated layer of fibers and the clay layer provides a flame-resistant product which will meet United States Department of Commerce commercial standard specification CS-42-49, effective November 15, 1949, and entitled "Structural Fiber Insulating Board", fourth edition, page 3, class F. The same product with the boric acid coating alone or the same product with a clay coating alone will not meet the specifications. Also, if the clay is combined with the boric acid solution and the two are applied as a single coating, the product does not meet the applicable specifications. In these respects, a dried fiberboard sheet treated with a boric acid solution, so as to secure adequate penetration behaves in the same fashion as the wet-end treated material.

Anhydrous Borax

E.A. Lauring (U.S. Patent 2,849,316; August 26, 1958; assigned to Minnesota and Ontario Paper Company) describes a practical method whereby integral treated fire resistant fiber composition board can be produced without substantial loss of the chemical or chemicals. The board produced by this method requires less fire resistant material to be added to the slurry and also results in improved fire resistant qualities.

The method requires the use of a chemical that initially has low solubility in water normally used in the makeup of a pulp slurry. Such material must be of a character that during drying an effective fire resistant chemical is produced in situ. Anhydrous borax is such a material and in its original or natural state it is a poor fire retarding agent. The natural mined "Rasorite" which is essentially anhydrous borax is the ideal material. In accordance with the process a chemical agent such as anhydrous borax is reduced to a size that at least 75% of the material will fall within the range of 150 to 250 mesh with substantially none of the material coarser than 100 mesh and substantially none of the material finer than 300 mesh. The particle size of the material is important in obtaining the desired results and in economical manufacture of internally heated fiberboard and will vary with individual machine setups since the time element involved between chemical addition and formation of the fiber product is dependent on the formation of the product before substantial solvation and hydration of the anhydrous borax occurs. Anhydrous borax which has been subjected to water and heat, as occurs in drying wet fiberboard, dissolves then crystallizes out as a solid which is a hydrated borax compound and this material is an excellent fire retarding agent.

In the process pulverized anhydrous borax is added to a pulp slurry just prior to the formation of the board or mat. It is preferred that it be added within a period of time not greater than three minutes before formation of the board or mat. The formed board containing the chemical agent is subjected to drying at the time it has 50% or more water content. The initial drying temperature should be within the range of around 325° to 400° F. so that the temperature of the wet board is at the boiling point, approximately 212° F. The temperature of drying may, after the initial application, be lowered to 200° to 350° F., the total drying time at these temperatures being two hours or more. Composition board produced in this manner should have a modulus of rupture of 500 in the machine direction formation.

For example, if a fiber composition board is to be formed as to provide approximately a one-half inch thick board 18 pounds per cubic foot density and the finished product is to meet class F of commercial standards as relates to flame resistance, anhydrous borax is reduced to the size described. As the particle size of the treating agent has a definite effect on the efficiency of the product, the pulverized anhydrous borax is fed to the pulp slurry just prior to the formation in an amount sufficient that the finished product will have 55 pounds per thousand square feet. The formed board is then subjected to various methods of reducing surplus moisture and then dried at a temperature within the range of 325° to 400° F. and a time not less than two hours. If the fiberboard with the integral treatment is to meet the requirements of class C of the Commercial Standards CS42-49 there must be incorporated anhydrous borax in the pulp slurry in an amount which will result in 65 pounds of material in the finished board, per thousand square feet and one-half inch in thickness, 18 pounds per cubic foot density.

Boric Acid, Ammonium Sulfate

R.S. Moffett (U.S. Patent 3,027,326; March 27, 1962; assigned to Insul Seal Products, Inc.) describes the treatment of shredded wood fiber material with ammonium sulfate, boric acid, potassium alum, a dye from the family of dyes consisting of the diamino derivatives of triphenyl methane, and water. In addition, a liquid aromatic agent is utilized to give the end product insulating material an acceptable aroma. These constituents are then chemically combined. After chemical processing, the insulation material is dried by being fluidized by air and is then packed into appropriate containers. A moisture absorption test of the material shows the following typical characteristics:

Time of Exposure, Hours	Percent Loss in Weight, 50% Relative Humidity	Percent Gain in Weight, 75% Relative Humidity	Percent Gain in Weight, 100% Relative Humidity
50	1.15	1.81	5.90
100	1.55	2.49	10.80
200	1.78	3.37	32.5

The fire resistant quality of the insulation material is illustrated by the following test. After being maintained at a constant temperature of 135°F. for 39 days, the insulation material is subjected directly to a flame from a Bunsen burner for a period of fifteen minutes. After removal of the flame, smoldering of the insulation material is visible for no longer than two and one-half minutes. The maximum penetration of charring into the insulation material is five inches. The dielectric strength of the insulation material is illustrated by the fact that no appreciable difference exists between the dielectric strength of air and the dielectric strength of the end product insulation material at relative humidities less than 50%. At relative humidities in the range of 75 to 100%, the dielectric strength of the insulation material is approximately two-thirds that of free air.

The treatment of the wood fiber fluff formed from used newsprint with boric acid and ammonium sulfate does not remove ink and other impurities from the fluff which existed in the newsprint. These impurities have been found to be attractive to rodents and roaches of various types. It has been found that the addition of potassium alum, $Al_2(SO_4)_3K_2SO_4 \cdot 24H_2O$, to the fluff renders the end product unattractive to such rodents and roaches. The use of used newsprint, even when treated to the extent previously described, results in a gray material which lacks any attractiveness in its color. A dye is therefore added to the material in order to give the end product insulating material a satisfactory color appearance for commercial use. A dye from the family of dyes consisting of diamino derivatives of triphenyl methane also adds fire retardant qualities to the insulating material. An example of such a dye is Brilliant Green, which has the formula shown below.

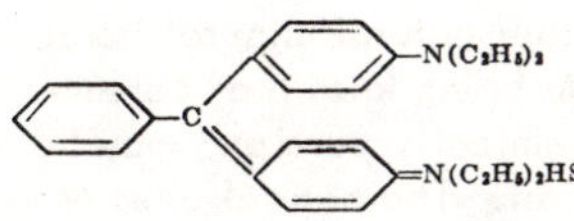

The preferred insulating material is produced by a process utilizing the previously named constituents in the following proportions by weight:

	Parts
Shredded wood fiber as used newsprint	68
Ammonium sulfate	4
Boric acid (powdered 99 1/2% pure)	6.5
Potassium alum	4
Brilliant Green dye	4
Alamask CPM	2
Water (liquid form)	8

thereby giving a total of 96.5 parts. The remaining 3.5 parts required to constitute 100 parts of the finished product consist of water taken in during processing in the form of vapor from the air.

Hydrated Borates

A process described by O.R. Videen (U.S. Patent 3,202,570; August 24, 1965; assigned to Wood Conversion Company) involves the manufacture of fire retardant panel material containing combustible vegetable fiber, such as the fiberboard useful in walls and ceilings, and in particular, to fiberboard having sound-absorbing porosity. It has been a practice to render such panel material fire retardant at the exposed surface by the use of special coatings or treatments of the surface layers of the panel. With the advent of punching or drilling such board material for acoustical uses, the surface treatments are insufficient for adequate fire retardancy.

Such panel board is commonly made by dewatering an aqueous slurry of fibers. To incorporate fire retardant additive well distributed throughout panel material, it must be present in the slurry and it must resist to a high degree a tendency to be carried away from the fibers by the water being removed from the fibers. To this end, it must be relatively insoluble in the slurry before and during the dewatering, and it must have a particle size distribution and surface properties favoring its retention.

According to this process, certain hydrated borate compositions having the desired properties for adequate retention in forming a wet mat to be dried have been found to impart fire retardancy when distributed within a fiberboard, and as a group they have an adequate content of water of constitution retainable in drying the wet mat, but releasable when the dried board is exposed to the heat of a flame. The borates involved are so poorly soluble in water that in a process incorporating them into a fiber slurry to form a wet mat for drying to a board, the time and temperature of subjection to water are ineffective to dissolve any appreciable portion of the borate present. However, the process, as described, of dewatering the slurry on a screen, as in a Fourdrinier machine, removes much more of the borate particles than is lost by dissolution.

Among the borates described is the preferred one known as Gerstley borate, a natural mineral. Its resistance to loss of water of constitution is representative of the properties of other borates. A specimen of Gerstley borate was heated for successive periods of time at increasing temperatures, with loss of weight determined for each period. As tabulated later, the mineral may have from 15 to 28% water and organics.

Period	Temperature, °F.	Time, hours	Equilibrium loss, percent
No. 1	210	24	6.4
No. 2	275	90	5.4
No. 3	325	13	5.6
Total			17.4

The loss in period No. 1 is no doubt free moisture of the natural mineral. In drying to form board, the temperature and exposure time are never so great as encountered in period No. 2, and the time is never so great as in period No. 1, although the entire mat reaches a temperature of at least 210°F., with the surface layers only attaining a higher temperature. The kind of vegetable fiber, the density of the board, the content of borate, the kind of borate, and the standard of fire retardancy, are all factors controlling that content of a particular borate which just meets a standard test. Consequently, it is not possible to prescribe a range of content or proportion of borate to fiber. The amount to be retained in a board is to be determined by empirical testing.

Gerstley borate is a natural borate salt. There are other natural mineral borates having a suitably high content of water of constitution for use as a fire retardant, two of these being known as colemanite and ulexite. The natural borates are preferred for economic reasons. They need not be chemically pure and may be used in association with minor quantities of naturally associated minerals of different composition. The natural ones predominate in borate-forms of CaO, but not to the exclusion of other metal oxides as illustrated by ulexite having the formula:

$$NaCaB_5O_9 \cdot 8H_2O \quad \text{or} \quad Na_2O \cdot 2CaO \cdot 5B_2O_3 \cdot 16H_2O$$

The following table shows known forms as identified in the literature and described as poorly soluble, some being natural and some synthetic:

	Name	Composition	Mols of H_2O per 30 mols B_2O_3
A	Monoborates	$CaO \cdot B_2O_3 \cdot 6H_2O$	180
		$CaO \cdot B_2O_3 \cdot 4H_2O$	120
		$CaO \cdot B_2O_3 \cdot 2H_2O$	60
B	Tetraborate	$CaO \cdot 2B_2O_3 \cdot 5H_2O$	75
		$CaO \cdot 2B_2O_3 \cdot 12H_2O$	180
C	Triborates	$CaO \cdot 3B_2O_3 \cdot 8H_2O$	80
		$CaO \cdot 3B_2O_3 \cdot 4H_2O$	40
D	Colemanite Series	$2CaO \cdot 3B_2O_3 \cdot 9H_2O$	90
		$2CaO \cdot 3B_2O_3 \cdot 7H_2O$	70
		$2CaO \cdot 3B_2O_3 \cdot 5H_2O$	[1] 50
E	Pentaborate	$3CaO \cdot 5B_2O_3 \cdot 9H_2O$	54
F	Decaborate	$4CaO \cdot 5B_2O_3 \cdot 9H_2O$	[2] 54

[1] (Colemanite).
[2] (Pandermite).

The naturally occurring Gerstley borate is preferred. It is a mixture of the colemanite which has $5H_2O$ in the table shown, and of ulexite Two typical analytical compositions of borates are shown below.

	Colemanite, percent	Gerstley borate, percent
B_2O_3	32.6	33.0
CaO	33.8	15.9
Na_2O	.5	4.9
Water and organics	15.2	28.2
SiO_2	3.0	9.5
MgO	1.2	3.5
Al_2O_3	.9	1.1
Fe_2O_3	.2	.3
CO_2	12.4	3.2
Undetermined	.2	0.4
Solubility in water as percent B_2O_3:		
15 Minutes Agitation:		
60° F	----------	0.04
70° F	0.06	----------
82° F	0.12	----------
180° F	----------	0.4

Example: A slurry is formed containing the following materials:

	Parts by weight
Board-forming wood fibers	100
Gerstley borate	23
Polyacrylamide flocculating agent	0.02
Ferric sulfate flocculating agent	1.0
Water	4450

The retention is slightly better using the flocculating agent than without it, so its use is optional. The borate mineral is added after dispersing the fibers, and then the flocculating agents are added either together, or first adding the ferric sulfate and then the polyacrylamide. The ferric sulfate may be replaced by other coagulants, soluble polyvalent metal compounds or salts being suitable, such as alum, calcium chloride, lime, and insoluble ones, such as bentonite. However, these are not 100% efficient in flocculating the mineral particles onto the fibers, so that a more efficient flocculating agent is used, such as polyacrylamide. The latter may be used as the sole flocculating agent, but being more costly than the inorganic ones, it is used to flocculate at least some of the particles which are not deposited by the inorganic flocculating agent.

The slurry is dewatered on a Fourdrinier machine to a mat which when dried yields a porous fiberboard of 17 to 20 pounds per cu. ft., and of conventional wallboard thickness, such as one-half inch, in which approximately 87.5% or more of the borate is retained. For uses of the board which are not decorative, the dried board is fire retardant to a useful degree. For decorative uses, the board may be coated with a decorative coat, such as Composition A, given below. This may be applied to the wet mat before drying or to the dried board. When a highly fire retardant coat is desired, intumescent Composition B is employed. For economic and other reasons, it is preferred to apply to the wet mat Composition A in the amount of 20 to 25 pounds per M sq. ft., and then to apply to the dried coated face Composition B in the amount of 10 to 20 pounds per M sq. ft. For acoustic purposes, the coated board may be drilled or punched to form acoustic holes. The holes preferably approach but do not meet the opposite face, not only for acoustical purposes, but also to avoid channelling hot gases of a flame through the board where it may be unbacked.

Composition A

	Parts by weight
Vegetable protein binder	6
Pigments:	
Titanium calcium pigment ($30\%\ TiO_2 - 70\%\ CaSO_4$)	36
Clay	33
Diatomaceous earth	5
Whiting	5
Lithopone	15

Composition B

Vinylidene chloride copolymer	15
Mono-ammonium phosphate	45
Pentaerythritol	16
Dicyandiamide	11
Titanium dioxide	13

The effect of the borate in an uncoated board is evident from the following comparative tests. A conventional aqueous slurry of wood fiber for forming a board has been used at a fiber concentration of 1.75 to 2.25 parts per 100 of slurry. Particles of the borate mineral are incorporated at recorded proportions relative to fibers. The slurry is dewatered by a standardized procedure comparable to the action of a Fourdrinier machine, to a wet mat and dried to a board of 17 pounds per cu. ft., and 1/2 inch thick. Test strips 1 x 8 inches are then cut.

The bottom end of the 1 x 8 inch strip suspended vertically is ignited with a Tirrill gas burner. The burner flame is adjusted to a height of four inches above the tip. The tip is then placed exactly three inches below the sample strip for a 30 second flame exposure on the sample. The burner is then removed and the duration of flaming recorded. The loss of weight by combustion is determined and the length of char height measured. The percent of the added borate mineral retained in forming the board is also determined. The following table gives these results:

Additive	Am't. per 100 fiber	Amount retained	Percent loss by burning	Duration of flame in seconds	Length of char
Ulexite (100-200 mesh) (No flocculating agent)	10	5.2	31.8	90+	6.2
	20	12.4	21.1	90+	3.5
	30	17.4	17.7	50	2.4
	40	24.3	13.1	20	1.5
	50	32.6	10.5	14	1
	100	55	9.5	5	1.4
Gerstley borate (100-200 mesh)	[1] 25	18.1	20.6	90+	3.5
	[2] 25	18.1	18.9	90+	3.5

[1] No flocculating agent.
[2] With flocculating agent.

Disodium Octaborate

H.A. Chase (U.S. Patent 3,438,847; April 15, 1969; assigned to Weyerhaeuser Company) describes the treatment of composite boards with borate chemicals. Boric acid and borate salts have been applied to the composite boards in a number of different ways. One method that has been practiced is to impregnate the outer surface of the composite board with a requisite combination of boric acid and borate salts by a pressure medium as is done in the impregnation of lumber and plywood. This method of impregnation has not been effective to achieve a fire retardant composite board since often only the outer surfaces are impregnated with the fire retardant chemical and the board does not measure up to fire retardant properties. In addition, redrying of the finished board is required, causing a loss in appearance and reduction in physical properties. Another method of adding the boric acid and borate salt combination to the composite board has been to mix the requisite combination of boric acid and borate salts with the adhesive and other materials that bind the wood particles together, and then press the same into its final form.

However, the composite board binding adhesives are intolerant to borate salt which limits the amount that can be used. By using only a small amount of the borate salts, some fire retardancy of the board can be achieved. The amount of chemicals that can be tolerated by the adhesive falls short of that required for gaining the ultimate fire retardancy that can be obtained with the optimum combination system. Accordinglyly, if an increased amount of borate salt is added to the adhesive to be applied to the wood particles, the finished compressed panel composite board falls far short of the requisite properties as to physical strength and bonding characteristics of the adhesive since the presence of the borate salts interferes with the bonding properties of the particular adhesives used.

A process has been found for forming composite boards using a combination of an acid material and borate salts during the manufacture of the boards which does not interfere with the bonding properties of the adhesive and has physical properties equivalent to an untreated board. In the process comminuted wood is formed by conventional means such as a hammer mill or flaker. A borate salt is then added to the comminuted wood furnish selected from a group consisting of sodium borate, potassium borate, ammonium borate, magnesium borate, preferably sodium borate. The amount of borate salt that may be added to the comminuted wood furnish may range from 3 to 25 parts by weight. The borate salt is added in solution form wherein the borate salt is dissolved in warm water ranging in temperature from 50° to 200°F. and is thoroughly mixed within the wood furnish.

After the borate salt is added to the wood furnish the borate salt is allowed to diffuse into the particles of the wood furnish. This diffusion is normally very rapid, being essentially complete in several seconds. In this manner, the sodium borate salt is buried within the wood furnish by diffusing into the wood cellular structure and becomes dissolved in the moisture within the wood itself. After the borate salt is completely diffused within the wood furnish, the wood furnish is then dried to from 2 to 9% moisture content. The wood furnish is next treated with an acidic material selected from the group consisting of boric acid, ammonium phosphate, ammonium sulfate, zinc chloride, or mixtures, and a suitable binder in a conventional manner. Boric acid is preferred. The amount of boric acid used to be mixed with the binder may range from 3 to 15 parts by weight. The particular binders usable in the process are the aminoplasts such as urea formaldehyde or melamine formaldehyde. The amount of binder used may range from 3 to 15 parts by weight. The wood furnish is now processed through a conventional mat forming apparatus and laid down in mat form and conveyed to heated platens to be pressed and cured at the desired pressure and temperature.

It has been found that the above procedure in forming the composite board structure provides a unique panel which is highly effective in fire retardant properties and also has excellent physical properties since the bonding properties of the adhesives used are not adversely affected by the borate salts. The following examples illustrate the process:

Example 1: A fire retardant panel was formed by mixing 100 parts of comminuted wood at a moisture content of 35% and 10 parts by weight of disodium octaborate in a 20% water solution. The comminuted wood was allowed to diffuse for one hour and then dried to 6% moisture content. A mixture of 10 parts granulated boric acid, 10 parts, solid basis, of 65% urea resin and 1/2 part, solid basis, 40% wax emulsion, and 2 parts water was sprayed on the wood particles in a blender and then followed by forming and hot pressing to a density of 47.5 lbs./cu. ft. The above board was tested and found to have excellent flame spread rating and high physical strength properties.

Example 2: A three layer board was prepared as follows: 100 parts of dry basis green comminuted wood was formed having 40% moisture content. Ten parts disodium octaborate in 20% water solution was sprayed onto the comminuted wood furnish and immediately dried. Next, 10 parts, dry wood basis, of 65% solids urea resin was mixed with 10 parts powdered boric acid and 1/2 part wax emulsion with 7 parts of water. This solution was spray mixed with the treated comminuted wood. The above furnish was then divided into two equal parts. An additional furnish was formed for the core portion of the three layer board. This consisted of 100 parts of dry wood particles having mixed therewith 7 parts of urea resin mixed with 6.3 parts of boric acid and 0.7 part sodium octaborate with 1/2 part wax emulsion and 1 part water for dilution. The above mixture was sprayed on the dried particles. Using 1 part surface to 2 parts core, the above prepared furnishes were formed into a panel with the surfaces sandwiching the core material therebetween. The panels were hot pressed to a density of 45 lbs./cu. ft. The strength and moisture properties were excellent as were the fire retardant properties.

Example 3: A water solution of disodium octaborate was sprayed on green wood particles to a level of 25% based on the bone dry wood content present. The wood particles were then dried. The wood particles were then sprayed with a mixture of boric acid containing 15% by weight based on the bone dry content of the wood and a urea formaldehyde resin containing 10% by weight based on the weight of the bone dry wood content. The wood particles were formed into a mat and pressed by standard methods. The board showed high strength characteristics and excellent fire retardant properties. Boards formed in accordance with this process were comparatively tested with known fire retardant boards. All the boards were tested according to ASTM E-162 standards to indicate the flame spread index. The results of these tests are shown in the following table:

Sample	Chemicals added	Average flame spread index	Type of board	Smoke deposited, (mg.)
1	Boric acid, 16% by wt.	18.2	Single-layer	
2	Boric acid, 25% in surface, 21% in core.	10.5	Three-layer	
3	Boric acid and sodium borate as in Example 1	3.0	Single-layer	
4	Boric acid and sodium borate as in Example 2	4.4	Three-layer	0.1
5	Mono ammonium phosphate, 18% in surface, 16% in core.		do	1.8

Urea Resin Binder and Aluminum Hydrate

H.L. Becher (U.S. Patent 2,611,694; September 24, 1952; assigned to Homasote Company, Incorporated) describes the use of an amine type resin binder with a chlorinated resin and aluminum hydrate. A typical product according to the process has the following composition:

	Percent
Fiber (news fiber)	48
Chlorinated paraffin (70% chlorine)	5.5
Urea resin	30
Aluminum hydrate	16

When produced as described, the product is in the form of a felted fibrous sheet which ceases to burn or glow within 30 seconds after removal of the burner in carrying out the "crib test". The product has a specific gravity between 0.6 and 0.7 and it may be sanded or tooled to present a very attractive finish. Its tensile strength and modulus of rupture are high and the sheet possesses excellent nail holding properties and an impact resistance exceeding that of cement asbestos sheets of like thickness.

In the preferred form of producing the sheets, the chlorinated resin and metal hydrate are mixed with the fibrous pulp. The material is then wet pressed until it contains not over 60% of water. The material is then redispersed in a solution containing the amine resin forming constituents and is then sheeted, pressed and dried. In this way, the fire resisting binder is distributed substantially uniformly throughout the sheet and a strong fire resistant product is produced. The following example illustrates the process.

Example: 72 pounds of paper such as newsprint are mixed with 3,600 pounds of water in a pulper and an emulsion containing 9 pounds of chlorinated paraffin (70% chlorine) are added together with 24 pounds of aluminum hydrate. The pH value of the resulting mix is between 5 and 6 and the solids content is 2.8%. After thorough pulping, the mixture is passed to a thickener and to a dehydrator or press where the material is subjected to a pressure of 210 pounds per square inch or until it contains 50% by weight of water. The liquid extracted is returned to the pulper to recover solids removed in the thickening and pressing operation.

The wet pulp from the pressing operation is passed to a mixer where it is disintegrated and redispersed in 2,895 lbs. of a solution containing the resin forming chemicals so as to produce a batch of 3,000 lbs. of mix containing 3 1/2% of solids. The solution is produced by adding 500 lbs. of dimethylolurea and 450 lbs. of urea to 1,945 lbs. of water. The mixture is agitated thoroughly and passed to sheet forming and pressing equipment where the product is subjected to a pressure of 50 lbs. per square inch. The product weighs 288 lbs. and contains approximately 50% by weight of water, whereas approximately 2,700 lbs. of the liquid pressed out of the sheet is returned for use in the treatment of an additional batch of material. The pressed sheet is then passed to a drier where it is subjected to temperatures about 212°F. until the dry weight of the finished product is 150 lbs. During the drying operation the resin forming constituents added after redispersion of the material are caused to polymerize forming a urea resin which is distributed with the chlorinated paraffin and hydrated alumina throughout the sheet.

The product when subjected to the "crib test" ceases to flame or glow in less than 30 seconds with a loss in weight of less than 30%. The product has a density of 0.6, a tensile strength of 1,000 to 1,300 lbs. per square inch, a modulus of rupture of 1,800 to 1,825 lbs. per square inch and a water absorption of from 12 to 20%.

Dicyandiamide-Formaldehyde-Metaphosphoric Acid Condensation Products

A process described by K.W. Müller and D. Deifs (U.S. Patent 2,872,355; February 3, 1959; assigned to Farbenfabriken Bayer Aktiengesselschaft, Germany) concerns salts of phosphoric acids, which have a smaller content of water than orthophosphoric acid, and basic condensation products formed by contacting aldehydes and nitrogen compounds having at least once the grouping

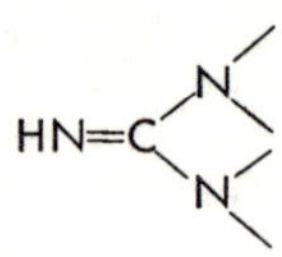

in an acyclic configuration. The salts of the process can be obtained by any suitable method, for instance by reacting the acids and the condensation products or their water soluble salts in aqueous medium and recovering the precipitated reaction products.

The salts of the process are useful products for imparting flame resistance to materials of all kinds, for instance fiber base material such as wood, fiber boards, paper and textiles. In the flameproofing treatment the salts may be applied in various manners, for instance in the form of a dispersion in water or organic liquids as well as in the form of a fine powder. It is also possible to produce the salts in situ, that means in the material to be rendered flameproof. In this case the materials are impregnated successively, in either order, with a solution containing the acid component and with a solution containing the basic component.

The application of basic condensation products containing alkylol groups gives especially good results. By heating materials impregnated with the phosphorus salts of these condensation products a further condensation takes place which leads to improvements of the mechanical and water resistant properties of the materials. By reacting wood pulp, as used in the manufacture of fiber boards, in the beater with pyro- or metaphosphates of the basic condensation products, containing free alkylol groups such as methylol groups, more of the customary sizing agents may be dispensed with to a more or less degree as the plates obtained by the usual method after pressing at elevated temperatures are not only flame resistant but also extremely stable against mechanical influences. The following examples illustrates the process.

Example 1: 84 parts of dicyandiamide are suspended with stirring in 160 parts of formaldehyde (30%), the suspension is transformed to a clear solution by heating to 90°C. and 47 parts of concentrated hydrochloric acid are dropped into the solution within 30 minutes. The temperature, which rises quickly, is prevented from exceeding 100°C. by cooling the reaction mixture. After the reaction is complete, the mixture is kept at 100°C. for a further 10 minutes and then cooled to room temperature. A clear viscous solution is obtained which, if desired, can be diluted with water with the addition of some hydrochloric acid. A solution of 58 parts of metaphosphoric acid in 200 parts of water is added to the aqueous solution of the dicyandiamide-formaldehyde condensation product. The precipitating metaphosphate of the condensation product is filtered off and dried. The yield amounts to 150 parts. By carefully adjusting the pH of the solution to 2.5 with ammonia the yield is increased to 170 parts.

Example 2: The viscous solution of the dicyandiamide-formaldehyde condensation product of Example 1 is mixed with 89 parts of pyrophosphoric acid in 200 parts of methanol; 185 parts of the pyrophosphate of the dicyandiamide-formaldehyde condensation product, which contains 12.1% of phosphorus, are then obtained.

Example 3: 84 parts of dicyandiamide are dissolved in 120 parts of hot formaldehyde (30%) and mixed with 88 parts of concentrated hydrochloric acid within 30 minutes, the temperature being kept at 100°C. After the reaction is complete, the mixture is kept at 100°C. for a further 10 minutes and it is cooled to room temperature and stirred into a solution of 82 parts of sodium metaphosphate in 1,500 parts of water. The precipitated metaphosphate of the dicyandiamide-formaldehyde condensation product is filtered off and dried. 180 parts of the product, which contains 11.75% phosphorus, are obtained.

Example 4: A pulp as it is used in the production of fiberboards, which consists of 20 parts of fibrous material in 800 parts of water, is mixed in a beater with 1 part of the finely powdered metaphosphate of the dicyandiamide-formaldehyde condensation product prepared as described in Example 1. The pulp is then worked up to fiberboards in the usual process during which the material is heated to a temperature of 150°C. for some time and pressed at this temperature. The difficultly soluble metaphosphate taken up by the fibrous material melts at this temperature, is absorbed by the fibers and condenses completely. The fiber plates obtained are highly resistant to flames and possess excellent mechanical properties. If two or three times the amount of the metaphosphate is added to the pulp, the fiber plates obtained are completely resistant to flames.

Example 5: A pulp of 10 parts of wood or cellulose pulp in 500 parts of water is mixed with 1 part of the metaphosphate of the dicyandiamide-formaldehyde condensation product described in Example 3 and worked up in the usual manner. The paper boards and paper produced from this material are flame resistant and possess a high stability to mechanical stresses.

Magnesium Oxide, Aluminum Sulfate

In a process described by W.S. Shisko (U.S. Patent 3,220,918; November 30, 1965; assigned to Domtar Limited, Canada) an incombustible molded fibrous board is produced containing principally cellulosic fibers, a certain amount of mineral fibers and, as other major ingredients, magnesium oxide and aluminum sulfate. While acceptable incombustible board meeting the requirements of United States Federal Specification SS-A-118b can be produced in accordance with this process without aluminum sulfate, it is necessary for best results to include aluminum sulfate. In a preferred example, cellulosic fibers constitute 50 to 65% of the total fiber content and mineral fibers 35 to 50%.

The mineral fibers may be glass fibers, rock wool, asbestos, etc. Another necessary ingredient of the board is magnesium oxide. It has been found that a quantity of between 40 and 100% by weight of magnesium oxide (based on total weight of fibers) is required to render the board incombustible. However, the quantity of magnesium oxide will vary within wider limits depending on the type of fiber, the desired physical characteristics of the board and other factors, but particularly the ratio of cellulosic fibers to total fiber content, a higher ratio of cellulosic fibers generally necessitating the addition of larger amounts of magnesium oxide.

In the process, very much larger quantities of aluminum sulfate than is customary are used, the aluminum sulfate apparently entering into a reaction with the magnesium oxide and this reaction product being incorporated with the fibers into the board along with unreacted magnesium oxide and unreacted aluminum sulfate. Quantities of aluminum sulfate, between 50 and 150% of the weight of total fibers, are added to the fiber suspension.

The process is carried out as follows: a slurry of mixed cellulosic and mineral fibers is prepared according to known methods, e.g., two separate slurries may be prepared and blended or dry mineral fiber may be added directly to a prepared wood fiber slurry. It is often necessary to incorporate small quantities of a binder such as starch with the mineral fibers. To this slurry aluminum sulfate is added either in solution or in solid form. The fiber slurry with the dissolved aluminum sulfate is thoroughly mixed by usual means of agitation and the magnesium oxide is added either in suspension or in solid form. The final slurry containing mixed mineral and cellulosic fibers, aluminum sulfate and magnesium oxide is made up to the usual consistency of board stock and fed to the machine, on which it is drained and formed into a mat. The mat is then pressed, dried and cut or shaped. It is a distinct advantage of the process that it can be carried out using conventional equipment such as the Oliver board-former or a Fourdrinier machine.

Sliming during the process may present a problem, particularly when certain equipment is used. Using an Oliver board-former, for example, this sliming problem is overcome with thorough agitation throughout the process and with the addition of magnesium oxide directly into the Oliver board-former head box instead of at an earlier stage, thereby reducing to a minimum the elapsed time between the addition of the magnesium oxide and the pickup of the fibers by the board-former. The early addition of sodium aluminate in solution, in minor quantities (up to 10% by weight of the fibers), will also assist in ensuring uniform distribution of the gelatinous slime and the fibers, thereby assisting draining. The resultant board is similar in appearance to ordinary cellulosic fiberboard, has also similar insulating, acoustical and strength characteristics, is readily machinable and may be drilled, planed, sawn and cut with ease. While it resembles, in so many respects, ordinary fiberboard, it differs from it primarily in that it is incombustible in that it passes the Class A requirements of United States Federal Specifications SS-A-118b and in this respect it resembles some of the existing mineral tiles.

The boards produced using magnesium oxide without aluminum sulfate meet the requirements of Federal Specification SS-A-118b, Class A, and may, therefore, be classified as incombustible. In the case of these boards, however, slow oxidation proceeds after the removal of the test flame from the test panel without flame or glow but with some evolution of heat. This continued slow oxidation was not observed in board containing both magnesium oxide and aluminum sulfate. Examples of compositions from which incombustible boards were satisfactorily made are given below:

(1)	Wood fiber	175 parts
	Fiber glass	50 parts
	Magnesium oxide	175 parts
	Aluminum sulfate	250 parts
(2)	Wood fiber	125 parts
	Fiber glass	100 parts
	Magnesium oxide	100 parts
	Aluminum sulfate	250 parts

The following examples illustrate the process.

Example 1: An incombustible fiberboard, where 55% of the fiber was wood fiber and 45% was glass fiber, was made by the following method: to a 3% glass fiber slurry containing 100 parts of glass fiber were added 10 parts of aluminum sulfate in 50% solution; to a 3% wood fiber slurry containing 125 parts of wood fiber were added 10 parts of sodium aluminate in solution. The two slurries were then blended together and thoroughly agitated. To the blended slurry were added a further 290 parts aluminum sulfate in solution. To the head box of the Oliver board-former were added 175 parts of magnesium oxide, 91% pure, in the form of a 50% slurry. To the resultant slurry containing all these constituents, recirculated "white water" was added until the final slurry had a consistency of 1.5%. The timing of the process was such that the Oliver board-former picked up the fiber mat within five minutes after the addition of magnesium oxide. The final slurry was drained, pressed, formed and cut to produce a fibrous board-like product meeting the requirements of Class A according to Federal Specification SS-A-118b.

Example 2: An incombustible board, where 60% of the fiber was wood fiber and 40% was glass fiber, was made by the following method: to a wood fiber slurry of 135 parts wood fiber and 7,950 parts water, 90 parts of dry glass fiber were added and thoroughly mixed. To this slurry 10 parts of sodium aluminate in solution were added and agitated. 250 parts of aluminum sulfate were then added in 50% solution with continued agitation. The mixture which then had a consistency of 3% was pumped to a storage tank and then to the main Oliver board-former head box next to the Oliver mixer. At the Oliver mixer 150 parts of magnesium oxide (89% pure) in the form of a 60% slurry were added along with recirculated "white water" from the Oliver board-former. The consistency of this final slurry was approximately 1.5%. The final slurry was drained, pressed, formed and cut to produce a fibrous board-like product meeting the requirements of Class A according to Federal Specification SS-A-118b.

Use of Urea as Decomposition Inhibitor with Ammonium Phosphates

R.P. Langguth and H.L. Vandersall (U.S. Patent 3,398,019; August 20, 1968; assigned to Monsanto Company) describe the use of decomposition inhibitors with ammonium phosphates. In general, for preparing wood fiber products, drying temperatures in excess of the decomposition temperatures of the ammonium phosphates are normally used. In order to prevent or minimize discoloration, therefore, it is necessary for the drying temperatures to be reduced below the decomposition temperatures of the ammonium phosphates thus requiring a longer drying time in order to produce a substantially "dry" product. Thus a process which permits normal drying temperatures, i.e., temperatures above the decomposition temperatures of the ammonium phosphates, and normal drying times, in making an ammonium phosphate fire retardant wood fiber product would allow the important advantage, among others, of greater economy in the process.

It has been found that drying temperatures in excess of the decomposition temperatures of the ammonium phosphates may be used to produce fire retarded wood fiber products which have an acceptable color by incorporating a minor amount of a decomposition inhibitor. Decomposition inhibitors suitable for use in the process are, in general, nitrogenous compounds which decompose to form ammonia below the decomposition temperatures of the ammonium phosphates and preferably decompose at a rate which allows relatively steady or uniform decomposition throughout the drying cycle with their substantially complete decomposition by the end of the completed drying cycle or time. In addition, for ease of application they preferably should be water-soluble. In particular, inorganic ammonium compounds, such as the ammonium salts containing nonoxidizing anions, i.e., halides, CO_3^{-2}, S^{-2}, and yielding ammonia on decomposition, as well as the hydroxy amine salts, such as the chloride salts, can, in many cases be used. In addition, organic nitrogenous compounds, such as the amides and amidines, and in particular, those having a low molecular weight, i.e., less than 100, and containing less than 4 carbon atoms, can, in many cases be used. Such nitrogenous organic compounds include urea, formamide, acetamidine and guanidine, with urea being preferred because of its relative low cost and availability.

Normally only small amounts of the decomposition inhibitor are required depending upon, inter alia, the temperature and time of the drying operation. It is preferred that weight ratios of 1:0.1 to 1:0.5 ammonium phosphate to decomposition inhibitor be used and in no event should a weight ratio greater than 1:1 be used because of the deleterious effects which the excess of the inhibitor has on the finished product, such as lowering the strength of the product and, with regard to such nitrogenous bases as urea which is hygroscopic, tends to draw moisture to the product which is highly objectionable, as well as the deleterious effects which the excess of the inhibitor might have on the process and drying equipment. In general, the method of producing the cellulosic material is as follows:

(1) A pulp is prepared from wood fibers such as southern pine, cotton wood, willow, Douglas fir, and hemlock, and in some instances these fibers are treated with chemicals for digestion. An aqueous slurry of pulp is then formed which can vary in solids content but is usually in the order of 1% by weight.

(2) The pulp is formed, most generally into sheets, and is then pressed to a desired density, usually 10 to 25 lb./cu. ft., which is most generally accomplished by pressure with the pressed tile containing a substantial amount of water, sometimes as high as 50% by weight.

(3) The pressed panel is then dried, usually in ovens, and is most generally carried out at drying temperatures from 270° to 500°F. for periods of time ranging from 30 to 120 minutes.

The degree of dryness is usally determined by the percent moisture remaining in the panel immediately after completion of the drying operation and the standard for a dry board, that is, a board considered to be substantially "dry" has generally between 1 to 9% remaining moisture. With respect to discoloration, in general the drying temperatures employed are proportional to the percent moisture in the product, i.e., high drying temperatures may be employed only on products containing a relatively high percent moisture otherwise discoloration of the product will occur. When using normal drying procedures and nonretarded products drying temperatures above 400°F. will discolor to an objectionable degree products containing less than 10% moisture. Therefore, for products which are to have a low percent moisture content and are to be considered substantially "dry", drying temperatures below 400°F. should be employed.

When fire retardant agents, such as the ammonium phosphates, are used they are normally incorporated into the board prior to the drying operation. Normally amounts necessary to give effective flame ratings are usually within the range of 5 to 30% by weight of the panel on a dry basis with an amount of 15% being adequate for a homogeneously impregnated product. The decomposition inhibitor can be incorporated by any manner into the product but it is preferably added to the product either in the solid form or dissolved in an aqueous medium during the slurrying of the pulp.

As being representative of the process, tiles 12 x 12 x 1/2 inches having a dry density of 17 lbs./cu. ft. were prepared from various representative pulps and dried in an oven, Despatch Model PLHD-2-5, containing ten drawers. Before the tiles were introduced for drying the oven and circulation fan were started with the fan providing an air velocity of 175 fpm and the oven brought to the desired drying temperature. The tiles were introduced with surface thermocouples on the upper tile surface and were dried at the indicated temperatures to the indicated dryness for periods of time varying between 90 and 140 minutes. In order to duplicate conditions existing in tunnel driers commonly used in commercial production the tiles were dried under high humidity conditions, i.e., above 80% relative humidity, for about the first 25% of the drying time. The tile were compared according to degree of discoloration based upon the following standards.

Color Number	Description
1	Clear surfaces, edges and interior. Some slight scorching allowed.
2	At least one side all tan, some light brown but mostly yellowish. No black or brown edges.
3	Clear interior, light brown to tan exterior. Usually charred edges.

(continued)

(continued)

Color Number	Description
4	Brown to dark brown surface. Edges charred but not soggy.
5	Dark brown to black even in tile interior. Charred, soggy, crumbly edges.

The following is the tabulated results of the discoloration comparisons:

	Retardant[1]	Decomposition Inhibitor (wt. ratio of retardant to inhibitor) (urea)	Drying Temperatures (° F.)	Moisture Remaining (Percent)	Color Number[2]
1	Monoammonium dihydrogen phosphate	- - -	365	9	3.5
2	Monoammonium dihydrogen phosphate	1:0.2	365	9	2.0
3	Monoammonium dihydrogen phosphate	1:0.5	365	9	1.5
4	Monoammonium dihydrogen phosphate	- - -	340	2	5.0
5	Diammonium hydrogen phosphate	- - -	340	2	5.0
6	Monoammonium dihydrogen phosphate	1:0.2	340	2	2.5
7	Monoammonium dihydrogen phosphate	1:0.5	340	2	2.0
8	Monoammonium dihydrogen phosphate	- - -	300	2	3.0
9	Diammonium hydrogen phosphate	- - -	300	2	3.5
10	Monoammonium dihydrogen phosphate	1:0.2	300	2	1.5
11	Monoammonium dihydrogen phosphate	1:0.5	300	2	1.5

[1]The monoammonium dihydrogen and diammonium hydrogen phosphate used was 15% by weight based on total dry weight of the tile.

[2]When color numbers are above 2.5 the degree of discoloration is considered "objectionable" and unacceptable.

As noted in the table when the ammonium phosphate retarded tiles (1), (4), (5), (8) and (9) were subjected in the drying operation to drying temperatures above the decomposition temperatures of the ammonium phosphates and within the range of 300° to 365° F. when producing a dry tile, i.e., one containing less than 10% remaining moisture, they exhibited discoloration to an objectionable degree. However, when the decomposition inhibitor, such as urea, was used to minor amounts under identical conditions significant improvement in the discoloration of the tiles were achieved and acceptable tiles were produced. This illustrates the effect of using minor amounts of the decomposition inhibitor to prevent or minimize the problem of discoloration when using fire retardants, such as the ammonium phosphates, under normal drying conditions.

Hydrophilic Colloids

In a process described by W.P. Fairchild (U.S. Patent 3,281,252; October 25, 1966; assigned to Kelco Company) an aqueous solution of any selected fire retardant chemical is prepared, generally making as concentrated a solution as possible, and incorporated with the said solution is a modicum, typically 1/10 to 1%, of a Xanthomonas colloid produced by the bacterium Xanthomonas campestris. This colloidal material is a polymer containing mannose, glucose, potassium glucuronate and acetyl radicals. In such a colloid, the potassium portion can be replaced by several other cations without substantial change in the property of the material for this purpose. The colloid, which is a high molecular weight, exocellular material, may be prepared by the bacterium Xanthomonas campestris by whole culture fermentation of a medium containing 2 to 5% commercial glucose, organic nitrogen source, dipotassium hydrogen phosphate and appropriate trace elements. The incubation time is approximately 96 hours at 28° C. aerobic conditions.

Treating compositions made up in accordance with the process have the property of exhibiting minimal penetration into porous compressed vegetable fiber structural panels, when applied to surfaces of the latter. The following example illustrates the process.

Example: To 183 parts water, one part Xanthomonas campestris hydrophilic colloid is added. This is heated to 140° F., and 18 parts boric acid is added to form a treating solution according to the process. This solution is applied to a porous compressed vegetable fiber structural panel made from bagasse, having a total thickness of 1/2 inch, and sold under the trade name "Celotex". The treated board was allowed to stand for 72 hours after which time it was dry. This treated structural panel was then subjected to scorch tests by applying a steel plate heated to 250°C. The scorch resistance of this panel was very good.

Boric acid solutions of 2%, 5%, and 10% containing 0, 0.1%, 0.3%, 0.75%, and 1% of a Xanthomonas campestris hydrophilic colloid were prepared in the manner described in the example. These solutions were stored for one week at 140°F. in tightly capped bottles. They were then used to coat portions of insulating board which had been formed but not dried on the machine during manufacture. That is, the board was in the same wet condition as is the same board on the machine ahead of the drying section. The coated portions of board were then placed under an infra-red lamp which supplied a surface temperature of 187°C. After 1.5 hours the surface of the board samples was dry. The degree of surface scorch of the various samples was compared with that of uncoated board. The degree of scorch and fiber toasting was found to be in direct relation to both the concentration of boric acid and the concentration of the Xanthomonas campestris hydrophilic colloid in the coating. For example, at each Xanthomonas campestris hydrophilic colloid level scorch decreased with increasing boric acid concentration. Likewise, for a given boric acid level, scorch and fiber toasting decreased with increasing Xanthomonas colloid concentration.

COATINGS

Latex Dispersions, Scrub Resistant Formulations

Processes described by E.K. Stilbert, I.J. Cummings and J.P. Talley (U.S. Patent 2,755,260; July 17, 1956) and E.K. Stilbert, I.J. Cummings and W.B. Cuerrant (U.S. Patent 2,684,593; July 27, 1954, both assigned to The Dow Chemical Company) provide an abrasion resistant, intumescent coating deposited from a 40 to 60% aqueous dispersion of a mixture consisting of nonfilm forming intumescent solids and an aqueous film forming latex-like dispersion of a polymeric material. The nonfilm forming intumescent solids consist of a foam forming substance, a nonresinous carbon yielding substance, and an organic nitrogen compound.

Materials suitable as nonresinous carbonifics are carbohydrates, modified starches and similar substances, a water dispersible protein such as gelatin or casein, or a polyhydric compound from the class consisting of the hexitols, such as mannitol, the pentitols such as arabitol, or the mono-, or di-tetritols such as mono- and dipentaerythritol. It is preferred to use pentaerythritol or the dimer of pentaerythritol, or mixtures of the two for the nonresinous carbonific. While an organic nitrogen compound is not entirely essential improved intumescence is obtained from a composition containing one. Amino compounds such as dicyandiamide, urea, dimethyl urea, guanylurea phosphate, and glycine are suitable organic nitrogen compounds. Of these, urea or dicyandiamide are preferred as the organic nitrogen compound. At temperatures at or near the temperature of combustion of wood, the nonresinous carbonific and organic nitrogen compound in combination with a foam producing agent, or spumific, forms a fire resistant carbonaceous foam. The spumifics that may be used in this process are preferably monoammonium phosphate, diammonium phosphate, phosphoric acid, and ammonium sulfate.

Such compositions, when employed without further modification lack resistance to abrasion. Therefore, a polymer latex, usually of 50% solids, is incorporated into the coating composition as a binder, so that the amount of the latex solids in the composition is from 5 to 35% of the total solids, but preferably from 15 to 30%. An aqueous film forming latex of a polymer may be employed. Examples of such latices are the polymer of vinyl chloride, and copolymers of vinyl chloride and other vinyl esters, the copolymers of styrene and butadiene, vinylidene chloride and acrylonitrile, and vinylidene chloride and vinyl chloride. The last three are preferred, as they give the greatest scrub resistance to the compositions. The following examples illustrate the process. All of the compositions were made by grinding the inert ingredients in a pebble mill for 3 hours with sufficient water to form a mixture of 55% solids content. To the resulting pigment slip containing the inert ingredients, the pentaerythritol, monoammonium phosphate, and dicyandiamide were added with stirring in the amounts shown in the tables shown on the following page. Subsequently, the latex was added and mixed with the other ingredients for 10 minutes.

Each of the compositions was sprayed evenly over the surface of 12 inch square panels (3/8 inch thick) of low density, previously uncoated fibrous wall board. The coated panels were dried at 158°F. for 10 minutes followed by a 5 minute period at 302°F. The dry coated panels were then aged 16 to 20 hours at 75°F. and 50% relative humidity. These panels were then exposed to flame in a standard fire test to determine the degree of intumescence and the area of surface char of each coating. The fire test consisted of exposing the coated surface of a panel, inclined at an angle of 45 degrees, to the complete burning of 1 cc of absolute alcohol, the alcohol flame being focused on the board at a point about 2 inches above the lower end of the panel. This fire test is described in Commercial Standard CS-42-49, U.S. Department of Commerce Bulletin, "Structural Fiber Insulating Board" (4th ed.). The amount of protection that each coating composition provides for each panel is rated according to amount of intumescence and according to the area of surface char.

Fiberboard

Example	1	2	3	4	5	6	7	8	9
Monoammonium phosphate	44	26.4	66	66	44	70	17.6	-	17.6
Dicyandiamide	-	30.8	22	-	44	9	13.2	44	57.2
Pentaerythritol	44	30.8	-	22	-	9	57.2	44	13.2
Titanium dioxide	12	12	12	12	12	12	12	12	12
Sodium alginate	0.6	0.6	0.6	0.6	0.6	0.6	0.6	0.6	0.6
Water	45	45	45	45	45	45	45	45	45
Plasticized latex[1]	17.5	17.5	17.5	17.5	17.5	17.5	17.5	17.5	17.5
Char area	11	12	12	10	11	11	16	20	15
Intumescent rating[2]	3	3	3	3	2.5	3	1.0	0	0.5

[1] Copolymer of 75% vinyl chloride 25% vinylidene chloride plasticized with 10% monooctyl diphenyl phosphate.

[2] 0 = no intumescence; 1 = very slight intumescence; 2 = poor intumescence; 3 = fair intumescence; 4 = good intumescence; 5 = excellent intumescence.

It is also possible to replace a minor amount of the pentaerythritol with starch. The effect of such a replacement may be seen by the Examples 10 to 14 tabulated below. These samples were prepared in the same manner described above. In addition, however, a 4 x 12 inch strip was cut from the uncharred portion of each of the coated panels and these were tested for abrasion resistance, according to Federal Specification TT-P38a using a 0.5% soap solution and 100 oscillations of the scrubbers. The "Gardner Straight Line Scrub Tester-Model 105" (obtained from Gardner Laboratory Inc.,) was used in the standard scrub test. At the end of 100 oscillations (200 strokes of the brush), a panel whose surface is still completely covered with coating, is one that exhibits maximum resistance to abrasion. A coated panel that exhibits no resistance to abrasion, is one whose board surface is completely exposed and the fire retardant coating is completely removed after 100 or fewer cysles under the Gardner tester.

Example	10	11	12	13	14
Monoammonium phosphate	56	56	56	56	56
Dicyandiamide	10	10	10	10	10
Pearl corn starch	21.2	15.9	11.2	5.3	-
Pentaerythritol	-	5.3	10	15.9	21.2
Titanium dioxide	12	12	12	12	12
Sodium alginate	0.6	0.6	0.6	0.6	0.6
Water	75	75	75	75	75
Plasticized latex[1]	25	21	21	21	21
Intumescent rating[2]	3.5	4.5	4.5	4.75	5.0
Wet scrub cycles[3]	350	400	420	800	550

[1] See above table.
[2] See above table.
[3] Number of wet scrub cycles to expose 10% of the board's surface.

As can be seen from the above data, as the amount of starch is increased both the intumescent rating and the scrub resistance decrease. It is preferred to maintain the amount of starch at less than 15% of the amount of spumific present.

The use of other latex binders in scrub resistant, intumescent, fire retardant coatings is described by E.K. Stilbert, Jr., I.J. Cummings and C.E. Clock. These pertinent patents all assigned to The Dow Chemical Company are as follows: U.S. Patent 2,912,392; November 10, 1959 (polyvinyl acetate), U.S. Patent 2,912,393; November 10, 1959 (polyacrylates) and U.S. Patent 2,912,394; November 10, 1959 (polystyrene).

In related work, N.R. Peterson and I.J. Cummings (U.S. Patent 2,917,476, December 15, 1959; assigned to The Dow Chemical Company) have found that certain solid plasticizers may be added to latex based intumescent coatings and treated as an inert filler. It is only necessary that the solid plasticizer be nonsolvating at room temperature and have a congealing point above room temperature and below a temperature at which the polymer will be degraded. Preferably these temperatures will be from 75° to 125°C. Typical examples of such solid plasticizers are solid chlorinated paraffins, such as Chlorowax 70S sold by The Diamond Alkali Company, diphenyl phthalate, tris (p-tert-butylphenyl) phosphate sold by The Dow Chemical Company as Plasticizer 77, p-toluene sulfonamide and triphenyl phosphate. With some of the plasticizers such as triphenyl phosphate it may be desirable to add a small amount of additional surface active agent to the composition to impart mechanical stability.

The solid plasticizers of this process may be treated as an inert solid and added prior to milling the pigment slip. This represents a substantial departure from prior methods employing liquid plasticizers where the plasticizer had to be incorporated into the latex. To the pigment slip so formed from the inert ingredients and plasticizer may be added the spumific and carbonific. The desired latex is finally stirred in for a short time. A smooth composition results, containing 50% solids, and having the stability of the unmodified latex. The stability remains with the composition until the temperature of the composition is raised above the congealing temperature of the plasticizer. When using liquid plasticizers and the prior plasticizing methods it was never possible to attain stability equal to that of the unmodified latex. The plasticizers exert their plasticizing action on the polymer during the curing step after the mechanical stability of the composition is no longer important. The compositions preferably consist of the amounts of materials previously described.

For example, several intumescent coating compositions were prepared by milling mixtures of 56 parts by weight monoammonium phosphate, 10 parts by weight of dicyandiamide, 22 parts by weight of pentaerythritol, 12 parts by weight of titanium dioxide, 5 parts of a plasticizer from those listed below, 1.25 parts of an anionic wetting agent, and 47 parts of water on a high speed colloid mill. The plasticizers used were a liquid alkyl aryl phosphate for comparative purposes and triphenyl phosphate, diphenyl phthalate, a chlorinated paraffin having 70% chlorine, tris-(p-tert-butylphenyl) phosphate and p-toluene sulfonamide. To the resultant pigment slip was added, with agitation, 40 parts (solids basis) of a latex prepared by the emulsion polymerization of comonomers consisting of 25% by weight of vinylidene chloride and 75% by weight of vinyl chloride, the latex having 55% evaporative solids.

The shelf life of the compositions was measured by means of a Brookfield viscometer at 60 rpm. It was found that the composition using the liquid alkyl aryl phosphate reached a viscosity of 10,000 cps. in less that 12 hours. The composition using triphenyl phosphate reached a viscosity of 10,000 cps. in 14 days; that using tris-(p-tert-butylphenyl) phosphate reached 10,000 cps. in 16 days; that using diphenyl phthalate leveled off at a viscosity of 900 cps. in 17 days; and the composition using the chlorinated paraffin leveled off at 375 cps. in 14 days. For comparative purposes an unplasticized composition leveled off at 450 cps. in 13 days but coatings made from the unplasticized composition were inferior, especially in their washability or scrub resistance.

As a further illustration of the advantages of these coatings the coating compositions prepared above were subjected to high speed agitation for 30 minutes. The coating having the liquid alkyl aryl phosphate plasticizer was completely coagulated. That coating composition using tris(p-tert-butylphenyl) phosphate had 5% coagulation. The composition having diphenyl phthalate had 2% coagulation. All the others had no coagulation. The coagulation was determined by filtering the latex following agitation and weighing the coagulum.

To illustrate the effect of curing on the above samples, the wall boards coated with the composition using triphenyl phosphate were cured at four temperatures. The samples were tested for scrub resistance using a 0.5% soap solution and the "Gardner Straight Line Scrub Tester Model 105" (obtained from Gardner Laboratory Inc.). The coating cured at 70°C. failed after 50 cycles; at 80°C. it failed at 300 cycles, at 90°C. it failed at 900 cycles; and at 100°C. it failed at 1500 cycles. The last three are considered excellent. Similar results were obtained when latexes of polyvinyl chloride, copolymers of vinyl chloride and vinyl acetate, vinylidene chloride and acrylonitrile, and styrene and butadiene were employed.

Polyvinyl Acetate Dispersions Containing a Borate

G.O. Orth (U.S. Patent 2,809,949; October 15, 1957; assigned to American Potash & Chemical Corporation) has found that if one uses a borate composition having a B_2O_3 to Na_2O ratio of between 4 and 5 to 1 on a weight basis, the resulting composition has greatly improved fire resistance and, in addition, is not susceptible to damage by moisture even under conditions far more severe than those encountered in any normal use. Such borate compositions are weakly alkaline; in a 10% by weight water solution, they will have a pH between 7 and 8 at 25°C.; the preferred borate compositions has a B_2O_3 to Na_2O ratio of 4.8 to 1 and a pH of 7.7 in a 10% by weight water solution at 25°C. These compositions are readily prepared by mixing boric oxide and sodium oxide in suitable proportions or by mixing suitable sodium and borate containing salts, as is described in U.S. Patent 2,096,266.

To demonstrate the utility of the compositions of this process, a typical water-base paint was made up as follows (the proportions are by weight):

Liquid portion:

780	polyvinyl acetate, 55% total solids
1170	water

Pigment portion:

126	titanium dioxide pigment
86	diatomaceous earth
210	coating grade clay
340	lithopone
86	mica.

One should use a polyvinyl acetate dispersion having a low polyvinyl alcohol content e.g., less than 5%. Also, the dispersing agent should be compatible with the borate composition; if the dispersing agent is not compatible, the borate breaks the dispersion and may cause the polyvinyl acetate to coagulate. A portion of the liquid was mixed with the pigment and ground, following which the balance of the liquid portion was added. To equal portions of the above paint composition, increments of a borate composition (having a B_2O_3 to Na_2O ratio of 4.8) were added in amounts corresponding to increments of 10% based on the total resin solids from 10 to 100%; that is, the paint formulation given above includes 429 parts by weight up to a total of 429 parts, corresponding to increments of 10% from 10 to 100% of resin solids.

In those formulations corresponding to 80, 90, and 100% of the borate composition based on the resin solids, the addition of further water was necessary to dissolve the borate. Each of the eleven compositions was then sprayed at the rate of 20 pounds of contained solids per 1,000 board feet over primed and unprimed hard board, primed and unprimed soft board, and on ground glass. After the lapse of sufficient time to permit of thorough drying, the respective panels were subjected to test in a Gardner wash test machine.

The specifications drawn around the use of this machine ordinarily require that the panel surface be unbroken after 300 complete strokes (see Federal Specification TTP-88a), and this is considered satisfactory. The results of this test are shown graphically in Figure 2.1a. From these, it will be apparent that the use of the borate composition of this process in a water-base paint does not markedly reduce water resistance until the composition is present in an amount in excess of 50% of the resin solid content.

FIGURE 2.1a: POLYVINYL ACETATE LATEX-BORATE COMPOSITIONS

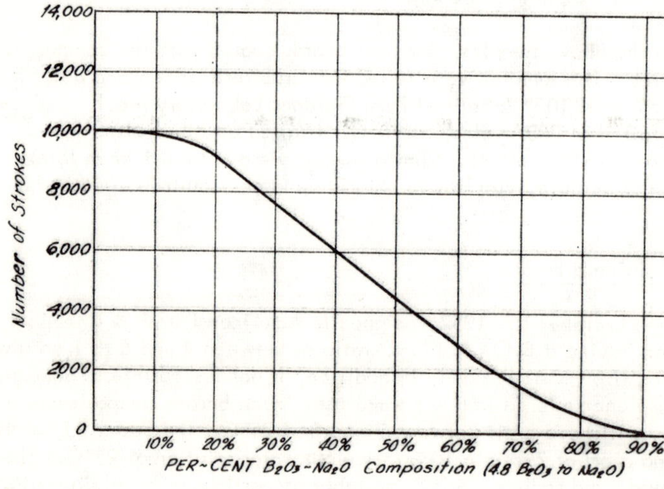

Scrub Resistance

Source: G.O. Orth, Jr., U.S. Patent 2,809,949; October 15, 1957

To illustrate the effectiveness of the borate composition in reducing char, soft board panels were prepared with each of the eleven paint formulations set forth above and in accordance with U.S. Department of Commerce Specification CS-42-49. The panels were then tested in accordance with this specification, the results being shown graphically in Figure 2.1b. From this, it will be noted that the presence of from 15 to 70% of the preferred borate composition will provide a surface having less than twelve square inches of char, the accepted test limit. The preferred quantity of the borate is from 15 to 70% of the resin solids with a more preferred range of from 40 to 60%.

FIGURE 2.1b: POLYVINYL ACETATE LATEX-BORATE COMPOSITIONS

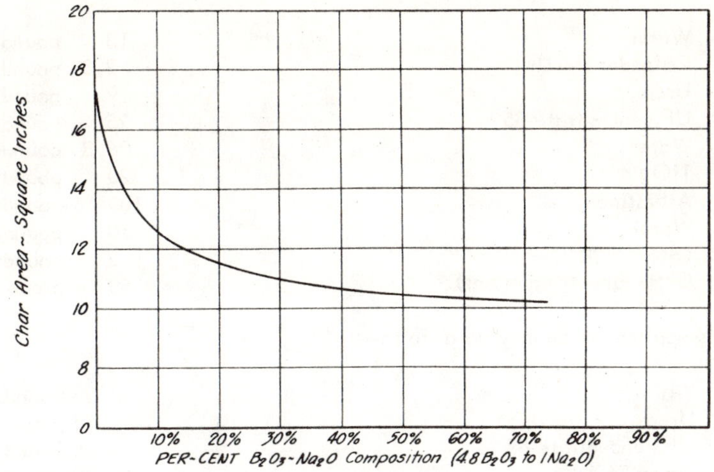

Fire Retardancy

Source: G.O. Orth, Jr., U.S. Patent 2,809,949; October 15, 1957

The composition was also applied to drilled acoustical cellulose units, each one-half inch thick, one foot square, and of 18 pounds per cubic foot density, at the rate of about ten pounds of solids per 1,000 square feet; one group contained 50% by weight of contained resin solids of the preferred borate composition, and the other group 70% by weight. These units were then subjected to the test outlined in Federal Specification SSA-118a. In this test, nine of the like coated units are assembled on a flat support and are subjected to a standard burner blast in which the edge flame temperature application follows the rate curve prescribed in the test and which attains a value of 800°C. after twenty minutes. After the specified twenty minute burn period, the results on both groups of units were considered satisfactory by an impartial referee laboratory.

Latex Applied over Intumescent Coating

E.A. Lauring (U.S. Patent 2,813,046; November 12, 1957) has found that by applying a flame resistant coating to the fiber wall board and then applying an additional coating over the first which contains a polymer latex in combination with urea formaldehyde resin in a definite relationship to the latex, that a washable flame resistant coating product is produced. It is preferred that the polymer latex of 50% solids be used in the composition in 8 to 12 parts to 4 to 6 parts of the urea formaldehyde resin.

Any polymer latex dispersion capable of forming a film is satisfactory for the purpose. Examples of latices are Dow Latex 762K or a polymeric material such as the polymers of vinyl acetate, copolymers of styrene butadiene, copolymers of vinylidene chloride and acrylonitrile. The fiberboard or acoustical tile has applied by suitable means an intumescent flame resistant coating of which an example is shown on the following page.

Water	60	pounds
Calender starch	2.3	pounds
Urea	9	pounds
UF concentrate 85	25	pounds
Water	82	pounds
TiO_2	25	pounds
Asbestine	60	pounds
Monoammonium phosphate	100	pounds
Water	10	pounds
Latex 762K	2	pounds
CPH (wetting agent)	80	grams

Generally, it is preferred that a second coat be applied of which the following is an example:

Water	13	pounds
Calender starch	2.3	pounds
Urea	9	pounds
UF concentrate 85	25	pounds
Water	106.3	pounds
TiO_2	20	pounds
Asbestine	100	pounds
Water	10	pounds
Latex 762K	2	pounds
CPH-30(wetting agent)	80	grams

Then the finished coating is applied which may be as follows:

Water	7.53	pounds
Urea	1.54	pounds
UF concentrate 85	4.26	pounds
Latex 762K	10.32	pounds
Water	1.50	pounds
TiO_2	1.67	pounds

The coating applied according to this process when tested in accordance with Commercial Standards 42-49, after being washed exhibited no smoldering and a char area of less than 3 square inches.

Kaolin Clay, Starch Prime Coat, Latex Top Coat

J.F. Campbell and E.M. Cagle (U.S. Patent 2,889,235; June 2, 1959) describe a process for producing a flame resistant, decorative insulating board. In the process, there is first formed, in any conventional manner, a wetlap or felted web of wood or other vegetable fiber. The wetlap is then, in accordance with the process coated on its upper surface, with a composition which serves to provide a priming coat for the decorative finish coatings to be subsequently applied, and to provide a surface that is substantially flame resistant, and of a color approximating that of the finish coating. This base coating is applied to the board substantially immediately prior to its entry into the ovens through which the board is passed for drying. As illustrative of a composition that may be employed for the prime coat in the process, is the following:

	Parts by Weight
Clay	800
Pearl starch	50
Anti-vermin and anti-mildew agent	5
Water in sufficient quantities to make approximately 250 gallons	

This composition serves well when the subsequently applied decorative finish coating is to be white in color. Where the decorative finish, however, is to be other than white, suitable pigments may be incorporated in sufficient quantity so that the prime coat will be of a color approximately that of the finish coat. When applied at the rate of 9 to 12 gallons per thousand square feet, viz., to deposit approximately 30 to 40 pounds of clay per thousand square feet, this composition provides a prime coat which is considerably resistant to the spread of flame, by reason of the presence of the high amount of clay, especially when the latter is one which is of the kaolin type. The composition is best applied by means of felt covered rolls in pressure contact with the surface of the wetlap immediately after it emerges from the conventional wet presses and prior to entrance of the wetlap boards into the drying ovens. By reason of the expanding action which the wetlap undergoes upon emerging from the wet presses, the composition tends to be more or less dewatered thereby, and hence may, if desired, be applied in a plurality of coats.

The expanding action, in conjunction with the pressure exerted by the applicator roll or rolls, serves not only to provide a film of the composition on treated surface, but also to permeate the sub surface portions of the relatively thick sheet of wetlap with the clay and other inert components of the composition. The decorative finish, according to the process, is provided by an aqueous composition which contains, essentially, in addition to pigment and a flame retarding material, a water-dispersible relatively inexpensive binder which serves also as a gloss producing agent, and a binder which serves also to minimize the brittleness of the film caused by the gloss producing agent and to impart a sufficient degree of water-proofness to the film to render it washable. Preferably, also, there is included a preservative.

The decorative coating is applied to the surface previously prime coated as described, in quantities to provide 1.5 to 2.0 pounds of the flame retarding agent per thousand square feet of the surface. When applied at that rate, the finished board tested according to the conditions of test specified in Commercial Standard CS-42-49, exhibits an area of char ranging from 9.0 to 11.0 square inches, with no glow or smoldering continuing 60 seconds after removal of the igniting flame. Thus, it meets the requirements of those as well as other accepted standards for flame resistant fiber insulating board. As illustrative of a composition for producing the decorative finish, the following example is given:

	Parts by Weight
Water	1000
Pigment	250
Disodium octaborate tetrahydrate	200
Starch	150
Acrylic resin dispersion	200
Preservative (sodium pentachlorphenate)	9

When the coating was applied at the rate indicated, and then dried in the conventional manner, the finished board not only exhibited the fire resistant properties above set forth, but also exhibited a uniform gloss on the surface, the dried film of the coating being also washable and sufficiently flexible to permit flexing of the board without cracking the film.

Kaolin Clay Coatings

A.J. Page and D.R.P. Haig (U.S. Patent 2,708,643; May 17, 1955; assigned to Johns-Manville Corporation) describe a coating to provide insulating board flame resistant characteristics which essentially comprises either a synthetic or a natural mixture of from 75 to 95% by weight of a nonswelling clay mineral and from 5 to 25% of a swellable clay mineral of fine particle size. Exemplary of such nonswelling clays are moderate particle size clays of the kaolinite group such as kaolinite, endellite, etc.; clays of the micaceous clay group such as illite, attapulgite, etc.; and other equivalent clay minerals.

Examples of the swellable natural minerals of fine particle size are swelling colloidal clay minerals which can expand along the C crystal axis without altering the other two major crystal axis (A or B), such as montmorillinite or bentonite, hectorite, beidellite, etc. The synthetic or natural mixture of the above materials in the indicated proportions should also have the physical characteristic of an average bonding modulus of at least approximately 200 lbs./sq. in. as determined by standard test conditions outlined in the Report of the Committee on Standards, The American Ceramic Society, Year Book 1921-1922. An example of a natural mineral mixture and the preferred essential mineral for fabricating a flame resistant fiberboard in the process is a natural ball clay mined at Crenshaw, Mississippi, and known as Mississippi M & D clay. This clay essentially comprises a mixture of kaolinite and montmorillinite, has a bonding modulus of 210 lbs./sq. in., and leaves upon ignition a lignitish residue.

A fiber insulating board may be rendered flame resistant by coating the board with an aqueous dispersion of the above synthetic or natural mineral mixture alone, in amount sufficient to deposit at least 14 lbs. of clay per 1000 sq. ft. of board surface. The maximum amount of such a mineral mixture which may be utilized is limited only by the amount capable of being dispersed in the desired vehicle to a workable coating consistency, and in turn, the amount of dispersion applied to the board. In normal practice the amount of aqueous dispersion applied when utilizing such a natural or synthetic mineral mixture alone will not exceed that required to deposit 30 lbs. of clay mineral solids per 1000 sq. ft. of insulating board surface. While an insulating board can be rendered flame resistant to a degree sufficient to exhibit a char area of less than 9 sq. in. by employing the synthetic or natural mineral mixture alone, various ingredients may be, and preferably are, added to the coating to improve the appearance and characteristics of the finally obtained product. The prime additive preferably utilized in conjunction with a natural mineral clay mixture such as M & D clay, is a light colored clay have high viscosity reduction when treated with a deflocculating agent, for example, a kaolin such as Tako or ball clay such as Cooley. An additive clay having such properties is utilized in conjunction with the natural clay mixture in order to enable the formation of a relatively high solids concentration dispersion for ease in application.

The additive clay should have a relatively light color which will serve to provide as light colored flameproofing clay primer coat as possible, obviously desirable when conventional finishes such as light colored casein paints are utilized as decorative top coatings. Since the additive clay is a substantially inorganic material, it is nonflammable and will also serve, in some degree, to render the insulating board flame resistant. The additive clay in itself, however, will not render the board flame resistant to a sufficient degree, and actual test measurements have indicated that a clay such as Tako, found suitable for this process as an additive and as a nonswelling mineral of moderate particle size, when used alone does not diminish the char area below 12 square inches.

When an additive clay, such as Tako clay, is utilized in conjunction with the natural clay mixture, such as Mississippi M & D clay, the amount of clay dispersion utilized should be sufficient to deposit at least 14, and preferably 20 to 25 pounds of clay solids per 1000 sq. ft. of board surface. Again the maximum solids content which may be deposited is dependent only upon the amount of solids capable of being dispersed to a workable viscosity, and the total amount of clay solids deposited will not generally exceed 40 pounds per 1000 sq. ft. of board surface. When using such a clay dispersion, the natural clay mixture of the above characteristics should be present in amount sufficient to deposit at least 10, and preferably 11 to 13 pounds of its clay solids alone per 1000 sq. ft. of board surface treated, and in the optimum compositions equal amounts of additive clay and the natural clay mixture are used.

The coating serves a dual function in the formation of decorative surfaced structural fiberboard products. Primarily, the coating serves to render a fiber insulating board flame resistant and, in particular, sufficiently flame resistant to meet the requirements set forth in various commercial standards such as Department of Commerce Standards CS-42-49 and the United States Government Federal Specifications SS-A-118a, F-3c. The flameproofing clay coating also serves as an excellent primer for conventional decorative paints utilized in the manufacture of insulating board products, such as a pigmented casein paint. The flameproofing coating of this process strongly adheres to the fibers of the insulating board and forms a good sealant for the decorative top coating. The coating is sufficiently flexible to allow the insulating board to be flexed without cracking the flameproofing coating and in turn, the decorative top coating.

The flameproofing coating also has the important characteristic of being capable of application without causing warpage of the board to which it is applied during the manufacturing procedure. The following are examples of various flame resistant cellulosic fiberboards and their general method of preparation.

Example 1: A 1/2 inch thick cellulosic fiber, wet felted insulating board having a density of approximately 15 to 18 lbs./cu. ft. was coated with two felts of a slurry having an approximate composition of:

	Percent
Starch	1.5
M & D ball clay	24.0
Tako kaolin	24.0
Clay dispersing agent	0.3
Anti-mildew agent	0.2
Water	50.0

The slurry was coated on the board in amount sufficient to deposit a total of approximately 25 pounds of mineral solids per 1000 sq. ft. of treated surface. The coated board was then painted with four felts of a conventional casein paint in amount sufficient to deposit approximately 18 pounds of paint solids per 1000 sq. ft. The adhesion of the priming mineral slurry to the treated board, and of the decorative coating to the prime coating was excellent, and no cracking of the finish appeared. When tested in accordance with Commercial Standards 42-49, the board exhibited a char area of only 8.3 square inches and exhibited no smoldering or after glow.

Example 2: A wood fiber insulating board was coated in the same manner as defined in Example 1 by employing, in place of the indicated primer, an amount of coating sufficient to deposit approximately 25 pounds of mineral solids per 1000 sq. ft. of board surface, having the following approximate composition:

	Percent
Starch	1.0
Peerless coating clay	34.0
Bentonite	5.0
Dispersing agent	0.2
Anti-mildew agent	0.2
Water	59.6

The coating exhibited no cracking and, as in Example 1, the adhesion of the coatings to the board and to each other was excellent. When tested in accordance with Commercial Standards 42-49, the insulating board exhibited no smoldering and a char area of approximately 8.8 square inches.

Monocalcium Phosphate, Starch Coating

L.W. Eckert (U.S. Patent 2,632,743; March 24, 1953; assigned to Armstrong Cork Company) describes a coating composition which comprises monocalcium phosphate, starch, and a binder. Preferably, a portion at least of the starch is in the form of a gelled solution, the remaining starch being present as an ungelled dispersion of starch particles, commonly referred to as dispersion of dry starch, the gelled starch solution serving as a binder. Other binders which may be used include urea-aldehyde resins, phenol-aldehyde resins, phenol modified resorcinol-aldehyde resins, melamine-aldehyde resins, and dicyandiamide-aldehyde resins.

For uses such as on perforated acoustical material, there may be combined with the monocalcium phosphate and starch a dicyandiamide-aldehyde resin and phosphoric acid, such a resin and acid combination providing an additional intumescent component in the coating composition, which also serves as a supplemental binder for the gelled starch or may serve as the binder if no gelled starch is used. The monocalcium phosphate is a water-insoluble salt, which fireproofs the starch solution and minimizes the possibility of paint failure which might occur under high humidity conditions. The following is an example of fire resistant coating composition which may be applied to fiberboard:

Example 1:

	Parts by Weight
15% starch solution	[1]35.0
Dry ungelled starch	18.7
Monocalcium phosphate	12
Phosphoric acid	3.0
Pigment	3.2
Filler	3.5
Water	24.6

[1]5.25 parts starch

In the preparation of the coating composition, the 15% starch solution is prepared by heating the starch in water until a temperature of 190°F. has been reached. This produces a starch gel which has binding properties. The temperature of the gel is reduced to 140°F., and the remaining ingredients are stirred in. The reduction of temperature is necessary in order to avoid the gelling of the dry starch, which would increase the viscosity of the coating composition to the point where application with conventional spraying equipment would be impossible. The amount of water added will vary depending upon the mode of application of the coating composition to the fiberboard. The quantity given above is for the preparation of a coating for pressure spray application.

The above coating composition when applied at the rate of 20 grams per square foot to an insulating fiberboard formed, for example, from southern pine wood fibers will render the product sufficiently fire retardant to meet Class F of "Federal Specification CS42-49, dated November, 1949, and entitled "Structural Fiber Insulating Board", when tested in accordance with the procedure outlined in paragraph 6.2.9 of that specification. For acoustical materials having perforations at the surface which materials are to be rendered sufficiently fire retardant to meet Federal Specification SSA-118a, dated February 12, 1948, and entitled "Acoustical Units: Prefabricated", it is preferred to add to the starch monocalcium phosphate coating an intumescent resin phosphoric acid component which serves to increase the fire resistant characteristics of the coating composition and provides added intumescence which will cause the coating composition to bridge effectively the perforations in the material when intumescence and coking result from the application of flame to the coating. The following is an example of a coating composition suitable for application to perforated acoustical fiberboard units:

Example 2:

	Parts by Weight
15% starch solution	20.9
Dry ungelled starch	17.7
Monocalcium phosphate	10.4
Water	19.8
Resin solution	13.5
Pigment	2.6
Filler	3.5
Water	11.6

In this example, the starch solution or gel is prepared in the manner described above; and the dry starch, monocalcium phosphate, and water are added after the starch solution has been reduced below 140°F. The resin solution is prepared by mixing the ingredients as shown on the following page.

	Parts by Weight
Formaldehyde (37% solution)	71
Dicyandiamide	22
Phosphoric acid	7

Preferred practice is to mix the formaldehyde and dicyandiamide together for two minutes and then add the phosphoric acid. It serves as a catalyst for the reaction. The reaction between the formaldehyde and dicyandiamide is continued until a clear solution is obtained. The reaction vessel is jacketed, and cooling water is applied to maintain the temperature of the reaction below 120°F., and preferably at a temperature of 90° to 105°F. With a batch such as referred to above, the reaction will be effected in about one hour's time.

The resin solution is mixed in with the starch monocalcium phosphate mixture, and the filler and pigments are added. The necessary amount of water to provide the desired viscosity for application is incorporated. In the above example, the quantity of water added has been sufficient to produce a composition suitable for pressure spray application. When 25 to 27 grams of the coating composition of Example 2 are applied per square foot to a perforated fiberboard acoustical unit, the same will meet the requirements of Section E-3c of Federal Specifications SSA-118a when tested in accordance with the procedure outlined in F-3c of the specification. The coating does not substantially reduce the sound-absorption efficiency of the product; and when dried at a temperature in the order of 325°F., the surface is tack free, smooth, and free of checks or cracks.

Phenolic Binder, Sodium Tetraborate, Boric Acid

A process described by E.A. Lauring (U.S. Patent 2,594,937; April 29, 1952; assigned to Minnesota and Ontario Paper Company) involves a composition which comprises a thermoplastic varnish, driers, solvents, sodium tetraborate compounds, boric acid, a supporting filler pigment, and a high hiding pigment. The ratio of the pigment to the binder (varnish) should be between 6 to 1 and 10 to 1, preferably within a range of 8 to 1. The filler and hiding pigment should not constitute more than 50% of the total pigment. About 1/3 of the filler and hiding gives the maximum flame retarding qualities. The thermoplastic binder, or varnish, must be such that it retains a thermoplastic character after aging and complete oxidation and polymerization of its drying constituents. It has been found that this varnish may be either a heat bodied resin oil combination, an alkyd, or a cold cut varnish of a thermoplastic resin and a drying or semidrying oil. It is preferred to use a thermoplastic resin of a melt point (ring and ball) of 140° to 150°C., preferably 145°C., together with a bodied linseed oil cold cut in a solvent. The preferred resin is of a modified phenolic type. One resin that has worked exceedingly well has a melt point of 142° to 149°C., acid number within a range of 10 to 20, a specific gravity of 1.10 and viscosity in poise of 3.3 at 56% solids in toluene. A similar resin to this is sold under the name Amberol F-7. A bodied linseed oil that is satisfactory is sold under the name Linogel 85. Example of the fire retarding coating composition that gives excellent results:

	Parts
Modified phenolic resin (Amberol F-7)	25
Bodied linseed oil (Linogel 85)	25
Drier (cobalt naphthenate 6%)	0.45
Drier (lead naphthenate 16%)	1.35
Solvent (xylol)	115
Solvent (V.M.P. naphtha)	90
Sodium tetraborate $5H_2O$, powdered	214
Boric acid, powdered	53
Calcium carbonate (Atomite)	107
Lithopone medium oil absorption	27

The sodium tetraborate compound may be ordinary borax containing 10 mols of water (52.8% anhydrous equivalent). However, it has been discovered that the sodium tetraborate with 5 mols of water (69.3% anhydrous equivalent) is at least twice as effective as the 10 mols of water variety. It has been found that if the 10 mol variety (borax) is dried until it has a uniform anhydrous sodium tetraborate equivalent between 69.3 to 80%, this combination is equally as good as the 5 mol variety. Calcined borax, which has an anhydrous equivalent of 77% is satisfactory.

The boric acid basically is used to prevent afterglow; however, it does have the effect on the intumescent effect of the borax. It has been found that a mixture of four parts of sodium tetraborate of the 5 mol variety with one part of boric acid gives the best all around results. The boric acid employed is powdered so it can act as a pigment as well as the sodium tetraborate. For a filler pigment, in general, nearly all pigment works; however, it has been found that calcium carbonate is the best since it contributes to the intumescence of the coating. Calcium carbonate is actually a supporting filler pigment. On the other hand, calcium carbonate contributes very little to hiding, so for nonyellowing and high hiding coatings, lithopone or its equivalent is recommended.

Guanidine-Phosphoric Acid Resin Salts

W. Juda and G. Jones (U.S. Patent 2,628,946; February 17, 1953; assigned to Albi Manufacturing Co., Incorporated) have found that reaction products of basic nitrogen-containing anion exchange or acid adsorbent resins with nonoxidizing inorganic acids, which reaction products can be termed "resin salts", are substantially less inflammable than the basic resins themselves and are capable of forming a fire retardant and heat insulating carbonaceous mass when subjected to elevated temperatures, for example, of the order of temperatures of flaming wood, wall board and the like, burning in air. By "nonoxidizing" acids it is meant the monovalent hydrogen halides and the polyvalent acids. As basic resins use is made of condensation products of amines or other organic bases with aldehydes, phenols or other condensing agents.

Acid adsorbent resins are obtained in general when basic nitrogen-containing organic compounds such as amines, or guanidine are condensed with aldehydes, hydroxy aldehydes (sugars, starch, etc.), or with mixtures of aldehydes and ketones, under such conditions that not all of the basic groups are linked to aldehyde residues in the formation of the resinous structure; that is, only resins containing free basic groups exhibit anion exchanging properties. Typical acid adsorbent or anion exchange resins suitable for the formation of fire retardant and potentially heat insulating compositions with strong inorganic acids include the following classes: Resins made by condensing

(1) meta substituted aromatic amines or aromatic polyamines with aldehydes and/or carbohydrates with or without ketones;

(2) aliphatic polyalkylene polyamines with aldehydes and ketones;

(3) aliphatic polyalkylene polyamines with phenols and aldehydes;

(4) a strongly basic derivative of cyanamide, with a weakly basic derivative of cyanamide, and with an aldehyde.

In addition, a carbohydrate may be included.

While it has been found that the anion exchange and acid adsorbent resins in general are suitable for the preparation of fire retardant resin salts, it has been noted that the resins of class (4) form more stable resin salts which are generally better fire retardants. In this preferred class of resins it is necessary to use an aldehyde as one of the resin forming agents. While it is possible to use a carbohydrate in addition to the aldehyde, carbohydrate cannot be used in place of the aldehyde because the resulting mixture containing no aldehyde is no longer a fire retardant resin salt. The following example illustrates the process:

Example: (a) 119 grams of guanidine carbonate and 126 grams of melamine were added to 322 grams of aqueous formaldehyde (37%). The ratio of the resin forming ingredients was thus 4 mols of formaldehyde to 2 mols of combined guanidine and melamine and thus a 2:1 ratio between the formaldehyde and the cyanamide derivatives. The mixture was allowed to stand for 20 minutes. Then 42.5 cc of concentrated phosphoric acid (85% H_3PO_4) was added in small portions until the evolution of carbon dioxide stopped. The mixture was agitated continuously until an aqueous, slightly viscous dispersion was formed. This dispersion gelled suddenly when heated on a steambath for 15 minutes. The gel was cut into small pieces and dried at 105°C. The dried gel was then washed with water, to remove any excess of unbound phosphoric acid, until the wash water was phosphate free.

Thus was obtained a resin phosphate of excellent fire retardant and potentially heat insulating properties. The resin phosphate is water-insoluble, the phosphoric acid being chemically bound to the resin. Thus the solid product obtained is very different from a mere mixture of phosphoric acid, or one of its salts with a resin. It is substantially water-insoluble. It resists treatment by HCl, H_2SO_4, and HBr, showing that the resin phosphate is more stable than the sulfate, hydrochloride or hydrobromide. It is substantially irreversible in the sense that a substantial fraction of the phosphoric acid combined with the resin can no longer be removed by treatment with alkali. The product, when subjected for two minutes to a temperature of 750°C. in an electric furnace, expanded to a carbonaceous mass having a volume several times that of the original sample and this expanded mass remained substantially intact throughout the test.

(b) The slightly viscous aqueous dispersion obtained in (a) prior to gelling by heat was applied to one side of a wallboard by brushing. The wet board was heated at 90°C. for three hours, and the water was evaporated and a uniform heat cured resin salt coat formed on the board. The coated board showed a greatly reduced inflammability on the coated side.

(c) 25 parts by weight of guanidine carbonate, 25 parts by weight of melamine and 62 parts by weight of aqueous formaldehyde (37%) were added to 60 parts of water. The ratio of the resin forming ingredients was thus 1 2/3 mols of formaldehyde to 1 mol of combined guanidine carbonate and melamine: To this mixture was added, very gradually and with thorough stirring, 42.5 parts by weight of phosphoric acid (85%). In a typical factory test application, a fiberboard was sprayed with this resin salt solution. The treated wet fiberboard was then run through a hot air oven approximately 60 ft. long, heated to approximately 400°F., at a range of 56 ft. of board per minute.

In another factory test application an untreated fiberboard was preheated by passage through a drying oven, was then sprayed with the resin salt solution and immediately subjected to the action of a metal surfaced pressure roll, the surface temperature of which approximated 500°F. In both cases there was obtained a fiberboard coated with a firmly adhering fire retardant resin salt film of excellent appearance. Plasticizers (for example p-toluene sulfonamide) and pigments may be added to the resin salt solution if desired.

Dicyandiamide-Formaldehyde-Starch Reaction Product

In a process described by L.W. Eckert (U.S. Patent 2,632,741; March 24, 1953; assigned to Armstrong Cork Company), a dicyandiamide-aldehyde resin is reacted with starch by mixing starch and water with a dicyandiamide-aldehyde resin and heating the resulting mixture to effect the desired reaction. The reaction product thus formed is then combined with monocalcium phosphate and preferably also with phosphoric acid to produce a fire resistant coating composition which will coke and intumesce to a substantial degree upon the application of a flame. The starch resin combination has binding properties and serves to hold the coating firmly anchored to the surface to which it is applied. This may be supplemented by other binders such as gelled starch obtained by heating a 15% dispersion of starch in water to a temperature of 190°F. and cooling to 140°F. or lower prior to combination with the other ingredients. The dicyandiamide-aldehyde resin may be prepared as follows:

Example 1:

	Parts by Weight
Dicyandiamide	22
Formaldehyde (37% solution)	71
Phosphoric acid	7

The formaldehyde and dicyandiamide are mixed together for two minutes, and then the phosphoric acid is added. The acid serves as a catalyst for the reaction. Reaction between the formaldehyde and dicyandiamide is continued until a clear solution is obtained. The reaction vessel is jacketed, and cooling water is applied to maintain the temperature of the reaction below 120°F. and preferably at a temperature of 90° to 105°F. With a batch such as referred to above, the reaction will be effected in about one hour's time.

The proportioning of dicyandiamide to formaldehyde preferably should be in the range of 1 mol of dicyandiamide to 2 to 3 mols of formaldehyde. An excess of aldehyde may be present without deleterious results, but there should be adequate aldehyde in the reaction mixture to combine with substantially all of the dicyandiamide. The resin thus prepared is employed in the formation of a heat reaction product with starch, in the following manner:

Example 2:

	Parts by Weight
Dry ungelled starch	37.8
Dicyandiamide-aldehyde resin	40.0
Water	213.0

The ingredients are mixed together and the mass is heated with agitation to elevate the temperature to 190°F., and such temperature is maintained for thirty minutes. Improvement in the final produce will be noted with heating for as little as five minutes at 190°F. There is believed to be a reaction between the starch and the resin, and it is known that the final product possesses better washability than a similar coating composition prepared from a simple mixture of dicyandiamide-aldehyde resin and starch, particularly when the product is applied to fiberboard and dried for 2 1/2 to 4 minutes at 350°F. It appears that the curing of the coating in the drying oven is accelerated by initial reaction of the dicyandiamide-aldehyde resin with the starch. Also improved intumescence is provided in the finished coating over that obtained with a mixture of starch and dicyandiamide-aldehyde resin. A coating composition may be prepared from the material of Example 2 with other ingredients as shown in Example 3 on the following page.

In the preparation of the coating composition, ingredients 1 to 4 are mixed together to produce a dispersion of the monocalcium phosphate and other ingredients in water. A dispersion of titanium dioxide in water is also prepared, and the remaining ingredients are added to the monocalcium phosphate dispersion and mixing is continued until a uniform product is obtained. In the composition of Example 3 the monocalcium phosphate is a water-insoluble salt which fireproofs the starch of the composition; being water-insoluble, it minimized the possibility of paint failure which might occur under high humidity conditions. The monocalcium phosphate is unreacted and is chemically stable in the dispersion. Because of the reactivity of the monocalcium phosphate with alkaline materials to produce gases, the composition must be substantially free of any such alkaline ingredients which would so react with the monocalcium phosphate, for when the monocalcium phosphate is thus reacted, the effectiveness of the fire resistant coating composition is lost.

When 25 grams of the coating composition of Example 3 are applied per square foot to a perforated fiberboard acoustical unit, the same will meet the requirements of Section F-3c of Federal Specification SSA 118a when tested in accordance with the procedure outlined in F-3c of that specification. The coating does not substantially reduce the sound absorption efficiency of the product, and when dried at a temperature in the order of 350° F. for 2 1/2 to 4 minutes, the surface is tack free, smooth, and free of checks and cracks and meets the washability requirements of Federal Specification TTP 88a.

Example 3:

		Parts by Weight	Resin Solids	Starch Solids
1.	Monocalcium phosphate	72.0	---	---
2.	Phosphoric acid	17.5	---	---
3.	Sodium pentachlorophenate	1.6	---	---
4.	Water	70.0	---	---
5.	Reaction product of Example 2	96.0	7.3	12.0
6.	Titanium dioxide dispersion (50% solids)	15.0	---	---
7.	Iceberg clay	22.5	---	---
8.	Pine oil	1.6	---	---
9.	Quadrofos (sodium tetraphosphate)	1.2	---	---
10.	Mica filler	13.0	---	---
11.	Dry ungelled starch	73.0	---	73.0
12.	Resin of Example 1	40.5	22.0	---
13.	Water	30.0	---	---

Aminoplast, Dicyandiamide and Carbohydrate

J.E. Robison (U.S. Patent 2,648,641; August 11, 1953; assigned to Fir-Tex Insulating Board Co.) describes a retardant coating which comprises a reacted aminoplast-aldehyde resin, an amido compound and a carbohydrate. To these active ingredients may be added pigments, preservatives, thickening agents, plasticizers, etc., as desired to meet the conditions of application. Either an organic solvent system or an aqueous solvent system may be used to carry the ingredients. The applied coating may be cured with heat or with a catalyst, depending on the circumstances of application.

The amido compound may be any derivative of cyanamide, such as are used normally in the production of an aminoplast. Examples are melamine and dicyandiamide. The carbohydrate can be any carbohydrate, such as sugars, starches, gums and cellulose. The following examples illustrate the process:

Example 1:

Water	60 parts
Melamine resin	20 parts
Dicyandiamide	40 parts
Starch	20 parts

The fire retardant coating made as above was applied as a coating to a structural fiber insulating board and was then tested for fire retardant properties in accordance with the procedure given in Commercial Standard CS-42-49. When the material was applied with a coverage of 15 grams of solids per square foot, the char area on the fiberboard underneath the foam was less than six square inches. Black foam was created by the flame, which rose more than one inch above the surface being tested, and formed an insulating layer which prevented the intense heat from the alcohol flame from reaching the inflammable wood fibers. Washability of this coating was tested in accordance with the procedure given in Federal Specification TT-P-88a. The coating withstood over 175 strokes of the brush before any failure of the film occurred.

Example 2:

Water	60 parts
Titanium dioxide	10 parts
Melamine resin	20 parts
Melamine	40 parts
Sugar	20 parts
Sodium pentachlorophenate	0.3 parts

This pigmented coating, when applied to structural fiber insulating board with a coverage of 23 grams per square foot (15 grams of fundamental ingredients per square foot), gave exactly the same fire protection as did the unpigmented coating given in Example 1.

HALOGENATION OF CELLULOSIC FIBERS

In Situ Halogenation of Cellulosic Materials

A process described by T.G. Brandts (U.S. Patent 3,092,537; June 4, 1963; assigned to Consolidated Paper Corporation Limited, Canada) involves the halogenation (in situ) of the organic material to be flameproofed, follow, if necessary, by a washing out and/or a neutralization of those soluble acidic products of the halogenation reaction which might later have a harmful effect. While flameproofing by bromination is often more effective than by chlorination, the latter, mainly because of cost, is the preferred form of halogenation. The preferred neutralizing agent is ammonia; lime can, under certain conditions, be found to be especially effective. The process has the following advantages: (a) it is relatively simple; (b) the cost of treatment is low; (c) the main flameproofing components remain an insoluble integral part of the material being flameproofed, such that subsequent water leaching does not seriously impair its fire resistant properties; (d) by a proper choice of neutralizing agent, the overall fire resistant properties can be further improved, for example, use of ammonia or lime markedly improves glow resistance; (e) in some applications, other properties of the material are improved, for example, after flameproofing insulating boards, the modulus of rupture is improved substantially.

Many cellulosic fibrous products and related materials contain phenolic substances such as tannins, phlobaphenes, and lignin. It has been found that, under suitable conditions, chlorine or bromine reacts rapidly with the phenolic material in these materials to give substitution and possible addition products which act as fire retardants. For example, when a piece of wood (white pine) was wetted, chlorinated at room temperature, and exposed to the open flame of a Bunsen burner, it did not catch fire; on the other hand, when a similar piece of wood at the same moisture content, but untreated, was exposed to the flame for the same period of time, it caught fire and burned down to ashes. Similar observations were made for such materials as wood pulp, straw and bark pulps, bark cork and the like; in these cases, all the materials showed a remarkable fire resistance when exposed to the open flame. The same effect was noticed when some of the reaction products, mainly hydrochloric acid or hydrobromic acid were neutralized with an alkali such as sodium hydroxide, ammonia and calcium hydroxide, etc. In the case of ammonia and calcium hydroxide, however, it was found that the glow resistance of the halogenated material was improved.

To evaluate the fire retardant properties quantitatively the following test was devised as other known tests were too elaborate or too cumbersome for a rapid evaluation of these properties: since the density of a material has a great influence on its fire retardancy, samples were reduced to pellets, 3/8" thick by 1/4" in diameter with a density of 60 lb. per cubic foot, by pressing one gram samples (10% moisture) in a die at 40,000 psi in a hydraulic press. For each fire test, four pellets were oven dried for 2 hours at 105°C. and the dry weights recorded. The pellets were placed longitudinally on two parallel connected and parallel placed spring loaded No. 22 Nichrome V resistance wires which were 0.20" apart and ignited by sending a 16±0.2 amp. 220 volt electrical current through the wires for a period of 3 minutes and 30 seconds. This current and ignition time guaranteed a total combustion of all untreated organic material. After burning and afterglow had stopped the pellets were dropped into aluminum dishes and the residual weight determined. The residual dry weight expressed as a percentage of the original weight was taken as a measure of the fire retardancy. Within each fire test these percentages varied remarkably little and only in some cases where the pellets were on the border line of catching fire or not could a larger spread be noticed. The process is illustrated by the following examples.

Example 1: A rectangular piece of pine wood of 1" x 1" x 3", moisture content approximately 20%, was kept in a chlorine atmosphere for thirty minutes. The color of the wood changed to an orange yellow. The residual chlorine in the container was removed and gaseous ammonia allowed to enter. The color of the wood changed to brown. After the neutralization was completed, after about three minutes, the sample was dried and conditioned to 7% moisture content. This sample was exposed to the flame of a Bunsen burner for a period of 2 minutes. When the burner was removed, flaming stopped within 5 seconds; the duration of afterglow was about ten seconds. An untreated, but similar piece of wood kept burning for four minutes; the duration of the after glow was one minute.

Example 2: 100 gram samples of the following material were chlorinated at a moisture content of 50%

(a) Straw — Oat straw was treated in an attrition mill equipped with spiked tooth plates. The average length of the straw particles was 1/4".

(b) Wood particles — Spruce chips were processed in a hammermill to obtain particles with an average length of 1/4".

(c) Groundwood — A coarse spruce balsam groundwood with a Canadian standard freeness of 430 cc.

(d) Bark fibers — Spruce balsam fibers, containing a small amount of wood fibers and cork.

The chlorinations were performed in a rotating, 5 liter, round bottom flask. The chlorine was introduced at a known constant rate through a Teflon gland with neoprene O ring. The round bottom flask was equipped with a calcium chlorine and sodium hydroxide tube through which the displaced air and possible hydrochloric acid and water vapor had to pass. The amount of introduced chlorine was determined by weighing the flask plus material before and after chlorination. The round bottom flask was cooled to keep the chlorination temperature constant at 20° ±2° C. The chlorinated products were divided into two equal portions; one part was dried at 105°C.; the other part was treated with ammonia gas to neutralize the hydrochloric acid formed during chlorination. One gram samples of air dry material were pressed into pellets and fire tested on electrically heated wires as described above. The results are given in the table below.

Raw material	Percent chlorine introduced B.D. basis	Percent residue after fire testing	
		Not neutralized	Neutralized with ammonia
Straw	0	2	2
Do	8	65	56
Groundwood	0	0.5	0.5
Do	4	21	25
Wood particles	0	0.5	1
Do	4	56	57
Do	10	65	62
Bark fibers	0	4	4
Do	4	41	54
Do	8	65	60

Example 3: 200 gram samples of the following materials were chlorinated at a moisture content of 40%. (a) Redwood bark fibers. (b) Douglas fir wood fibers. (c) Spruce-balsam bark fibers. All these fiber samples were shredded and defibered so as to make them suitable for insulating purposes (insulating wool, insulating pads, etc.). The chlorinations were performed at room temperature in a rotating, 5 liter round bottom flask. The chlorine was introduced through a Teflon gland and neoprene O-ring by means of a large calibrated gas burette. By raising a levelling bottle filled with a chlorine saturated calcium chloride solution a known amount of chlorine was measured into the reaction flask at a constant rate from the burette. The chlorinated bark fibers were neutralized with known amounts of gaseous ammonia in the same way (using paraffin oil in place of the calcium chlorine solution).

After chlorination and neutralization all samples were oven dried at 300° F. and conditioned at 50% R.H. and 72° F. Uncompacted samples of 3/4" cube at a density of approximately 4 to 5 lbs./cu. ft. were placed on two parallel, spring loaded, No. 22 Nichrome V resistance wires, which were 1/2" apart. The samples were ignited by sending a 22 ± 0.2 amp. 220 v. electrical current through the wires for ten seconds. Treated insulating wool samples were also leached in water for 3 hours, dried and conditioned before testing. The results of these tests are given in the table below. If the sample had not caught fire after ten seconds the current was kept on until the sample started burning. Recorded were: (a) Ignition time in seconds — The length of time elapsed from the moment the current was switched on to the moment the sample started flaming. (b) Flame time in seconds — The length of time the sample was flaming. (c) Glow time in seconds — The length of time elapsed from the moment flaming stopped to the moment glowing stopped.

Material	Chlorine added (B.D. basis)																	
	0%			5%			7%			9%			11%			13%		
	a	b	c	a	b	c	a	b	c	a	b	c	a	b	c	a	b	c
Redwood bark wool	6	14	44	23	2	13	[1]	0	10	[1]	0	10	[1]	0	3	[1]	0	5
Soaked	----	----	----	8	3	7	8	2	7	11	1	8	24	3	13	----	----	----
Douglas fir wood wool	6	27	13	7	25	6	8	4	9	15	5	12	26	6	2	37	3	0
Soaked	----	----	----	8	35	15	7	33	6	7	18	6	9	15	4	9	11	7
Spruce-balsam bark wool	6	20	42	----	----	----	16	5	10	30	7	5	[1]	0	10	[1]	0	5
Soaked	----	----	----	----	----	----	----	----	----	8	6	12	10	4	7	----	----	----

[1] Samples did not catch fire but were carbonized after approximately 60 seconds.

Halogenation of Lignocellulosic Structures

M. Lewin (U.S. Patent 3,150,919; September 29, 1964; assigned to the Government of Israel) has found that by the introduction of halogen into the lignin molecules the combustion resistance of the lignocellulose can be enhanced. This has the advantage that it is the inflammable material itself whose properties are modified, that is, the combustion resistance conferred in this way is not dependent upon the presence of an extraneous material which can be leached out. Since the lignin surrounds the fibrous cells, the process makes provision for a protective coating around each fiber.

For the purposes of the process the use of bromine has been found to be particularly useful. The proportion of halogen used for the treatment of the lignocellulosic material depends on many factors, e.g., the halogen used, the nature of the material to be flameproofed, the degree of flame resistance which is required. For a given degree of halogenation, the time required to effect the halogenation will in general be less where the lignocellulose to be treated is finely comminuted. This applies for example in the case of groundwood or wood pulp, and in other cases where a large specific surface is available for example in the case of yarn and rope, as contrasted with chips for the fabrication of chip board, or with wooden veneers or structural wood.

For example, groundwood treated with 4 to 8% by weight of bromine will produce a paper which does not burn when ignited with a match. Smaller additions of bromine will confer a smaller degree of noninflammability while nevertheless representing a considerable improvement as compared with paper produced from untreated groundwood. The general properties of the materials treated in this way are not substantially changed, especially if care is taken to avoid degradation of the cellulose by the acid formed as a by-product of the halogenation process, which can be done, for example, by neutralization of the reaction mixture or by the use of buffers. The process is illustrated by the following examples:

Example 1: 80.4 grams of air dry beech veneer sheets of 1 mm. thickness were placed in a closed desiccator-like vessel of 6 liters' capacity. The air was evacuated from the vessel, then 12.9 grams of gaseous bromine were introduced from an ampule into the evacuated vessel. After 30 minutes the bromination was completed and the vessel was flushed with air. The veneer sheets were rinsed in cold water until neutral, and dried in air. The veneers were then analyzed by the Carius method for bromine and were found to contain 8% by weight of bromine.

Specimens of the brominated beech veneer of a size of 20 x 20 cm. were tested in the Inclined Panel Inflammability tester according to British Standard 476/1932-Fire Resistance, Incombustibility and Non-Inflammability of Building Materials and Structures. The veneers did not inflame and were classified as "not inflammable". While no fire appeared or spread on the sample, only an area of 10 $cm.^2$ which was in contact with the ethanol flame used in the test became charred and underwent dry distillation without flame. A similar sample of non-brominated beech veneer burnt completely when tested under the same conditions. 19 specimens of the brominated beech veneer of 5 x 5 cm. size and 19 specimens of the non-brominated beech veneer were tested for biological stability according to the T.A.P.P.I. (Technical Association of the Pulp and Paper Industry, U.S.A.) Specification T 487M55 against *Aspergillus niger*, *Chaetomium globosum* and *Aspergillus terreus*. The microorganisms did not develop after 2 weeks on the brominated beech veneer specimens but covered completely the non-brominated specimens. This test showed that the brominated specimens were completely resistant to the attack of microorganisms according to the specification aforesaid.

Example 2: 24 grams of Okumè wood veneers of 1.5 mm. thickness were brominated with 2.35 grams of gaseous bromine for 15 minutes similarly as in Example 1. After the bromination the sample contained 4.1% of bromine. The sample was rinsed in water and immersed for 3 minutes in 300 ml. of a solution containing per liter 18 grams of borax and 1 gram of boric acid, and air dried. Specimens of the brominated Okumè veneers were then tested in the Inclined-Panel Inflammability tester in the manner described in Example 1. The specimens were classified as noninflammable. Non-brominated specimens of the same Okumè veneer burnt out completely when tested under the same conditions.

Example 3: 520 grams of commercial spruce groundwood were suspended in 4.0 liters of water at room temperature. A solution of 40 grams of bromine in 4.0 liters of water was added to the suspension with vigorous stirring in the course of 3 minutes, then the stirring was continued for another 2 minutes. The mixture was filtered and the amount of hydrobromic acid, determined in the filtrate by potentiometric titration with 0.1 N sodium hydroxide, corresponded to 20 grams of brimine, i.e., half of the bromine originally added to the suspension was recovered as hydrobromic acid. The other half had reacted with lignin of the groundwood structure.

The filtered mass was returned to the vessel and neutralized with a dilute ammonia solution so as to adjust the pH of the suspension to 7 as measured with a Beckmann model pH meter. The pulp was then filtered again. The lignocellulosic material contained 4% of bromine calculated on the over dry weight of the groundwood, as determined by Carius determination. A part of the pulp was made into standard sheets of a basic weight of 250 $g./m.^2$ in the Standard T.A.P.P.I. Sheetmaking Apparatus. The sheets were tested by the Inclined-Panel Test as described in Example 1. The sheets did not inflame and were classified as noninflammable. Several sheets were stored for 210 days at 21°C. and 70% relative humidity. After that time the sheets were analyzed for bromine by the Carius determination and were found to contain 4%. The stored sheets were tested by the Inclined-Panel Test and were noninflammable.

Phosphorylation of Halogenated Lignocellulosics

A process described by R.A. Davis (U.S. Patent 3,459,588; August 5, 1969; assigned to The Dow Chemical Company) involves improving the fire retardant properties of lignocellulosic articles in which the lignin component is halogenated. In the process lignocellulosic articles having improved fire retardant properties are obtained by halogenating, i.e., brominating or chlorinating the lignocellulosic material and then reacting the resulting product with trialkyl or triaryl phosphites, or phosphonates where the alkyl groups preferably contain from 1 to 4 carbon atoms, and the aryl groups may be halogenated, if desired. Phosphites and phosphonates that are particularly useful include, for example, triethyl phosphite, triphenyl phosphite, tripropyl phosphite, ethyl diphenyl phosphite, tris-(2-bromoethyl) phosphite, tris-(bromopropyl) phosphite, bis-(2-bromoethyl) 2-bromoethyl phosphonate and bis-(bromopropyl) bromopropyl phosphonate.

The temperature of reaction is from 100° to 150°C. and is advantageously 130° to 150°C. At these temperatures the alkyl bromines undergo a Michaelis-Arbuzov reaction with the trialkyl or aryl phosphite which results in the attachment of the phosphorus to the lignin in the form of an alkyl or aryl phosphonate and the loss of an alkyl or aryl bromide. Since the aryl and secondary alkyl bromines will not undergo this reaction, they will be left to furnish the halogen necessary for the synergistic effect on the fire retardancy. The lignocellulosic articles are advantageously halogenated to a content of from 2 to 12 weight percent, and then phosphorylated to a content of from 1 to 4 weight percent phosphorus. The extent of phosphorylation is controlled by the amount of primary alkyl bromine present, the reaction temperature and the time of reaction, which may be from a few minutes to several hours and is advantageously from 30 minutes to 2 hours. The process is illustrated by the following examples:

Example 1: Five grams of hardboard fiber (groundwood) which had been brominated by the method of U.S. Patent 3,150,919 and containing 3.86 weight percent bromine by neutron activation analysis was charged to a 500 cc flask. Triethyl phosphite, 99 g., was added and refluxed at 130°C for two hours. The treated fiber was then filtered on a vacuum filter and washed twice with 2B absolute ethanol using 50 ml. each washing. The treated, washed wood fiber was then placed in a 500 cc flask and held at 25°C. and 1 mm. pressure for two hours to remove traces of ethanol and remaining triethyl phosphite. After removal from the flask the material was analyzed by neutron activation and found to contain 2.9 percent by weight bromine and 2.6 weight percent phosphorus. The bromine content was thus reduced to 75% of its original value. A small amount of this treated fiber would not flame or smolder when ignited with a match flame and the flame was removed. A sample of the original brominated fiber would continue to burn.

Example 2: The experiment of Example 1 was repeated except that the phosphite employed was triphenyl phosphite (103 g.) and the temperature was 150°C. The treated wood was analyzed by neutron activation and found to contain 3.3 percent by weight bromine and 4.36 weight percent phosphorus. A small amount of this fiber was tested with a match as in Example 1 and did not flame or smolder.

PLASTER

Cementitious Plaster Composition

F. Ware (U.S. Patent 3,372,040; March 5, 1968) describes a method of preparing fire resistant cementitious plaster compositions by adding effective amounts of a naturally occurring phosphate rock and shale. Large amounts of exfoliated perlite and/or vermiculite are being incorporated in gypsum plaster or other cementitious mixtures in order to provide walls, partitions, ceilings, etc., with coatings which insulate these structures and thus provide fire protection. These fireproofed plasters may be sprayed directly to columns, beams, corrugated, flat, or cellular floors, with conventional plastering machines. It is important to insulate these members from extreme temperatures which cause them to weaken or fail under building loads. The critical temperature for structural steel is approximately 1000°F. and for walls, floors, and ceilings approximately 250°F. above room temperature. Structural members coated with insulating plasters have been able to withstand these temperatures for periods up to four hours by applying the plaster in thicknesses up to 2 inches.

While these insulating plasters are satisfactory for insulating structures from these temperatures for a period of time, they do not, however, retard the flame and are inadequate at extremely high temperatures. Both the gypsum and portland cement containing plasters when subjected to high temperatures have a tendency to lose water, which helps to retard the transmission of heat until all of the water is released, but then begins to harden and crack exposing the structural members to the open flames.

It has been found that cementitious or plaster compositions of the type used in the construction field, building churches, schools and hospitals, could be made fire resistant by adding to the composition effective amounts of a naturally occurring phosphate rock; the phosphate rock containing in addition to 0.25 to 40 percent by weight of P_2O_5 other associated elements which coact to function as a fire retardant, and thus prevent the transmission of heat. The presence of this P_2O_5-containing rock in these building materials prevents the transmission of heat at temperatures as high as 2000°C. and thus is effective in protecting the structure for periods acceptable by most building codes.

The naturally occurring phosphate rock and shale is obtained from phosphiria formations and deposits and has 0.25 to 40 percent by weight of P_2O_5. This rock and shale can be crushed to 10 to 350 mesh and added to the cementitious or plaster composition in the range of 1 to 50 percent by weight of the total mixture. Portland cement, calcined gypsum or calcium sulfate hemi-hydrate, and mixtures may be combined with up to 50 percent of this naturally occurring phosphate rock. In addition to the rock and cementitious material, other ingredients may be added depending on the ultimate use of the product. These may include, for example, 0 to 20 percent by weight of finished lime; 0 to 50 percent by weight of exfoliated insulating material, i.e., expanded perlite, vermiculite, and mixtures thereof; 0 to 2 percent by weight of an adhesive binder; and 0 to 1.0 percent by weight of a surface active agent or detergent. Either the portland cement, calcium sulfate, or mixtures of both of these materials in any proportions, may contain 1 to 50 percent by weight of the phosphate rock and shale, and small amounts of one or more of the other ingredients. Typical cementitious and plaster compositions of this process are illustrated in the following examples. The plaster composition may be mixed with water to obtain a wet composition having 40 to 60% by weight of water.

Example 1:		Percent by Weight
	Portland cement	0 to 90
	Calcium sulfate	0 to 90
	Lime	0 to 20
	Exfoliated insulating aggregates	0 to 50
	Adhesive binder	0 to 2
	Surfactant	0 to 1
	Phosphate rock and shale	5 to 50

Example 2:		
	Portland cement	50 to 90
	Lime	0 to 20
	Exfoliated aggregate (perlite and vermiculite)	0 to 50
	Adhesive binder	0 to 2
	Surface active agent	0 to 1
	Phosphate rock and shale	5 to 50

Example 3:		
	Calcium sulfate hemi-hydrate	50 to 90
	Lime	0 to 20
	Exfoliated aggregate (perlite and vermiculite)	0 to 50
	Adhesive binder	0 to 2
	Surface active agent	0 to 1
	Phosphate rock and shale	5 to 50

Example 4:		
	Calcium sulfate hemi-hydrate	50 to 90
	Lime	0 to 10
	Phosphate rock and shale	10 to 50

To illustrate the effectiveness of this rock and shale as a fire retardant in the cementitious or plaster compositions described, a specific petroliferous shale having a P_2O_5 content of approximately 9.7 percent, together with the associated elements, was used. This sample was taken from Cut A in section D, located in Section 15, Township 12, South Range 5, West Principal Meridian, Montana.

The phosphate shale, containing approximately 9.7 percent by weight of P_2O_5 was mixed with gypsum in ranges of 10, 25, and 50 percent by weight of the plaster, and subjected to a flame test to determine the effectiveness of the shale as a fire retardant. The test comprised the application of 3/8 of an inch of gypsum plaster, containing 10 percent of the shale on a 3/8 inch plaster board. The same procedure was followed in preparing 1/8 inch of plaster containing 25 percent of the shale on a 3/8 inch plaster board, and 1/16 inch of plaster containing 50 percent of the shale on wood and pressed board. The test was conducted by using a propane torch having a flame temperature in excess of 1900°C. on a 10 x 10 test area which was placed in front of the torch with the flame impinging the plastered surface. The temperature on the unexposed or reverse side of the plastered panels was determined by the time required for the heat to penetrate the plaster and char the substrate.

With the flame impinging the plaster surface, the substrate, i.e., backing of the plasterboard was charred after 15 minutes, whereas the panel containing none of the phosphate shale, or the blank, was charred within approximately one minute. Thus, the test indicates the effectiveness of the P_2O_5 and associated elements in the naturally occurring shale to inhibit the flame. Additional tests were conducted by using gypsum plaster and applying it in thicknesses from 3/8 inch to 1 1/2 inches onto an asbestos substrate. Temperatures were recorded on the reverse side of the substrate at a distance of 1/2 inch away while the plastered surface was subjected to a propane flame. The gypsum plaster contained 10 percent by weight of the phosphate shale having a P_2O_5 content of approximately 9.7 percent by weight. Plastered panels with and without the phosphate shale were flame tested with the results shown on the following page.

It should be noted from the described tests that the gypsum plaster in thicknesses of 3/8 to 1 1/2 inches gave temperature readings of 400°F. approximately 1/2 hour after the surface had been heated by the flame, whereas those samples which contained ten percent by weight of the phosphate shale did not exceed 176°F. The surface of the shale containing plaster, after the test, had a vitreous appearance but did not crack and crumble as did the samples which did not contain the phosphate shale. This indicates that even at extreme temperatures the P_2O_5 and associated elements in the phosphate shale effectively retard the transfer of heat and cause the plaster to harden and take on a stone-like appearance.

Plaster Thickness	Gypsum Plaster Temperature after ½ hr. (° F.)	Gypsum Plaster, 10% Phosphate Shale Temperature after 6 hrs. (° F.)
⅜ inch	400	176
¾ inch	400	139
1 inch	400	107
1½ inch	400	80

Use of Unexpanded Vermiculite of Specified Particle Size Prevents Spalling

G.S. Willey (U.S. Patent 3,454,456; July 8, 1969; assigned to United States Gypsum Company) describes a fire resistant, vermiculite based wallboard. In a study of the action of gypsum wallboard when exposed to fire, such as in a fire test, it was evident that there was a substantial shrinkage of the wallboard core at high temperature with consequent cracking which not only materially contributed to the passage of excessive heat and hot gases through the wall but also hastened the actual disintegration of the board under such adverse conditions. Hence, if this excessive shrinkage could be overcome or decreased the fire resistance would be substantially increased.

U.S. Patent 2,681,863 indicates that the even distribution of individual short textile glass fibers throughout a gypsum cast markedly reduces cracking when the cast, such as in the form of a core of a gypsum wallboard is subjected to fire. While the addition of the glass fibers to the board core did not markedly decrease the shrinkage, it enabled the calcined core to be held intact, and hence so materially increased the fire rating of the wallboard that a 5/8" thick board was acceptable to the Underwriters' Laboratories for a one hour fire rating when used as the ceiling under wooden joists with oak flooring placed thereover, i.e., their Design No. 1-1 hr. It has been established in a great number of building codes that such a construction must qualify for the above rating.

In U.S. Patents 2,526,066 and 2,744,022 there is set forth a further improvement in the fire resistance of a gypsum cast by the addition of minus 28 Tyler, 30 U.S. Standard, or minus 35 Tyler, 40 U.S. Standard, sieve sizes of unexpanded vermiculite particles to the calcined gypsum water mix prior to the setting. This addition substantially decreases the adverse contraction of a gypsum cast subjected to high temperature conditions because of the expansion of the vermiculite when it also becomes subjected to the high heat. In fact, the expanding vermiculite could cause an expansion of the core which, if excessive, would be undesirable as the board would buckle and become loosened. The expanding vermiculite not only aids in overcoming shrinkage, but also increases the heat insulation with consequent retardation of the flow of heat to the back of the board.

As mentioned U.S. Patent 2,744,022, the vermiculite as it expands causes the core of the board to disintegrate due to spalling which takes "place in a progressive manner so as to destroy the wall within a relatively short time...." This patent, as well as U.S. Patent 2,681,863 propose that the adverse effect of spalling be overcome by the addition of "bodying fibers", such as those made from glass. While ceramic fibers enable the unexpanded vermiculite to be used, they do not prevent the excessive expansion of the larger, unexpanded vermiculite particles which causes the core to become weak, limp and crumbly. This expansion is manifested by surface spalling when a gypsum cast containing such a vermiculite is subjected to high heat.

It has been found that very low fire shrinkage with practically zero spalling in a set calcined gypsum cast can be effected by a unique particle size distribution of unexpanded vermiculite in such cast. It has been found that unexpanded vermiculite of a particle size range which will pass through a No. 50 mesh sieve and be retained upon a 140 mesh sieve, preferably that passing a 70 mesh and retained from a 100 mesh, U.S. Standard sieve sizes, will not permit an adverse spalling condition but will still control shrinkage or contraction when a gypsum wallboard containing such particles is subjected to fire. It is possible, however, to have a small amoant of vermiculite particles slightly coarser than those which will pass a No. 50 mesh sieve, but such should be less than 10% of the whole and substantially all should pass a No. 30 mesh sieve. Particles substantially smaller than those retained upon a 100 mesh sieve can be used as they do not cause spalling; however, they are not as effective in controlling the fire contraction as the coarser sizes nor are they as effective in decreasing the flow of heat.

An unexpanded vermiculite ore, of a particle size distribution, similar to that known as BO5 from the Travellers' Rest deposit of the Zonolite Division of W.R. Grace & Company has been found to be very satisfactory in carrying out this process commercially. The importance of the particle size in preventing spalling, improving core integrity, and controlling expansion under fire conditions is clearly illustrated in the table on the next page. This table summarizes the results of a fire test of wallboards having set calcined gypsum cores. Each of these cores contained a different particle size range of unexpanded vermiculite from the Travellers' Rest deposit of the Zonolite Company.

These boards were all made as nearly the same as possible with the exception of the particle size range of the unexpanded vermiculite added to the calcined gypsum core composition prior to set. The composition used to make each board also contained substantially the same amount of certain of the well known set accelerators, consistency reducers and bonding aids usually used in the commercial manufacture of a gypsum wallboard in order to duplicate actual plant conditions as closely as possible. The same quantities of foam and water were used in all cases so as to obtain substantially the same core density. This along with the thickness was controlled so that the variations were no more than plus or minus 1%. No glass fiber or boric acid was used. Only unexpanded vermiculite was added to the gypsum core composition along with, of course, the usual board plant additives well known to the art and referred to above. All of the boards contained the same amount of unexpanded vermiculite, i.e., 6% based upon the weight of calcined gypsum used to form the boards. The core composition was enclosed between the same type of paper surfaces as used in regular commercial wallboards.

These boards were all thoroughly dried in the same manner so that none had burnt or under-dried cores. They were fire tested in a horizontal position as a ceiling under very closely controlled, uniform conditions in a small gas fired laboratory fire test furnace, which was designed so as to approximate on a small scale the fire test of the Underwriters' Laboratories. For each test the furnace temperature was held as closely as possible to the time temperature curve relationship shown on page 22 of Underwriters' Laboratories publication 263, dated May, 1959. A temperature of 1550°F. was reached in 1/2 hour, 1700°F. in one hour and 1850°F. in 2 hours at which time the test was stopped.

The linear contraction is obtained by spring actuated dial gauges attached to probes extending into the furnace and pushing against opposite edges of the wallboard under fire test. The thickness contraction was determined by measuring the core thickness when cold before and after the test, allowances being made for the thickness of the paper. The integrity is ascertained during the test by visual observation of the board and by a study of the core, such as the change in core thickness, i.e., the thickness contraction at the conclusion of the fire test. In actual construction under fire test, the linear contraction, if excessive, will cause undesirable cracks to form in the board and/or openings to form between the boards and permit the hot gases therethrough. The core integrity is also very important as it is indicative of the way the board will stand up under fire conditions.

The Effect of the Particle Size of Unexpanded Vermiculite (No Glass Fiber) in a 1/2" Thick Gypsum Wallboard Core upon Subjection to Fire

Board No.	Particle size range of unexpanded vermiculite: Passing	Retained on	Max. linear contraction during test, percent	Description of core after test
1	30 sieve	40 sieve		Disintegrated in 9 mins. 40 sec. and fell into furnace.
2	40 sieve	50 sieve		Disintegrated in 12 mins. 40 sec. and fell into furnace.
3	50 sieve	70 sieve	.025	Moderate spalling, core intact, good integrity.
4	70 sieve	100 sieve	1.8	Very fine shallow map cracks on surface.
5	100 sieve	200 sieve	.51	Surface smooth, some small map cracks.
6	200 sieve		.78	Similar to board No. 5, only surface smoother.
7	50 sieve	100 sieve	.17	Very fine shallow map cracks. Very slight spalling. Core intact.
8	BO5 see Table 3		.36	Very fine shallow map cracks. No spalling.
9	BO4 see Table 3			Center of board fell into the furnace in 13 mins.
10	No vermiculite		.73	Very deep fissure, some as wide as .090".

All tests were stopped after 120 minutes. All sieve sizes were U.S. Standard. The above data clearly illustrates the critical importance of the proper particle size range of the unexpanded vermiculite in order to have good integrity of the board core along with practically zero contraction such as shown in the minus 50 plus 70 mesh range (board No. 3); the minus 70 plus 100 mesh range (board No. 4); and the minus 50 plus 100 mesh range (board No. 7). Note the destructive spalling resulting from the use of the coarser size particles such as the minus 30 to plus 40 and the minus 40 to plus 50 ranges in the core. These are within the size ranges for the unexpanded vermiculite which the above identified patents predicted would cause spalling and disintegration of the core unless glass fibers are also used.

Board numbers 5 and 6 show that particle sizes smaller than 100 mesh are not as effective in preventing shrinkage, though they do not cause spalling. The effect of the sizes of the unexpanded vermiculite particles upon core integrity and shrinkage when boards containing different particle size ranges are subjected to fire is clearly set forth in a comparison of boards 8 and 9 made from two commercially available ores from Travellers' Rest, South Carolina, i.e. BO5, and BO4 respectively. The former board contained unexpanded vermiculite within the contemplated range and had good core integrity, while the vermiculite used in the latter was outside the size range of the process and hence the core disintegrated within 13 minutes.

In order to determine the effect of both glass fiber and unexpanded vermiculite upon the properties of the core of a gypsum board when subjected to fire, boards were made and tested similar to those referred to in the previously mentioned table. These boards contained 6 lbs. of chopped glass fiber or 0.4% by weight, per 1000 sq. ft. of 1/2" gypsum board evenly distributed throughout the core along with the 6% unexpanded vermiculite.

The Effect of the Particle Size of Unexpanded Vermiculite and Glass Fiber in 1/2" Thick Gypsum Wallboard Core upon Subjection to Fire

	Particle size range of unexpanded vermiculite		Max. linear contraction during test, percent	Description of core after test
	Passing	Retained on		
Board No.:				
11	30 sieve	40 sieve	.08	Very bad spalling but did not fall.
12	40 sieve	50 sieve	.058	Severe spalling and poor integrity.
13	50 sieve	70 sieve	.033	Slight spalling, core intact. Surface map cracks.
14	70 sieve	100 sieve	.025	Shallow surface map cracks.
15	50 sieve	100 sieve	.06	Very fine shallow map cracks. Very slight spalling.
16	BO5 Ore		.21	Very fine shallow map cracks. No spalling.
17	BO4 Ore		.19	Similar to board No. 12.
18	No vermiculite		.62	Deep fissure up to .050" wide. Core intact.

From the above, it is evident that the glass fibers aid in preventing a complete disintegration of the core due to spalling as mentioned in certain of the above identified patents. The glass fibers also reinforce the core after it has become weakened due to calcination. They also further aid in overcoming linear contraction. This process has been found to be most useful in the formation of gypsum wallboard cores whereby a markedly improved fire rating for constructions employing such wallboard is obtained. The process has made possible an Underwriters' Laboratories approval for a one hour fire rating on a construction referred to as Design No. 42-1 hr. This construction employs a one half inch thick gypsum wallboard containing chopped glass roving and unexpanded vermiculite having particle sizes within the range comprising this process. The Design No. 42-1 hr. comprises a gypsum wallboard ceiling under wooden joists with a wooden floor applied over the joists. This same board also qualifies for Design No. 41-1 hr., a construction similar to the above in which the board is furred from the joists with a resilient channel as set forth in U.S. Patent 3,090,164. This board also qualifies for Design No. 221-2 hrs.

In the past a 1/2" gypsum wallboard could not be used in constructions similar to Design No. 42-1 hr. A 5/8" thick gypsum wall board containing glass fiber but no vermiculite was required, i.e., Design No. 1-1 hr. referred to above. Except for minor details, such as the width of the finished floor and the size of nails, there is very little difference between Design No. 1-1 hr. and Design No. 42-1 hr. other than the former required a 5/8" board and the latter the 1/2" board of this process. It should be noted that the quantity of glass fiber in the 1/2" board of Design No. 42-1 hr. is the same quantity as is the 5/8" board of Design No. 1-1 hr.

The use of the thinner board not only saves material but also decreases the board weight with consequent savings in shipping cost and application labor. The gypsum wallboard used in the above Underwriters' Laboratories fire tests employed a 1/2" wallboard made with regular production equipment. The core of the 1/2" wallboard contained 6% of BO5 ore along with 6 lbs. per 1000 sq. ft., or about 0.4%, of chipped glass fiber roving of between 3/8" to 1/2" in length evenly distributed throughout the core, and 0.75% of boric acid. A bond aiding material, a foaming agent and a set accelerator, all well known in the art were also used. All percentages are based upon the weight of calcined gypsum used to form the core. The finished 1/2" wallboard had a minimum weight of 1925 lbs. per 1000 sq. ft. and a minimum average thickness across the field of the board of 0.508".

As evident from the foregoing, chopped glass roving is not necessary to offset any disintegration of the core due to spalling when vermiculite particles within the particle size range of this process are employed. However, glass fibers are always added to special fire resistant commercial gypsum wallboards so as to further improve core integrity thereby minimizing the possibility of failure due to stresses which the board may be subjected to as a result of fire. Good core integrity will enable the core to withstand the Underwriters' Laboratories hose stream test. The chopped glass roving also functions as a means to provide a further reduction in the fire or heat shrinkage. From the tabulated data the boric acid appears to decrease the fire contraction of the wallboard. The use of boric acid, however, is not necessary in all cases but when used at least 0.1% by weight of the calcined gypsum should be included.

CEILING TILE AND PANEL CONSTRUCTION

ACOUSTICAL CEILING TILE

Acoustical Correction Ceiling Panels

H.J. Sabine (U.S. Patent 2,838,806; June 17, 1958; assigned to The Celotex Corporation) describes a noncombustible or fireproof acoustical correction product. Due to the many factors which enter into the rating of a construction material as "fireproof," technical test specifications having been set up for the determination of such ratings. The generally accepted test specification for the rating of acoustical materials as to fireproofness is that of the U.S. Bureau of Standards, Federal Specification SS-A-118b, Acoustical Units Prefabricated.

To meet the strict requirements as to Class "A" (fireproof) products under the above test specification, the acoustical product is comprised of two parts; a base material and a surface coating. Referring to Figures 3.1b and 3.1c, unit 10 of the acoustical correction product comprises a sheet of gypsum board, the core of which is designated by numeral 11 which is provided with paper facings 12. To the outer surface of one facing 12 of the core 11 there is adhered a layer of porous paper 13, and a decorative coating of fire resistant paint 14 is applied to the outer face of the other facing sheet 12.

The complete combination described will pass the requirements for classification as type "A" (fireproof) under Federal Specification SS-A-118b. The base member of the product without face coating 14 will not meet the requirements for such classification. Obviously, face coating 14 so modifies and coacts with the base member 11 that its character is altered to the extent that the requirements for classification as fireproof are imparted. For this product the gypsum board base member is not an ordinary gypsum wallboard, although it is not different in appearance. The core portion 11 of the gypsum board is comprised of a special mixture imparting fire resistant properties.

Core portion 11, in addition to the usual gypsum wallboard core ingredients, contains Wollastonite finely divided, and relatively short mineral fibers. The gypsum board base member is prepared in the manner as described in Patent 2,080,009 or 2,172,076 except that in the preparation of the core portion 11 there is incorporated in the mix, finely divided Wollastonite to the extent of 10% by weight based on the weight of gypsum stucco. Additionally, there is incorporated in the mix about one-half percent, or within the range of about one-quarter to one percent based on the weight of gypsum stucco of mineral fiber of about one-half inch in length. These mineral fibers, of course, cannot all be cut to accurate length, and it is to be expected that such of one-half inch nominal length will probably range in length of the various fibers within the range of one-quarter inch to one inch.

A typical formulation for the core 11 of the gypsum wallboard sheet, for 1,000 square feet of nominal 3/8 ich thickness is:

	Pounds
Gypsum stucco	1070
Wollastonite, pulverized, 20 to 30 mesh	107
Mineral fibers, nominal 1/2 inch	2
Core adhesive	9
Accelerator	2 to 4
Rosin soap	1.8

Subsequent to the step of providing perforations 15 in the base sheet there is applied to one face a fabric sheet 13. The essential of sheet 13 is that it shall have porosity to air within definite limits and sheet 13 may be a felt or paper sheet formed of wood, glass, asbestos, rayon, or other synthetic fibers, or of hair, fur, or other fibrous material which may be felted to sheet form. Another alternative of the sheet 13 is a thin continuous sheet, such as a sheet of cellophane, saran,

FIGURE 3.1: FIREPROOF ACOUSTICAL CORRECTION PANELS

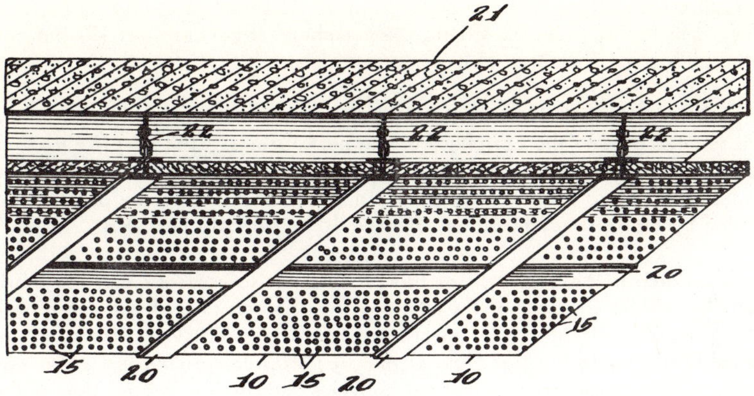

a.

b.

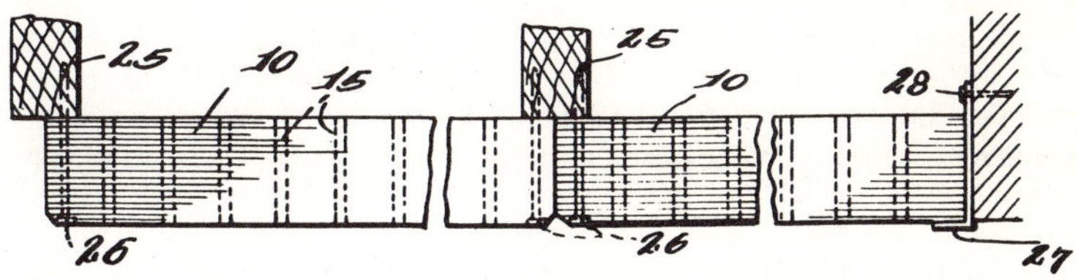

Figures 3.1a and 3.1b - Sound Absorbing Installations

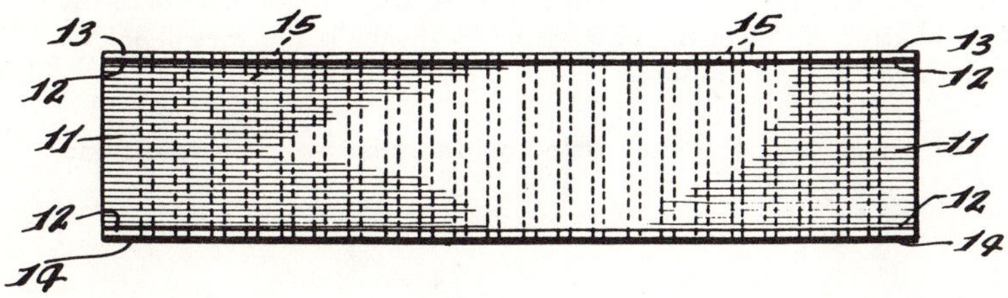

c.

d.

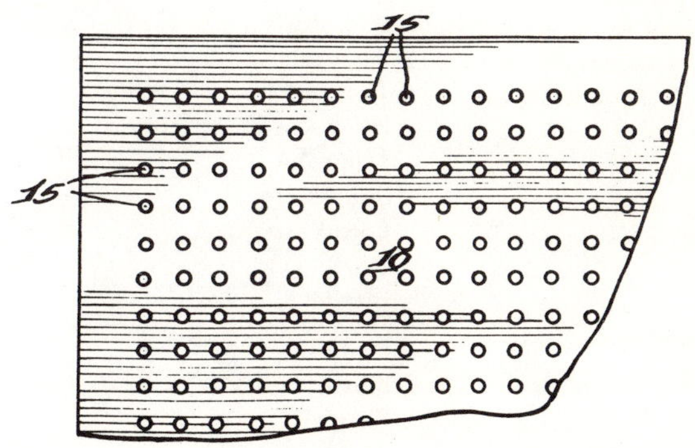

Figures 3.1c and 3.1d - Sound Absorbing Panel Material

Source: H.J. Sabine; U.S. Patent 2,838,806; June 17, 1958

or the like, which is very finely perforated to provide the requisite porosity. The fabric sheet 13 is adhered to the outer face of the paper facing 12 by suitable adhesive. Preferably, fabric sheet 13 is adhered to facing sheet 12 by adhesive applied to the outer face of the facing sheet 12. By such technique the fabric sheet 13 may be firmly and continuously adhered to facing sheet 12 without the adhesive getting on those portions of fabric sheet 13 overlying perforations 15. The fabric sheet 13 must be firmly adhered to facing sheet 12 around the periphery of perforations 15, but porosity to air of those areas of the sheet overlying the ends of perforations 15 must be retained.

In the application of the fabric sheet 13 it is critical that the material be firmly mounted to the rear surface of the board extending over the perforations 15 and secured around the periphery of the perforations. The specific manner of arriving at such results is relatively unimportant. The fabric 13, for example, might be in the form of small circles, or of other geometrical form, individually applied at each perforation 15 and each firmly secured around the periphery of the perforation. Best absorption would be provided by fabric sheet 13 alone, if it could be suitably mounted.

To secure the maximum absorption effect, which is the true basis of this process, it is necessary that the fabric sheet be sufficiently stiff that in effect it is nonvibratable on impact of the sound waves thereon so that the sheet, when impacted by the sound waves, at least does not appreciably vibrate. The porous sheet having effective stiffness, mounted so as to be nonvibrating, provides an air-permeable barrier to the free flow of the air pulses comprising the sound waves, and as a consequence the pulsing of the sound waves through the porous sheet converts and dissipates a portion of the energy of the sound waves to provide acoustical correction.

Due to the fact that practically, as a ceiling, the required porous sheet cannot be mounted so as to have the required stiffness, it results that resort must be had to the construction of this process. By exposing independent, relatively small areas of the porous sheet material 13 to the impact of the sound waves, each such area firmly secured around the periphery of its exposed portion, the portion overlying a perforation 15, the necessary effective stiffness of the porous sheet 13 is achieved.

The product so far described, because of the paper facings, will not, of course, pass the Class "A" fireproof rating of Federal Specification SS-A-118b. To enable the product to obtain the rating it is necessary to alter and protect the character of the facing sheet on the side of the unit from which the test procedure is applied. The requisite alteration or modification of the character of the facing sheet and provision of the necessary protection is accomplished as follows.

The modification of the facing sheet and protection is accomplished by a three-step application of coating. The three steps of such coating comprise the application of a first or base coat, the principal function of which is that of imparting flame resistance to the paper facing, and over which there is applied a second coat, the principal function of which is that of a protective coating, and finally a third coat, the principal function of which is to harden and consolidate the previously applied coatings so that the composite coating is integrated of itself and with the paper face sheet 12. This coating applied to a face sheet 12 is identified by numeral 14 in Figure 1.

The first or base coat of coating 14 is comprised of components which are listed below under two categories; one, "Pigment", and the other, "Binder":

Pigment		
Asbestos	15	lbs.
Aquasol	2.5	gal.
Ti Pure FF	25	lbs.
Hi White china clay	300	lbs.
Mica AA	250	lbs.
Santobrite	1	lb.
Borotherm	300	lbs.
Binder		
Protein	60	lbs.
Borax	10	lbs.
Pine oil	8	lbs.
Waxine	8	lbs.

In the formulation the asbestos is of the very fine or short fibered grade normally used as a paint filler. The Aquasol is a surface active agent product of the American Cyanamid Company, technically a sulfonated castor oil. Ti Pure FF is a white pigment available from E.I. du Pont de Nemours, being the anatase form of titanium dioxide. The Hi White china clay is a white clay of the grade normally used in paints, which is primarily a filler, although it has some pigmentation effect. Mica AA is a usual finely divided mica such as is commonly used as a paint filler ingredient. The Santobrite is a fungicide available from Monsanto Chemical Company, the active ingredient of which is sodium pentachlorophenate. The Borotherm is a flame retardent agent available from the American Potash Company, being a water-soluble complex borate salt containing 57% B_2O_3 by analysis. The protein is the so-called "A" or substantially pure protein which is available from many sources. The Waxine is an acid stable type wax emulsion of crystal and paraffin wax of melting point 122°F. which is comprised of 45% wax emulsified in water with 5% rosin based on the total composition.

The base coat described is applied to a surface sheet 12 of the base unit at the rate of approximately 2.75 to 3.25 gallons per 1,000 square feet and dried. The Aquasol wetting agent causes this coating to strike into the paper surface and in particular to carry the fungicide and flame retardant ingredients of the base coat, and thus to modify the sheet 12 to the extent of rendering it relatively flame and fungus resistant. Over the base coat there is subsequently applied a protective coat which is comprised of pigment and binder portions as follows:

Pigment	Pounds
Asbestos	10
Ti Pure FF	125
Hi White china clay	350
Mica AA	200
Santobrite	1
For tinting, raw sienna	0.2
Binder	
Carboxy methyl cellulose	1.5
Protein	60
Borax	8
Pine oil	8
Waxine	8

This second coating composition is applied over the dried base coat mixture at the rate of approximately 1.5 to 1.75 gallons per 1,000 square feet and suitably dried. Finally, to consolidate and integrate the paper facing 12 and protective coating, there is applied a third or "hardening" coat which insolubilizes the binder ingredients of the preceding coatings and anchors such to the paper facing 12, particularly through the integration or consolidation of the protein contents in which, as referred to, protein of the first coating has been caused to soak into or penetrate into the facing sheet 12. The such hardening of the coatings by the application of the third coat not only integrates and consolidates the paper facing 12 and the coatings, but it additionally so hardens the coating that it is protected from attack by roaches and is sufficiently hardened and waterproofed that a high degree of washability is imparted. The hardening coat comprises:

	Pounds
Dry alum	100
Borax	57
Sodium acetate	50
Formaldehyde, 35 to 40%	8

The above composition is made up with water to 150 gallons and is applied over the second or protective above described coat at the approximate rate of 1.5 to 2 gallons per 1,000 square feet and subjected to suitable drying. The composite coating 14, as described, combines with and alters the character of the paper facing sheet 12 so as to render it flame resistant, and additionally provides an integrated and mineralized protective layer thereon which together impart characteristics to constitute a combination which is capable of passing the Class "A" fire resistant rating of Federal Specification SS-A-118b. Additionally, the such coating integrated with the paper facing 12 provides a finished product, the surface of which is roach and fungus resistant and is sufficiently hard and insolubilized to the extent that it is quite washable, but it additionally comprises a decorative surfacing which in this instance, based on the compositions as above, is white and highly light-reflective, which is a normal requisite for a ceiling sound absorbing product.

Based on theoretical considerations and checked by actual test, it has been determined that the sound absorbing units, as described, function most efficiently when mounted in such manner that they are spaced approximately between 10 and 12 inches from a surface, as, for example, the ceiling of a room. A so suspended ceiling is shown in Figure 3.1a where the sound absorbing units are mounted on a grid of inverted T members 20 resting on the upwardly directed flanges thereof, and which inverted T members are suspended from a room ceiling or the like 21 by means of hangers 22. Merely as an example of a possible alternative mounting of the sound absorbing units 10, they may be mounted against the lower faces of spaced wood joists or the like 25 by nails or screws, or the like 26, and at the edge of an assemblage of such units 10, where they abut against a side wall, the extreme edge of a unit may be supported by means of an L shape support member 27, or an equivalent support member nailed or otherwise suitably secured to the side wall, as by beams 28.

Mineral Fiber Ceiling Tile

A process described by J.E. Cadotte and E.W. Juntti (U.S. Patent 3,248,257; April 26, 1966; assigned to Wood Conversion Company) involves the application of fiber-coating material to mineral fibers of a mineral fiber tile, to oppose the tendency of the tile to deform on exposure to flame.

The process involves the use of liquid containing chemical means, preferably aqueous suspensions, or solutions, or both, to be applied to the mineral fibers and to provide dry material on the surfaces of dry fibers which is a refractory deposit on the fiber surfaces, thus to stiffen them and to resist the normal tendency to flex when highly heated. The material to be applied may be so incorporated that fibers throughout the body are coated with refractory material, but this method with some materials offers difficulties. When a slurry is dewatered which contains the fibers and certain kinds of treating agents which are in solution or present as dispersions in water, some of the agent is lost in the removed water. Economy in use of that type of processing material is achieved by applying the liquid containing it to the surface of a fiber felt after dewatering, either while still wet from pressing a dewatered felt, or dried, or in two applications, one on the wet pressed felt and another after drying. When the felt is formed by an air-laying process, that type of treating agent may be introduced to the body of the fibers without such loss. Certain other materials which are applied and which react after application to produce a water-insoluble agent, are not so lost in dewatering.

When the refractory-forming agent is present throughout the body of the fiber felt, various methods of mounting the tile are available. But, when the processing is limited to the back face of a mounted tile, the method of mounting the tile is limited. Mineral tile or panel is commonly mounted at opposite edges by suspension from metal runners in such a way that the metal is hidden from view and protected from exposure to flame by a flange portion of the tile. The metal is more sensitive to heat than the mineral fiber body, and one objective in mounting is to protect the metal from exposure to direct heat of a flame. When the edge-mounted mineral body sags to an excessive extent, it may expose the metal runners to flame or direct heat of a flame, according to the design of the joint, which runners then warp or bend, resultingly permitting the body to drop or to be dislocated so that an opening in the tile ceiling is produced. Although a tile may remain integral on long exposure to heat or flame, it is the mounting of it that also must resist the heat and flame so that the tile remains mounted and its mounting means withstands the heat sufficiently to prevent dropping of the tile and prevent formation of an opening through the tile ceiling.

With respect to mineral fiber tile horizontally mounted as illustrated and described in this process, it has been found that when it lacks a treatment according to the process, its shrunken form will bend or sag downwardly when heated to redness to an excessive extent, and thus so shorten its span between metal runners that the metal could be exposed. Sometimes, the shortening pulls it away from at least one supporting metal flange, letting the tile drop. If this does not take place, the exposed metal runner so softens and warps that it permits the tile to droop without dropping, or to drop, in either case providing an opening through the tile ceiling for access of flame to the space above the tile.

In the Figure 3.2, the tiles are illustrated by conventional 12-inch squares. Suspended ceilings of such tile commonly have a plenum chamber above them and under a floor or roof above the tile in which space service equipment may be housed. The tile ceiling should confine flames to the space below it, at least for some predetermined required period of time. In Figure 3.2a, the numerals 10 designate parallel metal runners to which the tiles are mounted in a manner to protect the runners by at least a portion of the thickness of the tile.

One way to mount the tile to the runners 10 is to provide the runner with two lower flanges 11 and 12, in a T-formation with a vertical web 13, and to provide kerfs or grooves in side edges of tile to receive flanges. The tiles have interfitting tongue and groove structures in the two pairs of opposite edges, and these are constructed to provide continued protection to the metal flanges. A joint may open by limited sagging of tiles. In Figure 3.2a, a tile 15 is indicated as having a tongue 16 entering a groove in like adjacent tile 17, which groove duplicates groove 18 in tile 15, said groove 18 receiving a tongue of tile 19. The remaining tongue and groove of tile 15 for mounting on the runners 10 are shown in Figures 3.2b and 3.2c.

A tongue 20 extends to the right in Figure 3.2b, being about 1/3 as thick as the tile which is approximately 3/4-inch thick. The upper face 21 of the tongue extends into the body of the tile as a kerf 22, which kerf receives a flange 11 of a runner 10. The tile edge-face 23 above the tongue 20 is inwardly of the tile edge-face 24 below the tongue 20. The opposite grooved edge of tile 15 has a groove 26 in the central third of its thickness to receive a tongue 20 of another tile, thus forming two encompassing flanges 27 and 28.

Figure 3.2c shows tile 15 suspended on runners 10 alongside two similar tiles 15a and 15b of which the parts corresponding to the parts in tile 15 are similarly designated. In this relation tile flange 27 rests on runner flange 12 and forms a space 30 between it and the end of the opposite flange 23 for the web 13 of runner 10. The runner flange 11 fits into kerf 22. Figure 3.2d shows the normal joint of Figure 3.2c opened up to a degree, which happens when the tiles sag. In the normal joint the runner flanges 11 and 12 are protected by about two-thirds of the thickness of the tiles. As the joint opens, the extreme edge portion of runner flange 12 loses some of this protection from the heat and flame which

FIGURE 3.2: FLAME RESISTANT MINERAL FIBER TILE

a.

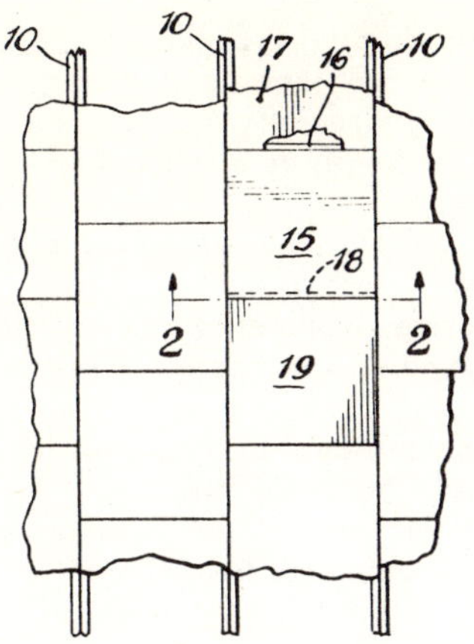

Suspended Tile Ceiling

b.

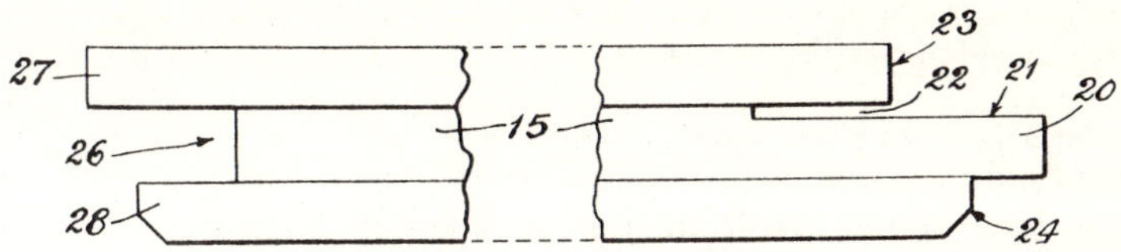

Broken Edge View of Tile

c.

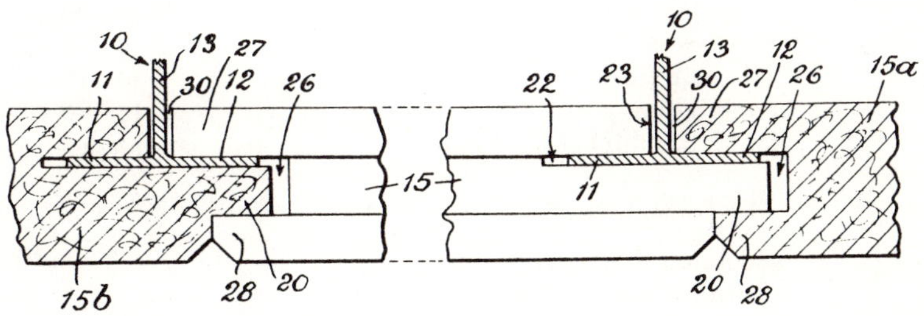

Mounted Tile

d.

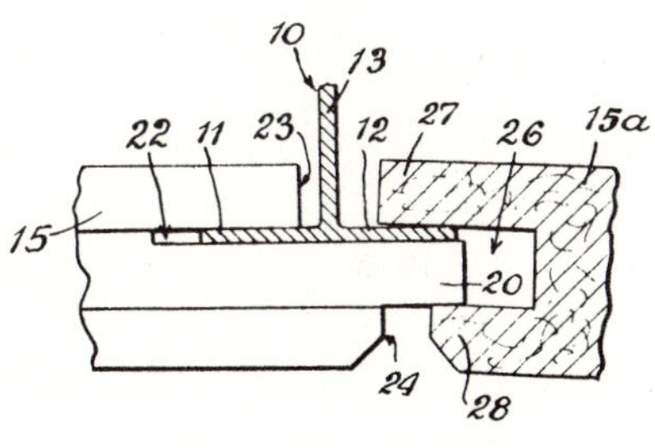

Mounted Tile Showing Sagging

e.

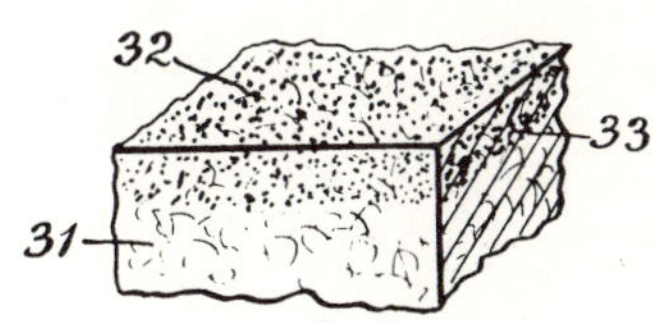

Portion of Tile Coated with Sodium Alginate

Source: J.E. Cadotte and E.W. Juntti; U.S. Patent 3,248,257; April 26, 1966

has caused the tiles to sag. The extent to which the joint may open before exposing the runner or dropping a tile is controlled by designing dimensions. To lengthen the tongue and groove to protect the runner results in weakening the tile at the joint and in likelihood of damage to a tile unit in handling. It has been found that when only the fibers at the back face and in the layer adjacent the back face have on their surfaces deposits of chemical means which is or which forms a refractory material, the treated fibers in their assembled relation in the tile form a rigidifying backing for the tile which stiffens it and lessens the degree of deformation. When the tile is mounted in the manner shown and when the chemical means is present only at the surface fibers, the entire back must be treated including especially the regions of the tile flanges which overlie the runner flanges 11 and 12. Otherwise, the tile flanges tend early to break and let the tile drop.

A variety of chemical means is available for selection to provide, when heated, refractory coverings on the fibers. Phosphate salts are suitable, such as sodium tripolyphosphate, and other chemical means, such as sodium chloride, sodium sulfate, ferric sulfate, ferric chloride, zinc hydroxide, zinc chloride, zinc oxychloride, mixtures of zinc chloride and zinc oxychloride, aluminum sulfate, sodium silicates, calcium aluminate, sodium aluminate, and magnesium sulfate.

However, the character of the mineral fiber is important when selecting the chemical means for the processing. When the fibers are reactive, as when they are alkaline, or contain sulfides, as does mineral fiber made from slag, the refractory material can be formed as a result of reaction of the applied chemical means with the fiber. Glass fibers are much less reactive to the chemical means, and in fact, practically inert in the process of producing the treated tile.

Sodium aluminate is the preferred agent for surface treatment, being alkaline and therefore nonreactive with glass fibers and slag wool. In practice, the preferred tile board is made to include the zinc oxychloride and the finished board is treated to back the tile with sodium aluminate. Suitable formulations for handleable fire-retardant mineral board having density in the range from 12 to 75 pounds per cubic foot, are given below.

Formulation 1	Parts by Weight
Feltable fibers	77 to 100
Mineral fibers	77 to 90
Synthetic mineral fibers	75 to 88
Asbestos fibers	2 to 15
Cellulosic fibers	0 to 10
Starch grains	5 to 20
Nonfibrous thickening agent	As needed

Formulation 2	
Nodulated slag wool	80
Sulfite cellulosic fiber	4.5
Amosite fiber	2.5
Tapioca flour	13.0
Wax-size suspension (aqueous)*	

*As described in Olson U.S. Patent 2,754,206

For forming the preferred board with inclusion of chemical means, the composition below may be used.

Formulation 3	Parts by Weight
Slag wool fibers	79.1
Amosite fibers	2.6
Sulfite cellulose fibers	2.0
Starch grains	13.5
Wax-size suspensions	0.8
Zinc chloride	3.3
Caustic soda	0.5
Coagulating agent	0.0175

A board of 23 1/2 pounds per cubic foot is formed as described. Then, the dry board in the form of cut tile units is coated over the entire back of the tile with sodium aluminate, at the usage of 17 pounds of solids per M square foot. Sodium aluminate has been found so far to be the most effective material for application to the back of a mineral fiber tile. It melts at 1650°C. or 3000°F. which is well above the temperature to which the back of the tile is exposed. It is soluble in water to a limited extent. There are numerous ways to locate sodium aluminate throughout the tile or at and adjacent the back face of the tile. It may be applied as such, or may be formed in situ. It may be sifted into the surface as a powder, or applied as a paste or suspension, either aqueous or nonaqueous. It is preferably used in an aqueous solution containing an excess of undissolved finely divided sodium aluminate. Confinement at and adjacent the back face is sufficient and preferred, thus permitting the exposed face to be decorated, for example. When the tiles are mounted in the manner shown, it is necessary that the location of the sodium aluminate be such as to lie in or on the flanges 23 and 27 to strengthen them for holding during exposure of the tile to flame. Penetration of the aqueous composition is minimized by the presence of undissolved particles and a thickening agent, such as methyl cellulose or starch paste. The thickening agent functions additionally to keep the undissolved content of sodium aluminate in suspension. Various methods of application require viscosity control for particular apparatus, or operation, or panel material, and varying the thickening agent is one way to control viscosity.

A tile meeting the American Society of Testing Materials Test E119-55 is made from Formulation 2, having a density of 20 pounds per cubic foot, and a thickness in the range from 1/2 to 3/4-inch. One face of the wet pressed mat to be dried is treated with the composition given below to apply 15 pounds of sodium aluminate per M square foot. Then, the mat is dried and to the same face when dry is applied the same composition in quantity to apply 10 pounds of sodium aluminate per M square foot. Then, the wet treated face is dried, by passing through an oven or under radiant lamps. The applied composition more easily penetrates the wet mat, so the first application carries the agent slightly inward from the surface and the second application confines it largely over the surface.

Formulation 4	Parts by Weight
Sodium aluminate	450
Water	550
Methyl cellulose	8

Figure 3.2e shows a tile 31 having fibers at its back surface coated with sodium aluminate as indicated at 32, and fibers adjacent the surface, likewise coated as indicated at 33. The following example shows the use of a furnish containing components of the fiberboard with and without chemical means.

Example:

Slag wool	145	parts by wt.
Asbestos	8	parts by wt.
Sulfite cellulose	8	parts by wt.
Tapioca starch	23.6	parts by wt.
Wax-size	1.6	parts by wt.
Water	5000	

A control tile of 18 pounds per cubic foot, was formed in the conventional manner. A test piece 3 x 8 inches was exposed to a flame at 2000°F. on test supports, and it sagged and dropped in 4 to 5 minutes. The same furnish was used to make a like tile with the addition of equal molar amounts of zinc chloride and of caustic soda, as follows:

	Parts
Zinc chloride	6.4
Caustic soda	1.85
Flocculating agent	0.032

Tested in the same manner as the control tile, the test tile remained in place for 15 minutes with a maximum deflection of 9 1/2 mm. Sodium zincate is effective when used like sodium aluminate, as described above. Usage of 10 to 20 lbs. of sodium zincate on the back of a tile decreased the sag in a laboratory test in 15 minutes from 12 to 13 mm. for a control tile to 7.5 to 9 mm., for the treated tile.

In the formula above, glass fibers have been substituted with results not quite as good as with slag wool tile. The difference is not explained, possibly resulting from the differences in diameter and length of fiber. The following table shows the comparative times of failure in a laboratory fire-test, with different chemical means in the furnish.

Chemical Means	Usage[1]	Glass Fiber Failure in Minutes	Comparable Time For Slag Wool	Sag, mm.
Control	-	4.5	7	-
Zinc Chloride	4	6.25	15[2]	12
Zinc Chloride	4	-	-	-
Caustic Soda	0.8	10.75	15[2]	4.5
Floc. Agent	0.02	-	-	-

[1]Expressed as parts per 100 parts of total fiber.
[2]Time not exact, as test stopped after 15 minutes.

Fibrous Acoustical Ceiling Panels

R.F. Shannon (U.S. Patent 3,333,656; August 1, 1967, assigned to Owens-Corning Fiberglas Corporation) describes a method for producing acoustical panels resistant to high temperature and flames. This is accomplished primarily by creating a continuous compressed web of bonded mineral fibers, splitting the web into panel blanks, applying over the upper surface of the blanks a slurry of fine mineral particles, the slurry having sufficient fluidity to soak into the blanks to a predetermined level.

In Figure 3.3a is shown a production line in which acoustical ceiling boards are being treated with a back application of mineral particles. As indicated, panel blanks 10 are received from a conveyor 12 preferably directly from blank producing apparatus including a fiber forming and collection station, a compressing pair of conveyer flights carrying the continuous mat of collected fibers through a binder curing oven, a surface sanding operation and transverse slicing means to cut the cured mat into blanks.

The density of these base panels or blanks of bonded fibrous glass may, for example, be approximately twelve pounds per cubic foot with an organic binder content of twelve percent by weight. However, the density may reside in the broad range from three to twenty five pounds and the binder content run from five to forty percent. As the panel blanks 10 travel along the conveyer 14 they receive on their rear sides, which usually have been lightly sanded and are here facing upwardly, a layer of impregnating material 22. This is in slurry form and flows from the weir 16. A constant flow from the weir is maintained by a continuous supply of the material arriving at the weir through piping 18. The material is spread in an even layer by the inclined flat chute 20.

The impregnating or filler material is of comparatively inert composition, preferably completely or predominantly of a mineral nature. Various materials including silicas and clays may be utilized. An example of a satisfactory formula has solid components of potassium silicate in the amount of sixteen percent by weight, forty-two percent of ball clay, and forty-two percent of colloidal silica. These ingredients are dispersed in water, with the latter constituting 30% by weight of the resulting slurry. The amount of water varies with the fluidity desired and the available drying facilities and may run over 50%. The slurry composition should not include ingredients which expand or contract on drying to an extent to cause warping of an impregnated panel.

To facilitate penetration and even distribution there may be need of fractional percentages of additives such as a wetting agent, a viscosity lowering ingredient, and an anti-forming agent. To effect a more complete sealing action effective for sound blocking by the applied layer of slurry, a minor amount of a film forming element may be added to the mixture. Examples of organic film formers are methyl celluloses and vegetable gums. For sound blocking purposes the sealing layer should be applied to the edges as well as the back of the panel.

The impregnating material soaks into the interior of the fibrous blanks to a depth determined by the porosity of the fibrous mass, the size and shape of the slurry particles and the fluidity of the slurry, facilitated if desired by air drawn at an effectual velocity into the suction chamber 23. The last portion of the slurry remaining on the surface of the blanks is wiped evenly across the blanks by the reciprocating brush 24, having bristles about one inch in length. The brush holder 25 rides in a guide 26. The brush may be approximately one and one half inches thick and be of a length to extend across the full width of the conveyer. A reciprocating stroke eight inches in length with a frequency of seventy strokes per minute has proved satisfactory.

About fifty ounces of the solid content of the slurry in this instance is spread over the back of a panel having nominal planar dimensions of 24 by 48 inches and intended to serve as a ceiling board supported directly along the edges of its exposed face by hanger bars. This amounts to one ounce of the solids applied to approximately twenty-three square inches of the rear surface of such a panel. This single application of material is generally sufficient for ceiling boards for fire and high heat resistance.

While fifty ounces of the solids is preferred for incorporation in the back of one example of ceiling board with the cited dimensions, protection of a more limited nature against fire and heat may be secured with as little as twenty-five ounces of the material. For sound blocking properties a still lower quantity of material may be used in forming a tight back and edge sealing layer. A final sealing coat of material may be deposited upon the upper surface of the panel blanks by the sprinkler pipe 28, particularly if they are intended for acoustical tiles. This material 30 applied in an amount depositing only five ounces of solids over an area of 24 by 48 inches is formulated so as not to so easily enter the body of the blank but to remain on or adjacent the surface.

Uniform distribution is secured through the action of a second reciprocating brush 32. The holder 33 of this brush rides in a guide 34. The application of the slurry 22 from weir 16 and the slurry 30 from the pipe 28 in this example, results in a total of fifty-five ounces of mineral particles being deposited for penetration into the rear surface of each panel having planar dimensions of 24 by 48 inches. The movement of the two brushes 24 and 32 is derived from motor 36. A rotating disc 38 is driven by the motor, as shown in Figure 3.3b. A connecting rod 40 is mounted at one end to an eccentric pin 41 projecting from disc 38. The other end of the connecting rod is joined to the pivoting arm 44 which turns

FIGURE 3.3: FIBROUS ACOUSTICAL CEILING TILE

a.

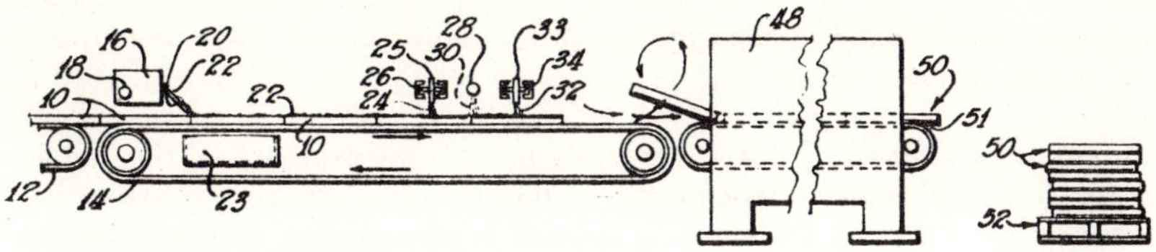

Figure 3.3a - Production Line

b.

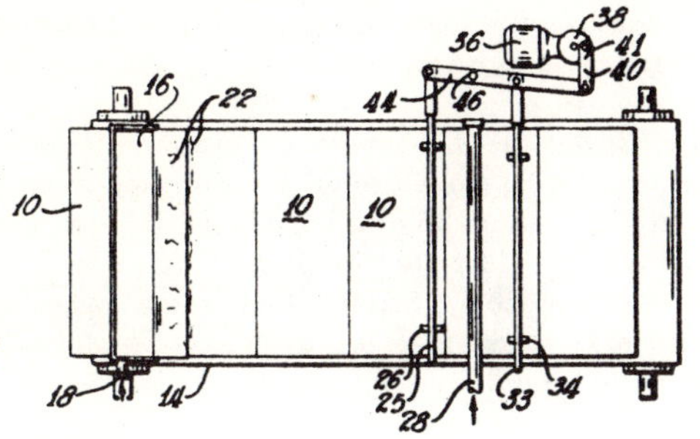

c.

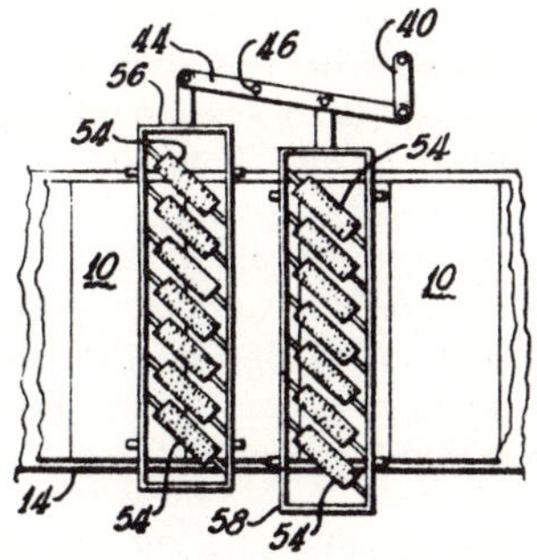

d.

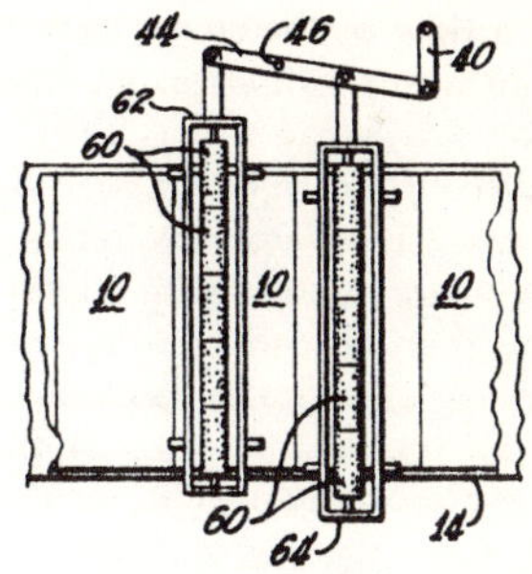

Figure 3.3b, 3.3c and 3.3d - Detail of Production Equipment

e.

g.

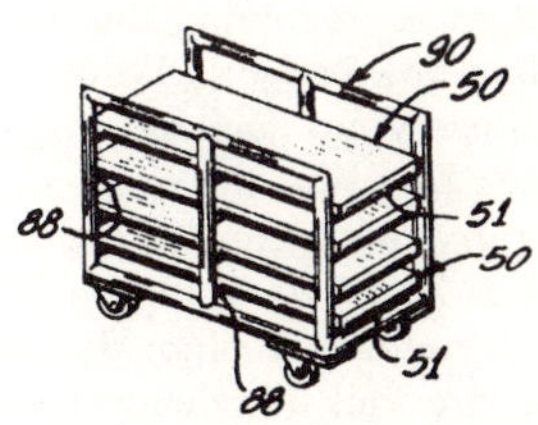

Figure 3.3g - Portable Rack

f.

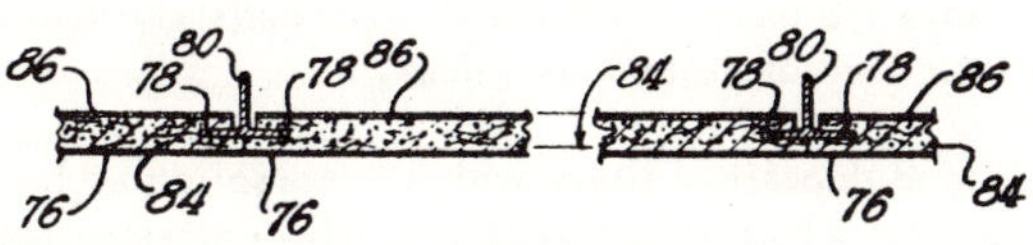

Figure 3.3e and 3.3f - Vertical Sections of Ceiling Tile

Source: R.F. Shannon; U.S. Patent 3,333,656; August 1, 1967

on the pivot pin 46. The ends of the brush holder 25 and 33 are reciprocated by the pivoting action of arm 44 to which they are both connected. The action of the brushes in some manner not completely understood consolidates the portion of the impregnating material disposed upon and adjacent to the upper surface of the panels. It is apparently a fiber spreading and light ramming action. If not so brushed the filler or impregnant has an objectionable porosity which not only permits too great a transmission of sound through the panels but which also detracts from the reinforcing strength available from the mineral impregnant.

A microscopic study of an unbrushed coating of slurry frequently reveals the presence of tiny bubbles. These may be due to an inherent frothing of the slurry, air caught beneath the curtain of slurry, resistance to wetting by the mineral fibers, or to a combination of such or other factors. The brushing appears to disperse any such bubbles and to thus produce a more complete sealing layer of the mineral particles.

The consolidation of the surface material into a fairly impervious layer is also promoted by turning over the panels after they have been back coated and before the coating has been dried. This reversal is indicated by the arrows immediately preceding the oven 48 of Figure 3.3a. Any further penetration of the slurry into the body of the panel is thus curtailed and a flow of the slurry back toward the surface is induced. With this procedure the layer of mineral particles imbedded in the surface fibers may be restricted to a compact thickness of three thirty seconds of an inch.

As discharged from the oven 48 the panels 50 carry the coating 51 on their lower faces. The panels are then loaded on a pallet 52 for transfer to subsequent treating or packaging stations. Instead of being moved through the drying oven 48 the panels 50, as shown in Figure 3.3g, may be laid in reversed position on the cross bars 88 of a hand truck 90, the latter being then placed in a drying oven. Instead of narrow brushes such as 24 and 32 illustrated in Figures 3.3a and 3.3b, rollers 54, as shown in Figure 3.3c, covered with bristles, nap or pile fabric, or sponge material may serve in consolidating the surface layer of mineral particles. The rollers are mounted for rotary movement within frames 56 and 58, which are reciprocated in the same manner as brushes 24 and 32 through a connecting rod 40 and an arm 44 pivoted on pin 46.

Another arrangement with aligned rollers 60 in frames 62 and 64 is illustrated in Figure 3.3d. Other means such as broad rotary brushes and flexible wiping blades may be used to accomplish the spreading and sealing of the mineral coating. When fire stopping panels intended for acoustical ceiling tiles with subsequently kerfed edges for receipt of hanger flanges are processed, deep penetration of the saturant of mineral particles is desired. The suction chamber 23 is then actuated to help draw the slurry from the weir 16 down into the panels.

Additional slurry 30 is then deposited by the sprinkler pipe 28. Such extra material is required to provide a rear sealing coat, consolidated by brush 32; with the slurry from the weir 16 drawn deeply into the panels, the mineral particles are deposited in a somewhat porous or dispersed relation. The fluidity of the slurry, particle size and shape, and the degree of vacuum of suction chamber 23 are selected to cause the slurry to descend into the panels to a predetermined point at least even with the lower walls of the kerfs to be later cut into the edges of the panels. The extra integrating strength under high temperature conditions provided by the mineral particles thus extends down to a point beyond the flanges of the hangers to within one eighth of an inch of the face of the panel, the region where planar splitting or delamination occurs when conventional kerfed ceiling tiles are subjected to high temperature flames. The depth of penetration should be approximately three eighths, one half and five eighths of an inch respectively for panels one half, five eighths and three quarters inch thick.

In Figure 3.3f is illustrated a portion of a suspended ceiling installation incorporating acoustical ceiling tiles 76 with kerfs 78 in their edges for receiving the flanges of hanger bars 80. These panels have the described impregnation of mineral particles extending down slightly below the kerfs 78 as indicated by the terminating line 84. The panels in addition have a substantially impervious back coating 86 provided by the slurry from sprinkler 28. In one example of the process in connection with ceiling tiles the deep penetration is secured with a slurry having a solid content of approximately 50% by weight, the entry of which into the body of the panels being facilitated by air drawn down through the panels into a suction chamber such as 23 shown in Figure 3.3a. A reciprocating wiping member such as the brush 24 of the apparatus of Figure 3.3a also contributes to a minor degree to the entry of the slurry into the body of the panels. About sixty-four ounces of solid particles are thus introduced into the panels for a back surface area of eight square feet.

The back coating 86 of the acoustical tile 76 secured from the second deposit of slurry which is consolidated by a wiping device such as the brush 32 of Figure 3.3a, is restricted to the surface area of the panels by slightly altering the composition of the slurry. Such composition alteration may involve increasing the solid content, omitting a wetting agent or the inclusion of fine mineral fibers or some other bulk providing constituent. In this instance sufficient slurry is applied to provide approximately nineteen ounces of solid mineral particles in a sealing layer for a surface area of approximately eight square feet. This is less than one third of the amount of material in the slurry from weir 16 with which the panel is saturated to the lower sides of the kerfs.

In Figure 3.3e is illustrated a portion of a suspended ceiling installation incorporating acoustical ceiling boards. Smaller acoustical tiles with unkerfed edges would present a similar appearance. As shown, the edges of the boards rest upon the lateral flanges of the hangers 72, which are of inverted T cross section. The hangers 72 are suspended from elements of the permanent ceiling by wires, straps or other suitable devices. The upper or rear side of the panels incorporate the consolidated layer 51 of mineral particles. This layer or coating may have a thickness of only 3/32 of an inch and for the most part lies beneath the surface of the panel. It is therefore, reinforced by fibers of the boards. Laboratory tests have proved that such panels will maintain themselves in position for over two hours with flames at a temperature of 1850°F. directed against their lower or room-facing surfaces. This stability is of great value in restraining the spread of a fire arising in an area of a building beneath such panels.

Tests have also shown that these panels with the consolidated back coating have such a low porosity that in addition to their sound absorbing capacity they serve very satisfactorily in blocking sound transmission therethrough. The panels are accordingly adapted for installation in suspended ceilings above adjoining rooms, where it is desired to keep sounds from one room from passing up through the panels and over the partition separating the rooms into the next room. While a thinner back coating of mineral particles may be produced according to this process and is effective in curtailing sound transmission, its strengthening effect for maintaining the panels in position under fire conditions is impaired.

For sound blocking purposes the layer of particles should be sufficiently compact that when air is passed through the coated panel by establishing a vacuum on one side, the pressure on the outlet side is preferably indicated by a measurement well above sixteen inches in an air pressure measuring water column. However, quite satisfactory performance in this regard is assured if the coating is sufficiently consolidated by wiping and by reversing the position of the panel immediately after it is wiped so that the pressure measurement is as low as sixteen inches in a water column.

Mineral material in an amount between forty-five and fifty-five ounces for every eight square feet of panel surface area has been found to serve satisfactorily for ceiling boards from a fire retarding and sound blocking standpoint. Where deep penetration is also required particularly for ceiling tiles the amount of mineral material is increased to as much as ninety ounces for an area of eight square feet.

Where maximum protection against the spreading of a fire is not required and sound transmission loss is the main consideration the back coating may contain considerably less than forty-five ounces of mineral material for eight square feet, the minimum limit of the specified recommended range, and be of a thickness from only one thirty second to three thirty seconds of an inch. However, the accompanying reduction in weight lowers the sound blocking capacity of the layer. To prevent sound energy traveling angularly into the borders of panels and laterally upwardly from the edges of the panels, the edge surfaces should also be sealed. The same sound blocking mineral composition may be applied to the edges by a roller or spray nozzle.

BUILDING PANELS

Structural Honeycombs for Curtain Wall Panels

A process described by B.M. Randall and R.S. Frobisher (U.S. Patent 2,910,396; October 27, 1959; assigned to The Bettinger Corporation) provides reticulated webs or honeycombs which are highly fire and flame resistant and which have superior strength for resisting stresses and strains, as compared with prior comparable web structures.

Reticulated webs or honeycombs of the general type to which the process relates have been made of treated paper and have constituted cores intervening between panel-facing elements secured to opposite sides of the webs or honeycombs. Previously, however, the paper stock of the webs or honeycombs has been impregnated with resinous substances to impart needed stiffness and strength to the cores, and the resinous substances have made the webs or honeycombs suitably moisture resistant. But, the resin-impregnated paper cores are highly inflammable, and the prior bonding agents for securing the facing elements to the cores, required to be the same as or compatible with the resinous substance with which the cores are impregnated, have not been capable of withstanding flame. Also, the use of resinous substances and adhesives, besides being relatively expensive, creates problems when the resin-impregnated cores must be bonded to various types of facing elements, such as, for example, facing elements of metal, commercial asbestos-cement board, and commercial gypsum board.

A fragment of a commercially available variety of a reticulated paper web or honeycomb structure 10 is represented in Figure 3.4a, in a partially opened state. The reticulated web or honeycomb structure 10 is available commercially in compacted strips of any desired width, which may be drawn out to provide the partially opened honeycomb structure of Figure 3.4a which is shown in fully opened coated condition in Figure 3.4b

This process employs conventional reticulated webs or honeycombs of relatively low cost sheet material which preferably is substantially free of resins and sizing substances, so that the sheet stock is in condition to absorb, or be penetrated by, a fluid cementitious substance, at least to a substantial degree. The honeycomb structure may be immersed in such a fluid cementitious material to thoroughly coat all exposed surfaces of the honeycomb structure, with the cementitious substance entering appreciably into the sheet stock for effecting a secure bond between the sheet stock and the coating when the cementitious material sets and ultimately dries.

FIGURE 3.4: STRUCTURAL RETICULATED WEBS AND HONEYCOMBS

a.

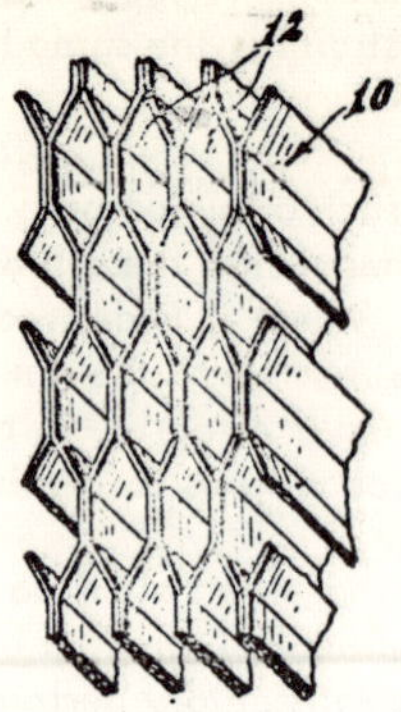

Partially Opened Paper Honeycomb Structure

b.

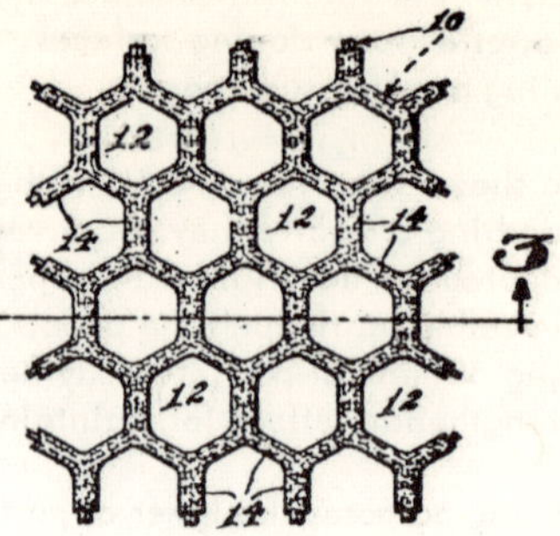

Fully Opened Structure with Cementitious Coating

c.

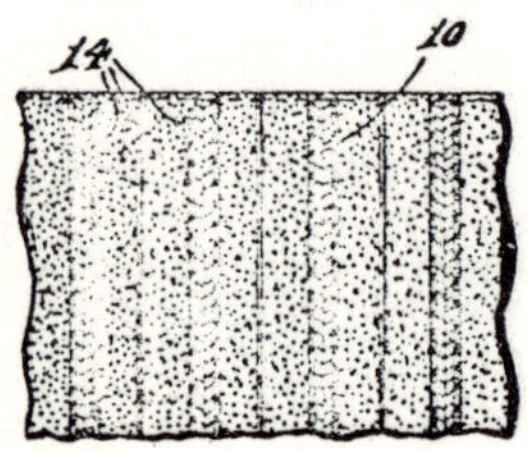

Cross-Sectional View Along Line 3 — 3 of Figure 3.4b

d.

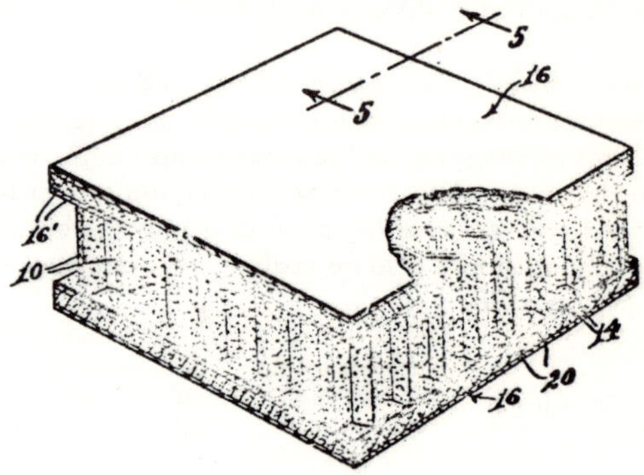

Isometric View of Section

e.

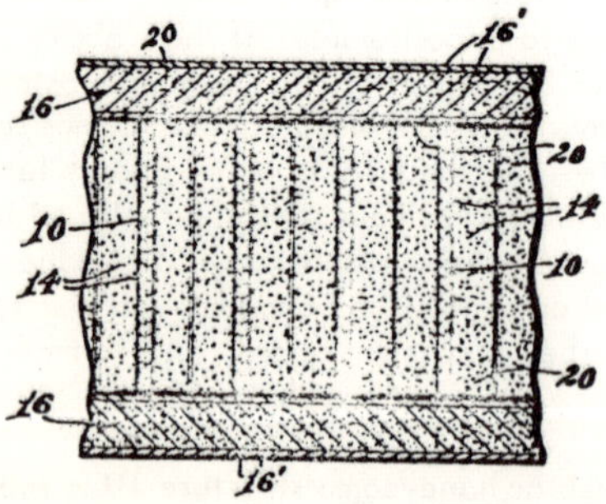

Cross-Sectional View Along Line 5 — 5 of Figure 3.4d

f.

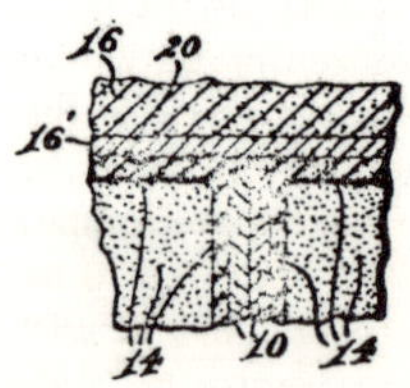

Detail of Panel Fragment

Source: B.M Randall and R.S. Frobisher; U.S. Patent 2,910,396; October 27, 1959

Paper stock which is substantially free of sizing substances and resins is considered to be a preferred sheet material from which to form the reticulated webs or honeycombs for use in the improved structural panels. The paper honeycomb structure, as represented at 10 in Figure 3.4a, may be purchased with its paper stock free or substantially free of resins and sizing substances, and such a paper honeycomb of suitable width, and of suitable area when fully expanded, has proven to be a highly acceptable reticulated web or honeycomb structure capable of being effectively coated with a cementitious material and effectively combined with any of a variety of panel-facing elements or skins in accordance with this process.

Examples of cementitious coating materials suitable for use in producing the improved reticulated webs or honeycombs, and the improved structural panels are Portland cement, gypsum plaster, Keene's cement, and other comparable relatively low cost cementitious materials of the class which may be more particularly described as mineral hydraulic cementitious materials. Portland cement, reduced by addition of water to a suitable consistency, is regarded as a preferred cementitious material for coating the reticulated web or honeycomb structures of panels which must be highly moisture resistant as well as highly flame and heat resistant, such as panels which will be exposed to the elements at exterior portions of buildings, for example. Usually, addition of a relatively small amount of an accelerator or a retarder will be advisable, for suitably controlling the setting of the Portland cement. On the other hand, a water and gypsum plaster, of suitable consistency, is considered to be a preferred cementitious material for coating the reticulated web or honeycomb structures of panels which are not required to be highly moisture resistant, such as panels for use interiorly of buildings, for example.

Either the Portland cement or the gypsum plaster, when coated on the reticulated web or honeycomb structure 10, as at 14 in Figures 3.4b-f, render the web or honeycomb structure highly heat resistant and substantially flameproof. When Portland cement is used, the coated structure is also highly moisture resistant. Commercial "Magnesite", and silicate of soda combined with calcium carbonate or whiting, are other materials falling within the general class of mineral hydraulic cementitious materials which may be used for coating the reticulated web or honeycomb structures.

The prior comparable structural panels which have employed resin-impregnated paper honeycomb structures between panel-facing elements have, to a considerable extent, been limited as to the varieties of facing elements which could be effectively combined with the honeycomb structures or cores. Problems have been involved to effectively and durably bond the facing elements to certain varieties of conventional panel-facing elements. It is a notable feature and advantage of these coated web or honeycomb structures that they are effectively and durably bondable to any of the conventional varieties of panel-facing elements by means of a bonding material the same as or compatible with the coating material.

Referring to Figures 3.4d-f, the facing elements 16, 16 represent sections of commercial plaster board, each of the facing elements or boards being of a commercial variety which is surfaced on both sides with paper 16'. The cementitious coating 14 on the web or honeycomb structure 10 may be either gypsum plaster, Portland cement, or the like. Assuming that it is gypsum plaster, each facing element 16 will be coated at 20 on one side with gypsum plaster, or a cementitious substance compatible therewith, preliminary to combining of the elements 16 with the coated honeycomb. One of the facing elements 16, coated on one side as at 20, may be arranged in a suitable box or form with its coated side up, and the web or honeycomb structure, in its coated condition of Figure 3.4b and while its coating continues in a plastic condition, is arranged on the element 16 in the box or form with the walls of the honeycomb perpendicular to the element 16. If the honeycomb is not fully spread or opened it may be fully opened as it is arranged on element 16, after which it is pressed against element 16, so that their plaster coatings merge as in Figure 3.4f. The second facing element 16 with its plaster coating on its under side, next is arranged on and pressed against the web or honeycomb 10 within the box or form.

The assembled panels may be stacked one upon another and be left to dry in atmosphere, or they may be subjected, according to known procedures, to conditions for hastening a complete curing and drying of the panels. The complete panels are exceptionally well suited for interior portions of buildings and for other uses where it may be important to have flameproof and highly fire and heat resistant paneling which also provides great strength and rigidity at relatively low cost.

Sandwich Building Panels

In a process described by A.I. Poms (U.S. Patent 3,462,339; August 19, 1969; assigned to Koppers Company, Inc.) a sandwich panel is provided having a foamed polymeric core and a face of denser material such as wood or metal and having a thermal capacitance disposed between the face and the polymeric core. The thermal capacitance is a material capable of dehydrating upon exposure to heat, preferably calcium dihydrate (gypsum).

The hydrated calcium sulfate material of the gypsum board, upon application of heat, loses water of hydration, and limits the temperature that heat can be passed onto the polymeric core. The gypsum passes first from the dihydrate form to the hemi-hydrate form and eventually becomes anhydrous calcium sulfate. In the losing of the waters of hydration, heat is absorbed and thus the surfaces immediately adjacent are cooled by the dehydration process. This feature enables the foam polymeric core structure to be cooled sufficiently to prevent melting of the core due to the transmission of the heat from the outer surface. Additionally, if the facing material be wood, the cooling, with respect to the outer facing, inhibits or retards the charring action by the fire. Furthermore, since the gypsum layer does not shrink as much as does the wood layer, the joint where two adjacent panels abut does not separate at the abutting gypsum faces to expose the interior and so the heat does not penetrate beyond the protecting gypsum layer as would occur if only a wood facing were used.

FIGURE 3.5: FIRE RETARDANT SANDWICH PANELS

a.

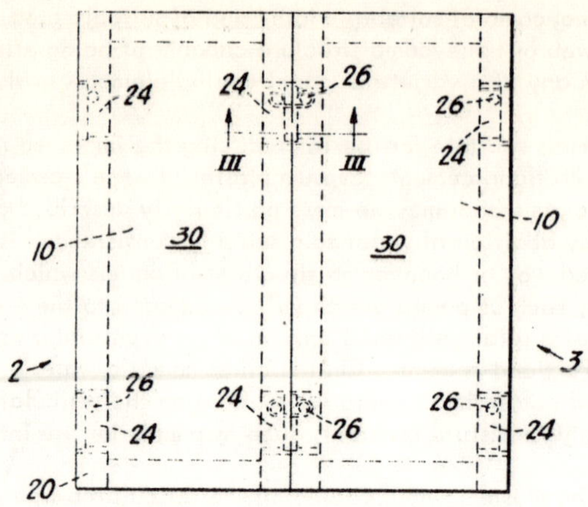

Sandwich Panel Construction — Side View

b.

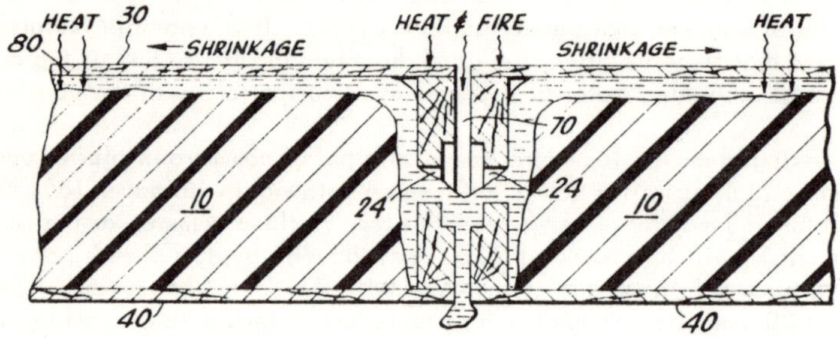

Effect of Heat on Prior Sandwich Panel

c.

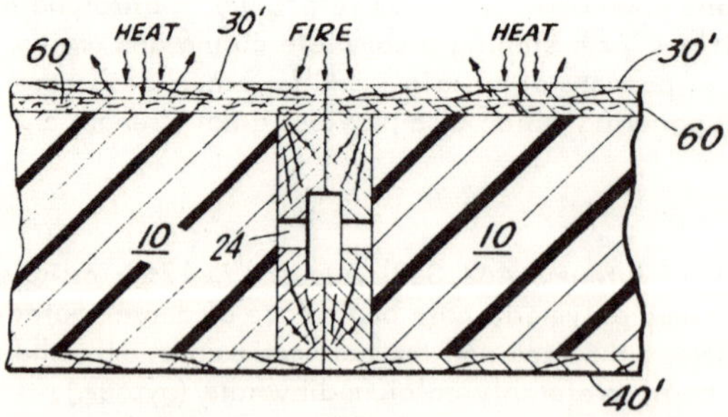

Effect of Heat on Sandwich Panel with Gypsum Thermal Barrier

Source: A.I. Poms; U.S. Patent 3,462,339; August 19, 1969

Referring to Figure 3.5a, the panels 2 and 3 each comprise a rectangular framework 20 which may be constructed of wooden beams. These wooden beams or rails have preferably been treated with fire retardant salts to render the wood noncombustible. The framework 20 of each panel is provided with openings or recesses 24 which house locking mechanisms 26 to secure adjacent panels as is shown in Figure 3.5a. These recesses or openings are necessary to provide a flush fit between adjacent panels; however they provide a communication between the interior of the panel and the joint which can have deleterious effects.

The interior of the panel, as seen in Figures 3.5b and 3.5c comprises insulating core of foamed polymeric material 10. The foamed insulation material, may comprise any of a number of foamable polymeric materials such as polystyrene, copolymers of styrene, polymerized alphamethylstyrene, polyethylene, polyurethane or polyester resins. These materials may be foamed in place utilizing, for example, the apparatus described in Patent 2,998,501.

The thickness of the foam insulating core may vary (as will the corresponding thickness of the wood framework) with the degree of insulation required. For example, the insulation of a foamed polystyrene core can be so great that a three-inch-thick core is sufficient for maintenance of a 40°F. temperature on one side of the panel while the opposite side is exposed to a 90°F. ambient temperature.

Previously panels have been made in the nature illustrated in Figure 3.5b. The panel has, for example, a front facing sheet 30 and a rear facing sheet 40. Facing sheets 30 and 40 are conventionally made of a plywood material and one or both sheets preferably are constructed of a plywood which has been treated with mineral salts to render it fire retardant similar to rail 20. However, as shown in Figure 3.5b, the facing member 30 of conventional core device panels, when exposed to excessive heat due to flame, may shrink as indicated by the arrow exposing the joint between two adjacent panels. The flame may then enter the joint 70, exposing the foam core to the flame and heat through the openings 24. The molten polymer then flows out openings 24 into the joints and is consumed by the fire providing more fuel and negating the effect of treating the rails 20 with materials to render them fire retardant.

Although the treated rails do not provide fuel for the fire, the molten plastic material flowing through the openings 24 in the rails do provide such fuel. The openings are necessary to house the locking mechanisms needed to fasten adjacent panels together. In addition to the fire entering the joints of Figure 3.5b due to the shrinkage of the wood face 30, face 30 may transmit enough heat as indicated at 80 to begin to melt the polymer at the interface adjacent the wall exposed to the heat and fire. The wall 30 is being exposed to heat at 80 and the heat is penetrating through the plywood wall 30 melting the polymer adjacent the wall as well as the polymer which is melting due to the entrance of the fire at the joint. Although wood face 30 may be treated with mineral salts to render it fire retardant, it is still subject to a charring action upon prolonged exposure to flame. This charring might well eventually expose the entire foam core to the flame.

The panel of this process as shown in Figure 3.5c remedies the defects of the panel of Figure 3.5b, and the difference in effect can be readily noted. With an outer panel or facing member 30' there is provided behind such facing a thermal capacitance 60. The thermal capacitance 60 has the ability to absorb heat and maintain an even temperature until the material is completely consumed or more properly chemically altered so that it no longer has the capacity to absorb heat. The particular material used is a gypsum material which has the chemical formula $CaSO_4 \cdot 2H_2O$ (calcium sulfate dihydrate). This gypsum layer 60 when exposed to excessive heat dehydrates but does not shrink appreciably. The lack of shrinking prevents the entrance of the fire at the joints as shown at 70 in Figure 3.5b. Thus no material melts and flows into the joint because no fire is transmitted into the joint to heat and melt the foam.

At the same time the gypsum prevents the heat from penetrating through the face of the panel into the foam. Any heat which does penetrate through layer 30' is absorbed by gypsum layer 60 as it dehydrates. This insures that the foamed material will not melt and run out of the panel to feed the fire for the amount of time otherwide necessary to bring a fire within control if no further fuel is added to the fire.

Although the opposite surface is not illustrated as of multilayer construction, side 40' may also be an outer wood facing member backed by a layer of wallboard if there is danger of exposure to fire from either side of the panel. Normally, however, only the fire side of the panel has the protective layer 60 of wallboard. For example, when used as the outer wall or the roof of a building only the interior side has the gypsum layer 60. However, for an interior wall both sides of the panel may be so constructed.

The rail portion 20 may be of conventional 2 x 2 construction or may be of a thicker width depending upon the amount of insulation required for the application which in turn will depend on the temperature differential desired between the two faces of the panel. The rail 20 is usually constructed as a rectangular frame and the gypsum and plywood layers are fastened circumferentially to this outer rail in any suitable manner as by cement or nails.

Veneer Panels

G.F. Wilhelm (U.S. Patent 3,480,494; November 25, 1969; assigned to R.S. Bacon Veneer Company) describes a method of preparing a fire retardant panel. The use of plywood, that is, a wood base or core to which are adhered on either side thin sheets of wood veneer, is widely used. A board comprised of Portland cement reinforced with asbestos fiber has also been used as a base or core upon which is adhered, on one or both sides, a sheet of decorative wood veneer so as to provide a panel having low combustibility. Such panels as have been produced previously have had the disadvantage of not being highly resistant to flame spread and of being stained owing to "bleed through" of the adhesive bonding the decorative face veneer to the core. Such staining occurs because the adhesive used must be of a nature to penetrate and seal the asbestos-like core and must overcome the alkaline problem inherent when a lime-containing cement-asbestos core is used. Such an adhesive will bleed through and stain the face veneer. In other instances, thick layers of soft inexpensive wood and thin sheets of asbestos paper have been interposed between a balsa wood core and a veneered panel for the purpose of providing a sound deadening board. This type of board is heavy and bulky and has little or no resistance to flame spread.

To overcome such objectional "bleed through" and staining of the decorative face veneer in a fire retardant panel, this process provides an intermediate layer of wood veneer cross banding between the cement-asbestos base and the decorative face veneer. In order to accomplish this without building up a panel of excessive weight and thickness, and to minimize flame spread, one adhesively applies over the cement-asbestos core, on one or both sides, a very thin layer of wood veneer cross-banding. After the adhesive, preferably phenol-resorcinol resin or its equivalent, has cured and the "bleeding" has, therefore, stopped, the thin layer of wood veneer cross-banding is further reduced in thickness to that of a thin skin without exposing the cement-asbestos core or the layer of adhesive bonding the wood veneer cross-banding to the core. This produces a well sealed core over which a second adhesive (such as melamine-fortified urea resin) is then applied and the sheet of decorative face veneer and a balancing back veneer (optional) are laid over opposed faces and the whole is pressed together in a hot plate press. This adhesive adheres to the wood surfaces and does not have "bleeding" characteristics, hence, the finished panel is devoid of stains, and is of minimum thickness while affording maximum strength and exposing a minimum amount of wood body to support flame spread.

FIGURE 3.6: FIRE RETARDANT PANEL

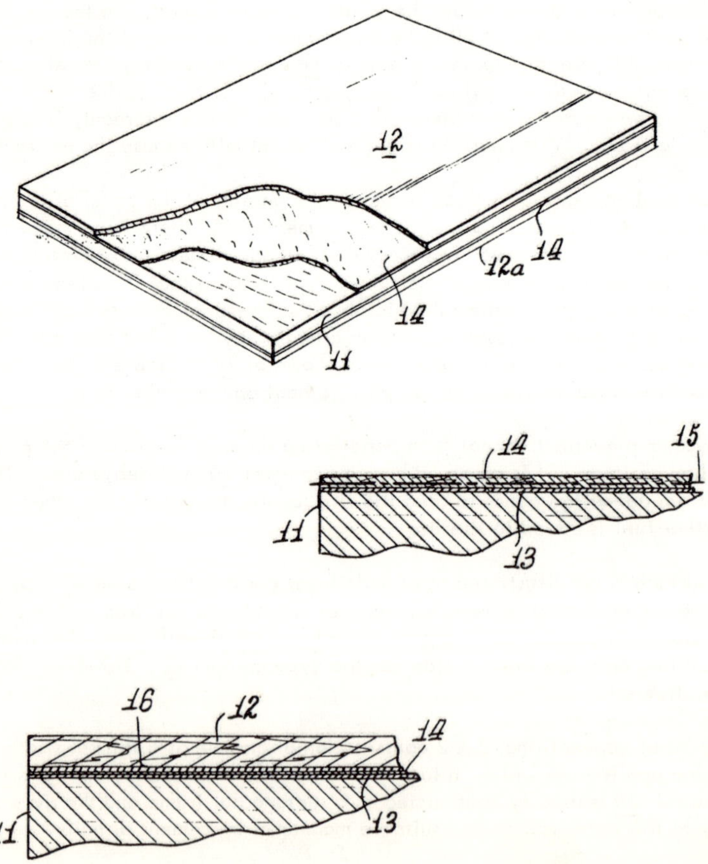

Source: G.F. Wilhelm; U.S. Patent 3,480,494; November 25, 1969

Referring to Figure 3.6, the laminated panel is comprised of a core 11, and a ply of decorative face veneer 12, adhered to one side with a balancing backing veneer 12a, on the opposite side (optional). The core 11 is comprised of a mixture of cement with asbestos fibers pressed into the desired thickness and set. This type of material affords a hard, rigid, dense, noncombustible core.

In the fabrication of the finished product, the cement-asbestos board or core 11 is coated on each surface with a layer 13 of adhesive, such as phenol-resorcinol resin or its equivalent, which has the capacity of soaking into the core so as to afford a good bond. Over each face is laid a sheet 14 (preferably not more than 1/16 inch in thickness) of wood veneer cross-banding, all of which is pressed firmly in place in a hot plate press and allowed to dry thoroughly. The nature of the adhesive used is such that it will bleed through the wood veneer cross-banding 14, thus staining same. However, since this is an innerply in the finished panel, the stain is of no consequence. The wood veneer cross-banding layer 14 then is reduced in thickness as by sanding. This sanding continues to the extent that the wood veneer cross-banding has been reduced to a minimum thickness without exposing core 11, or the adhesive bonding the wood veneer cross-banding to the core.

A layer 16 of another adhesive, such as melamine-fortified urea resin or its equivalent, which is also capable of withstanding fire testing without failure of bond, is applied to the sanded surface of the remaining thin film of wood veneer cross-banding 14. Over this, on one side, is laid the decorative face veneer 12 and on the other side, the balancing backing veneer 12a (optional). The adhesive 16 is of a character that does not bleed through, consequently there is no staining of the decorative face veneer 12 or veneer 12a. The whole is then pressed and cured in a hot plate press and subsequently trimmed to the required size. The cured product then is sanded and polished as may be required.

Air Permeable Fire Retarding Partitions and Doors

G. Rätzel (U.S. Patent 3,027,272; March 27, 1962; assigned to Alim Corporation) has found that fire retarding and flameproof partitions, doors, and windows while normally permitting a practically undisturbed passage of air can be produced from air permeable carriers, such as fine meshed nettings or grids of any desirable material, such as, for example, iron wire, or other metals or also of organic materials. According to the process, these carriers are coated with a fire protecting coating which is incombustible and expands or puffs up upon heating. Such coating is applied to the carrier so that the openings serving for passage of air are not sealed. When the carriers are subjected to the action of the heat the fire retarding coating expands to close the openings and thus prevents the penetration of flames. In view of its incombustibility and insulation, direct transfer of the heat of the fire is hindered.

Numerous fire retarding compositions suitable for the fire retarding coating have already been developed. Those which essentially consist of a synthetic resin and a foam forming material, and if desired a carbon forming component such as glue or starch have proved particularly effective. Compositions of this nature are fully described in U.S. Patent 2,523,626 and U.S. Patent 2,881,088.

The resin component of the fire retarding compositions according to the process, preferably is a urea-formaldehyde resin and/or a urea-formaldehyde-acrolein resin. Mixtures of 40 to 60% of urea-formaldehyde resin with 60 to 40% of urea-formaldehyde-acrolein resin have been found particularly suitable. Urea-formaldehyde resins in which the proportions of formaldehyde component to urea component is 1 to 2.5 mols of formaldehyde component per mol of urea component are particularly suited. Urea-formaldehyde-acrolein resins in which the proportions of the components are about 2 to 2.5 mols of formaldehyde component and about 0.12 mol of acrolein component per mol of urea component are preferred.

The fire retarding compositions employed are generally applied to the surfaces of the air permeable carrier as a solution or suspension or emulsified in a liquid medium. Drying oil or bituminous paint vehicles or other nonaqueous vehicles may also be employed. The method of application of fire retarding compositions to the air permeable carrier can be by dipping, spraying or painting. Instead of employing nettings or grids as the air permeable carrier, other air permeable structures can be employed, such as, for example, perforated materials in which the small perforation will seal off under the influence of heat as described above. Also, free hanging curtains of single wires, asbestos fibers and the like can be provided with the fire retarding coating.

The product, for example, renders it possible to provide a means for preventing spread of fire through cable shafts, mine shafts or other openings which cannot be closed off by fire walls or doors because of required air circulation, in that the described discontinuous surfaces which have been provided with a foam forming fire retarding coating are firmly affixed transverse to the axis of the shafts or openings. In order to provide certain effects, it is also possible to provide a plurality of nettings or grids in series in which the individual nettings or grids may contain different sized openings which can be staggered with respect to each other and which also may be provided with fire retarding coatings of different thicknesses. The following example illustrates the process.

Example: A netting having square meshes with a mesh opening of 5.5 mm. and formed of iron wire 1 mm. in diameter was coated with an aqueous suspension containing 200 parts by weight of urea-formaldehyde resin, 170 parts by weight of urea-formaldehyde-acrolein resin, 484 parts by weight monoammonium phosphate, 20 parts by weight of calcium phosphate,

76 parts by weight of soluble starch, 50 parts by weight of asbestos powder, all having a particle size passing through a screen of 1,600 meshes per cm.^{2}. The application of the coating was such that upon drying a coating weighing 900 g. per square meter of netting was obtained. Such netting, when employed in partitions, doors and the like, normally permitted good circulation of air, but upon access of fire prevented spread of fire, as the heat of the fire caused the coating to foam and swell up to seal off the mesh openings with an incombustible, stable, heat insulating foam.

Acoustical and Thermal Insulating Siliceous Fiber Boards

A process described by R.F. Shannon and J.L. Helser (U.S. Patent 3,286,785; November 22, 1966; assigned to Owens-Corning Fiberglas Corporation) is directed to the provision of low density decorative and structural elements formed from siliceous fibers, which exhibit structural integrity and dimensional stability at temperatures in excess of 1000°F., while retaining the acoustical and thermal insulating properties for which they are valued.

Structural integrity, as used here, is employed to define the ability to resist the transfer of heat and flames through apertures resulting from the delamination or disintegration of the structure, or through apertures resulting from the separation of the low density boards from their fastenings, suspension systems or from abutting, adjacent boards, upon exposure to temperatures in excess of 1000°F. Consequently, the term "structural integrity" contains an aspect of dimensional stability in respect to resistance to shrinkage, expansion and warping or sagging. It should also be noted that the acoustical properties of the structures include both sound absorption and resistance to sound transmission. The former is provided by the low density, void-filled region which is retained at one major surface of the board, while the latter is improved by virtue of the formation of a dense zone which is substantially impervious to sound at the opposite major surface. At the same time, the insulating values of the boards are enhanced by virtue of the opacifying effect of the dense layer.

In the process, an aqueous slurry containing a specific combination of inorganic ingredients formulated to provide the desired structural integrity and dimensional stability, is applied to one surface of the low density fibrous element and then deposited within the voids present in the fibrous medium and adjacent to the point of application. In determining the nature of a layer adequate to provide the prescribed structural integrity, and of the low density zone of the opposite surface adequate to yield the desired acoustical and thermal insulating properties, it has been found that a structural layer having a thickness of 15 mils is adequate, while the acoustical-thermal low density zone or layer must have a thickness of at least 50 mils. However, since the low density boards which are treated in the process normally have a thickness of at least 375 mils, the structural layer or impregnated zone normally has a thickness of at least 50 mils while an unimpregnated zone having a thickness of at least 125 mils is preferred.

After depositing and positioning the impregnant to the desired depth, the aqueous component is dispelled by drying to leave a strong rigid structural phase or layer. The impregnants contain both inorganic particulate matter, and a binder phase which is capable of interadhering the inorganic particles both at ambient conditions and during exposure to temperatures in excess of 1000°F. Specifically, the binder phase contains between 5 to 95% by weight of colloidal silica and between 5 to 95% by weight of bentonite, and comprises between 2 to 25% by weight of the total solids present in the impregnant. The impregnant contains between 1 to 20% by weight of colloidal silica and between 1 to 15% by weight of bentonite. It should be noted that references to quantities of colloidal silica relate to the quantities of silica solids, although these compositions are normally employed as dispersions in a liquid carrier medium such as water.

The inorganic particles which are interbonded by the binder phase include a combination of both materials possessing a melting point below 2000°F., and materials which melt above that temperature. The lower melting materials are the aluminosilicates of the Group I and II metals, sodium, potassium calcium, magnesium, and barium, and aluminosilicates containing two or more of these metals. The high melting materials comprise certain hydrous aluminum silicates which are subsequently described. This combination is based upon the discovery that the formation of a molten or ceramic phase is highly desirable in the attainment of structural integrity at temperatures experienced during a fire or a fire test.

It is believed that the lower melting materials provide a new or auxiliary adhesive phase upon exposure to temperatures in the range of 1000° to 2500°F., which serves to interadhere the higher melting particles and to maintain structural integrity. While structural integrity resulting from a "liquid" or molten phase may at first appear contradictory, the nature of this "liquid" must be considered. Specifically, the Group I and/or Group II metals function as fluxes which facilitate the melting of a portion of the ingredients of the impregnant and form an extremely viscous adhesive. For example, molten glass compositions commonly possess viscosities in the range of 10^4 to 10^7 poises and consequently are not highly fluid liquids in the ordinary sense of the word. The value of thermally inert particles interbonded by the described "liquid" adhesive, over thermally inert but structurally weak impregnants is aptly demonstrated by the fire ratings achieved with these impregnants.

The previously described binder phase, i.e., bentonite and colloidal silica, is employed to interbond between 75 to 98% by weight of a mixture of low and high melting materials. In turn, this mixture comprises between 10 to 90% by weight of the higher melting hydrated aluminum silicate particles and between 10 to 90% by weight of the lower melting sodium, potassium, calcium, magnesium and barium aluminosilicates. Consequently, the impregnant composition contains between 7.5 to 90% by weight of both the lower and the higher melting materials.

Thus, the impregnants generally comprise:

	Percent by Weight
Colloidal silica (solids)	1 to 20
Bentonite	1 to 15
Hydrated aluminum silicate	7.5 to 90
Na, K, Ca, Mg and Ba aluminosilicates	7.5 to 90

Preferably the impregnants contain:

	Percent by Weight
Colloidal silica (solids)	2 to 6
Bentonite	3 to 5
Ball clay	30 to 60
Feldspar	30 to 60

The hydrated aluminum silicates employed are ball clays consisting primarily of kaolinite, illite and montmorillonite, with minor quantities of ferric oxide, finely divided silica and titanium dioxide, and trace amounts of lime, magnesia, soda, and potash. More precisely, they contain between 50 to 75% by weight of SiO_2, between 15 to 35% by weight of Al_2O_3 and between 0.5 to 10% by weight of Fe_2O_3 with conventional silica, alumina and iron contents of approximately 57:27:1. Typical of these hydrated aluminum silicates are the Maryland and Tennessee ball clays generally, and such specific materials as Saxon clay, Yankee ball clay, Rex clay, XB ball clay and pyrophyllite. The examples in the following table provide a number of formulations suitable for use in the process, in which the quantities of ingredients are expressed as percentages by weight.

Ingredients	Examples								
	1	2	3	4	5	6	7	8	9
Feldspar I	47	73.3	13	—	—	36.75	—	76	—
Feldspar II	—	—	—	—	—	—	70	—	—
Feldspar III	—	—	—	—	—	—	—	—	70
Nepheline	—	—	—	73.5	—	—	—	—	—
Ball clay I	47	18.3	79	—	73.5	18.5	17	—	17
Ball clay II	—	—	—	18.5	—	—	—	19	—
Ball clay III	—	—	—	—	73.5	36.75	—	—	—
Colloidal silica (solids)	2.7	5.1	3	3	3	3	10	1.7	5
Bentonite	3.3	3.3	5	5	5	5	3	3.3	8

The impregnants of the above examples were prepared by mixing the ingredients with water to form a slurry. While a slurry containing 20 to 70%, and preferably 50% by weight of water is preferred in most cases, the ratio must be gauged to yield an impregnant of the viscosity desired for the specific applicator system and substrate employed. For example, in the case of low density fibrous substrates, the viscosity of the impregnant is normally increased by increasing the percentage of solids. This may be employed to the rate and depth of impregnation. However, the impregnation depth and rate may also be controlled by means independent of the impregnant. For example, vacuum or mechanically pressurized impregnation, e.g., a doctor blade, may be employed to force a highly viscous impregnant to the desired depth within the fibrous substrate. Alternatively, the fibrous substrate may be pretreated with a wetting agent, e.g., sprayed with a solution of an organosilicone fluid, to facilitate and control the rate and depth of impregnation by viscous slurries.

A wetting agent which has proved highly satisfactory for the above purposes is sodium dioctyl sulfosuccinate having a molecular weight of 444 and available from American Cyanamid under the tradename Aerosol OT. It should be noted that all of the impregnants of the above examples were applied to siliceous fiber acoustical boards and subjected to a simulated fire test to yield fire ratings in the range of 2 to 4 hours. These ratings represent a substantial improvement over untreated boards which yield a rating of approximately 15 to 45 minutes, or boards treated or impregnated with clays, silica, alumina, and the like, which normally provide a one hour rating.

The products of the process comprise an acoustical panel, board, or tile formed from siliceous fibers bonded or entangled into a mass having a density in the range of 1 to 30 pcf, and possessing two substantially parallel major surfaces, one of which is impregnated with the above compositions. The product retains its desirable acoustical and thermal insulating values, while yielding highly improved structural integrity, and resistance to heat or flame transfer when exposed to high temperatures. In order to preserve the acoustical absorption properties of the board, the impregnation of the board throughout its entire thickness, i.e., the filling of all of the voids present in the board, must obviously be avoided. However, satisfactory acoustical properties, and some degree of thermal insulation, are realized when a low density region having a depth of no more than 50 mils is left upon the unimpregnated surface of the board. In turn, an impregnated region having a thickness of no more than 15 mils is adequate to provide structural integrity which permits the board to endure temperatures as high as 2500°F., without failing. The acoustical boards comprising glass fibers have an apparent density of from 1 to 30 pounds per cubic foot and are from 250 to 3,000 mils thick. These boards are impregnated to a depth of at least 15 mils with the treating material. Preferred forms of the process comprise fibrous glass boards having a density in the range of 8 to 15 pcf, a thickness of from 350 to 1,500 mils, and an impregnated region having a thickness of about 50 to 750 mils and generally about 50 to 200 mils. Normally, the unimpregnated zone will comprise approximately one-half of the thickness of the board. Examples of such structures comprise a one inch thick board with an impregnated zone or structural layer having a thickness of 50 mils and a 5/8 inch thick board having an impregnated zone having a thickness of 500 mils.

In the case of boards having a density in the range of 9 to 13 pounds per cubic foot and a thickness of no more than one inch, impregnants constituting between 0.3 to 0.8 pounds per square foot of the impregnated board have proved generally satisfactory. Similarly, a preferred product comprises a board having a density in the range of 9 to 13 pcf, a thickness of between 350 to 1,500 mils, and containing between 0.3 and 0.8 pounds of the impregnant present in each square foot of the impregnated board. However, as previously noted, boards having densities of between 1 to 30 pcf, and a thickness as great as 3 inches, have been prepared and have yielded the desired properties.

A product prepared by the process is depicted by Figure 3.7a which provides a fragmentary, sectional view through a fibrous board. As may be seen, one surface 11 of the low density, fibrous board 12 has been impregnated to yield a dense layer 13 which serves as a structural member. As also may be observed, the dense layer 13 also contains the fibers 14 of which the board 12 is composed, and which provide an additional benefit in reinforcing the dense layer 13. In turn, a low density unimpregnated phase 15 is left intact at the opposite surface 16 of the board 12 to provide the desired thermal and acoustical properties. It should be noted that in the illustrated products the structural value of the dense layer 13 functions primarily after the fibers of the low density unimpregnated phase 15 have been weakened, softened or melted and following pyrolysis of the resin binder. At such time, the dense layer 13 becomes the primary or sole supporting, spanning, or suspending means and may provide the sole source of structural integrity for the board 12.

The impregnated structures are prepared by depositing a slurry of the impregnant upon one surface of the fibrous board, introducing the impregnant within the voids present upon one surface of the porous board without filling the voids present upon the opposite surface of the board, and drying the impregnant. A method for fabricating these products is depicted by Figure 3.7b which shows a schematic representation of apparatus suitable for the impregnation of low density siliceous fiber substrates.

FIGURE 3.7: HIGH TEMPERATURE RESISTANT ACOUSTICAL BOARD

a.

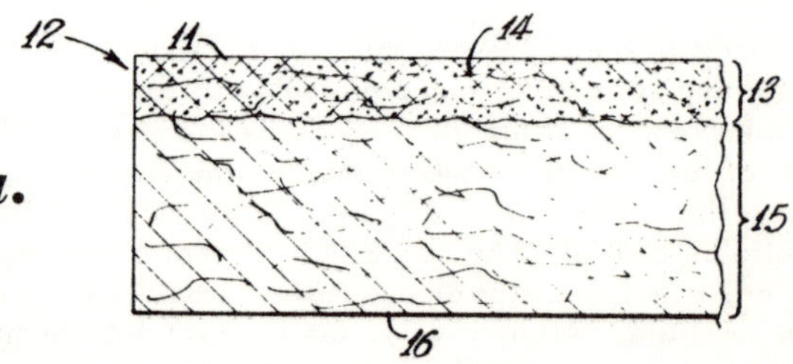

Sectional View of Fibrous Board

b.

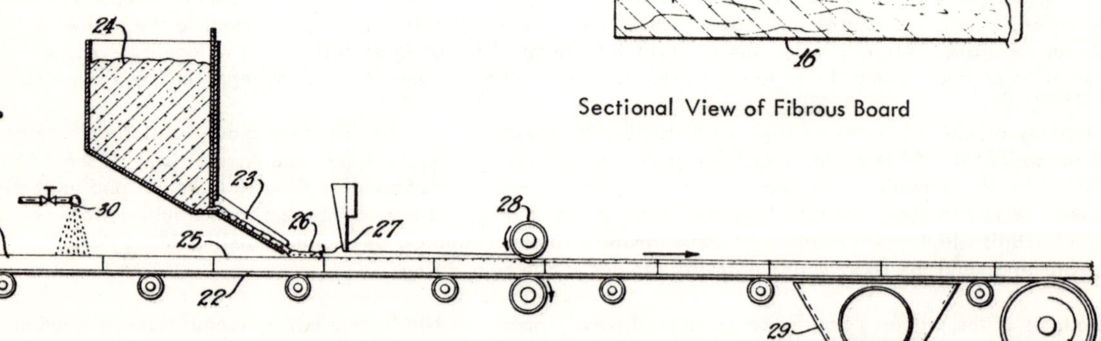

Apparatus for Impregnation of Fibrous Board

Source: R.F. Shannon and J.L. Helser; U.S. Patent 3,286,785; November 22, 1966

As shown, fibrous boards 21 are carried by means of a foraminous conveyor belt 22 beneath a trough 23 which dispenses an impregnating slurry 24 upon the upper surface 25 of the boards 21. The deposited impregnant 26 is then passed beneath a metering knife blade 27 which controls the thickness and consequently the quantity of the deposited impregnant 26. The boards 21 are then passed beneath and in contact with a pressure roll 28 which forces the deposited impregnant within the upper surface of the boards 21. To facilitate and control the depth of penetration of the boards 21 by the impregnant, the boards 21 and foraminous conveyor belt 22 may then be passed over a suction box 29 which serves to draw the impregnant within the boards 21 and may implement or replace the effect of the pressure roll 28. Alternatively, or in conjunction with the foregoing process, a sprinkler 30 may deposit a wetting agent such as a silicone fluid upon the upper surface 25 of the boards 21 prior to the deposition of the impregnant.

The use of such a wetting agent facilitates the rapidity and extent of the penetration of the boards 21 by the impregnant. The drying of the impregnant may be accomplished by conventional oven treatments. For example, boards having a density of approximately 11.5 pcf, a thickness of between 350 to 1,000 mils, and containing an impregnant layer weighing approximately 0.8 pounds (aqueous slurry, 50% solids) per square foot of the board, may be oven dried at a temperature of 400°F., in approximately one hour.

MISCELLANEOUS

Composite Metal Sheets

R. Gaeth, B. Schmitt and R. Breu (U.S. Patent 3,186,888; June 1, 1965; assigned to Badische Anilin- & Soda-Fabrik AG, Germany) describe metal sheets having heat insulating properties. U.S. Patent application 175,832 describes fireproofing compositions in the form of sheets which comprise an alkali metal silicate, preferably sodium silicate, which contains water in an amount of 20 to 70% by weight, preferably 40 to 60% by weight, and fibers, preferably synthetic silicate fibers, in an amount of 10 to 40% by weight, preferably 15 to 25% by weight, the percentages being with reference to anhydrous alkali metal silicate. The fibers may be glass fibers, rock wool and the like. The alkali silicate should have a molar ratio of alkali metal oxide to silicon dioxide of about 1:2 to 1:4, preferably of from 1:2.5 to 1:3.5. The fibers should have a minimum length of about 30 mm.

The composite sheets according to this process may be prepared simply by combining sheet metal with the fireproofing sheets with a suitable adhesive, for example, waterglass solution. Composite sheets of this type may be made particularly advantageously by pressing together the fireproofing sheets and the sheet metal, preferably consisting of iron, in a heated press at temperatures of 50° to 150°C., preferably of 80° to 120°C. The pressure to be maintained during the pressing depends on the temperature used. Coherent composite sheets are obtained at higher temperatures, for example, at about 140°C., with lower pressures than when pressing at lower temperatures, for example, at 80°C. Thus for example, at temperatures of about 140°C. a pressure of about 20 atmosphere is suitable with a pressing time of about five to ten minutes.

Composite sheets may also be produced continuously by passing the sheet metal and the fireproofing sheets between two heated pairs of synchronized rollers under pressure. The composite sheets may be provided with a fireproofing sheet of an alkali silicate containing water and fibers on one side or on both sides of the sheet metal. It is also possible, however, to provide the alkali silicate fireproofing sheets on both sides with a sheet of metal. Composite sheets according to this process may be processed directly into shaped articles by conventional deep drawing machines for working metals, and the operation may be carried out at elevated temperatures at which the fireproofing sheets are plastic but not yet expanded. This is the case at temperatures of 50° to 150°C., preferably of 80° to 120°C. Sheet metal combined with a fireproofing sheet having, with reference to anhydrous alkali metal silicate, a water content of 50% by weight and a fiber content of 17% by weight, may, for example, be extended at a temperature of 100°C. up to 60% of its original length without noticeable cracks appearing in the fireproofing sheet.

The side of the sheet metal which is not covered with a fireproofing sheet may be provided without difficulty with a conventional baking lacquer. It is advantageous to use lacquers which are baked at temperatures of about 100° to 120°C., because the fireproofing sheets begin to expand at higher temperatures. Composite sheets prepared according to this process not only have excellent heat insulating and flame insulating properties but also have a greatly diminished drumming effect and this makes them particularly suitable, for example, for the manufacture of body parts in motor vehicle construction. The process is illustrated by the following example.

Example: A fireproofing sheet 40 x 40 cm. in size and 1.2 mm. in thickness, which contains 17% of glass staple silk 5 cm. in length and 50% of water, both with reference to anhydrous alkali metal silicate, is laid on a piece of sheet iron of the same size and a thickness of 0.4 mm. and the composite sheet is pressed for five minutes at a pressure of 100 atmospheres and a temperature of 90°C. in a laboratory press and cooled under pressure. The composite sheet thus obtained is heated to 90°C. and shaped on a conventional deep drawing press to a ball segment having an upper diameter of 7.5 cm. and a height of 2.0 cm. A commercial lacquer based on urea formaldehyde and alkyd resin is spread on the metal face and baked for an hour at 100°C. The metal is not separate from the fireproofing sheet.

Insulative Structure for Roof Deck Construction

A process described by F.W. Curtis (U.S. Patent 3,466,222; September 9, 1969; assigned to Lexsuco, Inc.) involves an improved fire retardant insulative structure for a roof deck construction. As described in U.S. Patent 2,861,525, a conventional metal roof deck construction contains large quantities of combustible bituminous materials such as asphalt and coal tar pitch. It has been found that notwithstanding the heat shielding and fire resistant properties of the metal roof deck itself these combustible materials can and do contribute to the spread of a fire in a building under a metal roof deck construction.

A process has been found which provides a fire retardant, insulative structure which is capable of reducing the rate of heat transfer from a fire below, through a roof deck, to combustible materials above, to such a degree, that the combustible material is heated gradually and substantially uniformly through its thickness. Thus, by the time the combustible material closest to the roof deck has reached a vapor generating temperature, the combustible material further from the roof deck has at least softened to a point which permits the escape of vapors upwardly.

Figure 3.8 shows a fire retardant, insulative structure comprising a laminate having a core 10, formed of a substantially rigid body of synthetic, organic, polymeric foam; a lamina 12 on one surface of the core 10, formed of a particulate, inorganic material and a binder; and an optional lamina 14 on the opposite surface of core 10, providing a traffic and mopping surface 16. If desired, the laminate may be provided with a thin sheet of metal foil 18, adhered to the exposed surface of lamina 12. Metal foil 18 serves a useful function in the fabrication of the fire retardant, insulative structure.

Core 10 may be formed from a variety of synthetic organic polymeric materials, all of which produce low density, lightweight, low heat transmission, substantially rigid foams. The foam may be formed of a polyurethane resin, a vinyl resin (e.g., polyvinyl chloride, copolymers of polyvinyl chloride and polyvinyl acetate), epoxy resins, phenolic resins (e.g., phenol formaldehyde), polyethylene resins, silicone resins, urea formaldehyde resins, polystyrene, acrylic resins, and synthetic rubber.

A particularly suitable core is one formed of a foam having closed cells. This provides a core having extremely low permeability to moisture. However, cores formed of foam having open cells may also be used. The thickness of the core can vary over side limits, depending on the environment in which the insulative structure is to be used, the density of the foam, the nature of the synthetic organic polymeric material forming the foam, and the insulation requirements of the particular application for the insulative structure. In general, however, the thickness of the core for most applications will be within the range of about 1/2 inch to about 2 inches or more. In the case of an insulative structure intended for use as a component in a metal roof deck construction, a core formed of polyurethane foam will satisfy most commerical specifications if the core has a thickness with the range of about 1/2 to about 2 inches. This is based on the fact that a polyurethane foam having a K-factor [thermal conductivity expressed in Btu/(hr.) (sq. ft.) (deg. F./in.)] of approximately 0.15 will have commercially recommended C values [thermal conductance expressed in Btu/(hr.) (sq. ft.) (deg. F.)] in the range of about 0.06 to about 0.36.

FIGURE 3.8: FIRE RETARDANT INSULATIVE STRUCTURE

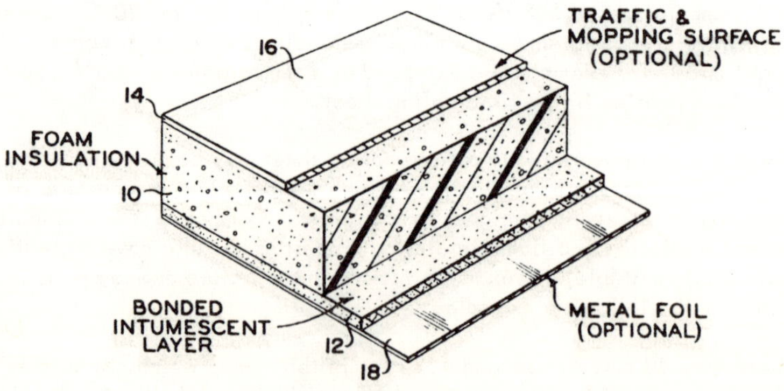

Source: F.W. Curtis; U.S. Patent 3,466,222; September 9, 1969

Lamina 12 is formed of a particulate, inorganic material and a binder where at least about 50 weight percent of the inorganic material is unexpanded vermiculite, the remainder, if any, being any other natural or synthetic inorganic, preferably low density, insulative material, such as expanded vermiculite, perlite or shale. Laminae containing higher concentrations of unexpanded vermiculite are generally preferred, and in some circumstances laminae containing 100% unexpanded vermiculite as the particulate inorganic material are especially preferred.

The advantages of unexpanded vermiculite are manifold. It is readily available at low cost, in a wide variety of sizes or grades. In addition, vermiculite expands at temperatures of about 300°F. and above, which means that the lamina 12 begins to expand and protect core 10 at temperatures below those which will begin to degrade the polymeric foam from which the core is formed. Another advantage of vermiculite, is that it releases water vapor upon expansion or exfoliation, and the latent heat of vaporization absorbed in the process of converting water to water vapor, helps to slow the rise of temperature in the vicinity of core 10.

The binder for the particulate, inorganic material may take the form of an inorganic adhesive such as sodium silicate, or an organic resinous material such as a urea formaldehyde or phenol formaldehyde resin. In general, organic binders are preferred because of their ease of handling and mixing with the particulate, inorganic material, and the speed with which they set up. Preferably, the amount of binder used is kept at a minimum, since it serves no other useful function than to bond the particles of inorganic material to each other to form a cohesive lamina. Thus, the binder may represent as little as about 1% by weight of the inorganic material, up to about 5% by weight.

The thickness of the bonded lamina of particulate inorganic material where 100% unexpanded vermiculite is used, may range from about 1/32 of an inch to about 1/4 inch, and preferably within the range of 3/32 of an inch to 1/8 inch. Thicker laminae may be used, but generally cannot be economically justified, since a 1/8 inch lamina of unexpanded vermiculite, will expand to a thickness of about 1 inch, which is generally adequate for most applications. Lamina 14, providing traffic and mopping surface 16, may be formed from a variety of materials, such as pressed hardboard, asbestos-cement sheet, and a sheet of synthetic elastomeric material, such as butadiene rubber. From the standpoint of economics, lamina 14 is preferably formed of a conventional roofing felt, which may be a cellulosic or synthetic fibrous body saturated with a bituminous material.

Referring to Figure 3.9, there will be seen a fire retardant insulative structure of the process, incorporated in a metal roof deck construction. Roof deck 20, is of conventional design, and is formed of a plurality of panels which are interlocked in any suitable fashion, and are supported by structural members forming a part of the building, such as purlin 22. Roof deck 20 may be formed from a variety of metals such as steel and galvanized steel. The fire retardant, insulative structure may be secured to metal roof deck 20 by any suitable means such as adhesives or metal fasteners. It will be noted that the fire retardant, insulative structure is placed on metal roof deck 20 so that lamina 12, formed of a particulate inorganic material and a binder, or metal foil 18, if such is used, is positioned between the metal roof deck and foam core 10. The metal roof deck construction is completed by applying to core 10, or to traffic and mopping surface 16 of lamina 14 if such is included in the structure, a "built-up roof" which comprises at least one covering of weather resistant which may be a rubbery substance applied as a coating or sheet, but in the form illustrated, is a coating of bituminous material 26, and preferably a plurality of such coatings separated by built-up lamina 28, which may be formed from a variety of materials.

FIGURE 3.9: METAL ROOF DECK CONSTRUCTION UTILIZING FIRE RETARDANT INSULATIVE STRUCTURE

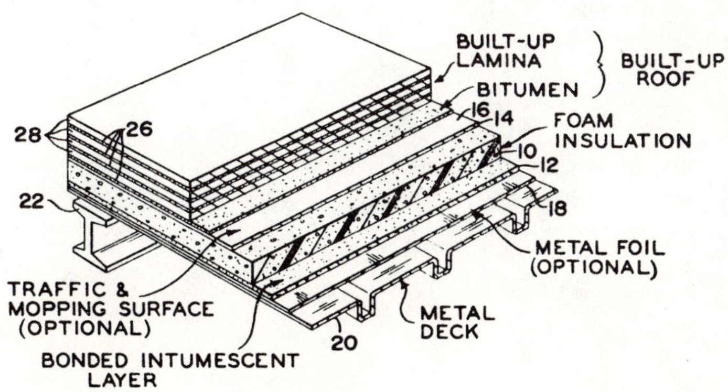

Source: F.W. Curtis; U.S. Patent 3,466,222; September 9, 1969

In general, the built-up roof portion of the metal roof deck construction is formed by mopping a bituminous material on the traffic and mopping surface, applying a layer of roofing felt, and alternating this procedure until 3 or 4 plies of roof felt have been built-up. The top layer is usually a coating of bituminous material, which may have embedded in it granules of a conventional mineral material, such as gravel, marble chips or volcanic rock.

The fire retardant, insulative structure of the process provides excellent thermal insulation at ambient temperatures and superior fire retardance at extreme temperature conditions. Should a fire develop in a building under a metal roof deck comprising a fire retardant insulative structure of the process, the following would occur. At temperatures well below those necessary to vaporize and/or liquefy the combustible material forming a part of the built-up roof portion of the metal roof deck construction, the unexpanded vermiculite in the lamina of particulate inorganic material, expands with the release of water vapor. This provides a substantial body of heat insulative material between the metal roof deck and the organic foam core. The expanded lamina together with moisture vapor released during the expansion of the vermiculite both would serve to protect the organic foam from heat degrading temperatures and slow the rate of heat transfer to the combustible material. Since the foam itself has low heat transmission properties, as long as it maintains its structural integrity, the foam in turn helps to slow the rate of heat transfer to the combustible material. In this manner, additional time is provided to bring the fire within the building under control before the combustible material vaporizes and/or melts, and thereby contributes to the spread of the fire along the under surface of the metal roof deck.

Should some vaporization and/or melting of the combustible material take place before the fire within the building is brought under control, the slow rate of heating of the combustible material would have by this time softened the entire mass of combustible material permitting the vapors to escape upwardly, so that they would not contribute to the spreading of the fire.

ASPHALTIC PRODUCTS

ROOFING MATERIALS AND SHINGLES

Asbestos and Vermiculite Fillers

F.B. Burns and M.S. Larrison (U.S. Patent 2,569,399; September 25, 1951; assigned to United States Gypsum Company) describe a coating composition which is so designed that it becomes a tough, cohesive, continuous inorganic covering after it has been once exposed to heat or direct flame for a period of time sufficiently long to completely burn out the organic constituents of the coating composition. If the heat or flame is applied for a sufficient length of time to burn out all or substantially all of the cementitious organic composition, there will still remain a continuous, cohesive ash, capable of retaining its shape and continuity even when exposed to a wind having a velocity up to 12 miles per hour, when it is supported on a roof deck having a slope of about 5 inches per lineal foot. This test is of particular significance since the ash protects the substructure. The composition is further designed to resist or accommodate thermal stresses or strains caused by the burning brand or fagot on the surface of the roofing so that substantially no cracks develop in the body of the roof coating.

Any material which will intumesce and form a crust during heating and decrease and prevent the flow of the superheated asphalt is suitable in the process. It has been found that a certain quality and quantity of asbestos serves the purpose very well when associated with materials such as ground unexpanded vermiculite, ground bituminous coal, capable of coking, though other equivalent materials may be used satisfactorily. Fillers such as powdered limestone, flaked slate dust, ground clay, calcined dolomite, etc., can be used. The active materials may include asbestos, ground bituminous coal, ground unexpanded vermiculite, etc. These materials are blended with the asphalt coating and applied to the surface which is to be protected, such as a saturated roofing felt sheet. The following examples illustrate the process.

Example 1:	Coating asphalt	61.5 parts by weight
	Asbestos fiber	15.4 parts by weight
	Ground bituminous coal	7.7 parts by weight
	Finely ground unexpanded vermiculite	7.7 parts by weight
	Diatomite	7.7 parts by weight
Example 2:	Coating asphalt	61.5 parts by weight
	Asbestos fiber	15.4 parts by weight
	Ground unexpanded vermiculite	7.7 parts by weight
	Finely ground clay	15.4 parts by weight
Example 3:	Coating asphalt	66.7 parts by weight
	Asbestos fiber	16.7 parts by weight
	Ground bituminous coal	8.3 parts by weight
	Limestone dust	8.3 parts by weight
Example 4:	Asphalt	50.0 to 61.5 parts by weight
	Asbestos fiber	8.4 to 20.0 parts by weight
	Ground bituminous coal	12.5 to 15.5 parts by weight
	Finely ground mineral fillers	7.5 to 25.0 parts by weight

In all of the examples of coatings given, it will be found that when they are applied to a saturated roofing felt, all will resist flowing and all will form a firm, strong crust when fired. It is not the intention to form a coating which has absolutely no flowability under heat, as it has been found desirable to have a slight amount of this action in order to close and seal the butt edge of the shingles against moisture, wind, etc. When the composition is burned the resulting crust is strongest before all of the organic matter has been burned out. However, the ash formed by any of the examples listed on the previous page will be found to be firm and strong to protect further the supporting structure. The coating asphalt used in these examples has a ring and ball softening point between 210° and 235°F. Other suitable varieties like the steep asphalt have a ring and ball softening point between 185° and 195°F., while the flux asphalt has a ring and ball softening point of 107°F. Other asphaltic materials such as Trinidad asphalt, coal tar pitch and stearine can be substituted for those given in these examples.

Mineralized Roofing Material

H.W. Greider and G.A. Fasold (U.S. Patent 2,610,928; September 16, 1952; assigned to The Philip Carey Manufacturing Company) have found that a roofing material which comprises an organic fiber felt base and a bituminous waterproofing coating can be made at the cost of ordinary bituminous roofing material. The material made by this process is so highly fire-retardant that a three-layer covering on a combustible roof deck will satisfactorily withstand exposure to the Class A fire-retardant tests prescribed by Underwriters' Laboratories, Inc. It has also been found that such roofing material, when applied so as to provide two layers, will satisfactorily pass the Class B fire-retardant tests prescribed by Underwriters' Laboratories, Inc.

In U.S. Patent 2,424,234, a bituminous composition was described which is especially adapted for use as the mopping asphalt in the construction of built-up roofings, the composition possessing high fire resistance by virtue of its formulation so as to comprise bitumen having a softening point between about 80° and 200°F. and of the range 35 to 60% of the composition, so as to comprise intimately commingled and distributed uniformly throughout the bitumen finely divided solid water-insoluble heat-resistant mineral filler which is of the range 40 to 65% by weight of the composition with at least 2/3 by weight passing a 100 mesh testing sieve, and so as to have a Wagner-Bowen plasticity value at 400°F. not substantially greater than 250 grams, the filler in the composition having a flow resistance coefficient of at least 75. This bituminous composition successfully solved the specific problem of providing a mopping asphalt suitable as the waterproofing adhesive between the plies of built-up roofing which is capable of application on the job using an implement such as a mop and which at the same time overcomes the low order of fire resistance that is an inherent attribute of conventional mopping asphalt compositions.

As distinguished from the mopping asphalt adhesive composition described in U.S. Patent 2,424,234, this process is directed to the provision of highly mineralized roofing products of very high fire resistance which comprise a base sheet and coating layer. The highly mineralized coating composition combined with a base sheet according to this process comprehends coating compositions whose Wagner-Bowen plasticity value at 400°F. is substantially greater than 250 grams while not being substantially greater than 1,500 grams. The roofing product involves coatings having the commercially important advantage of being highly blister-resistant, the coatings in such case utilizing bitumen having a softening point above 200°F.

It has been found that in the formulation of bituminous coating compositions suitable for roofing there is a critical point below which a finely divided mineral filler exerts little, if any, effect upon the combustibility and flow resistance of bituminous coating compositions but above which even slight further additions of the mineral filler exert a profound effect upon the combustibility and flow resistance of the bituminous coating composition. This critical point usually is in the neighborhood of the maximum amount of the filler in question that can be successfully mixed and spread as a satisfactory coating upon a base sheet such as a sheet of asphalt-saturated organic fiber felt. This critical point is different for different mineral fillers and usually is found to occur at a mineral filler content of about 60 to 80% by weight of the bituminous composition, depending on the particular mineral filler used. The critical point also is affected by the type of bitumen that is used as the waterproofing base for the bituminous composition.

In the exemplary form, the base or foundation sheet is ordinary roofing felt made of vegetable and animal fibers, e.g., a roofing felt made from organic fiber stock weighing about 6 pounds per 100 square feet. The felt is impregnated with a suitable impregnating material, e.g., a conventional bituminous roofing saturant having a softening point of about 120° to 170°F. The base sheet as composed has little fire resistance and, when exposed to fire, is consumed quite readily.

A typical example of the special coating composition may be made as follows. The bitumen used in the coating is preferably a residual asphalt flux from the refining of Mid-Continent petroleum which has been oxidized as by air blowing until it has a softening point of about 200° to 210°F. and constitutes about 32% by weight of the composition. With this asphalt there is intimately commingled and distributed uniformly throughout finely divided mineral filler which constitutes 68% by weight of the composition. The filler consists of powdered limestone which constitutes about 47% by weight of the composition and asbestos dust which constitutes about 21% by weight of the composition. The powdered limestone is limestone which has been reduced to a pulverulent condition, about 85% by weight of it passing a 200 mesh testing sieve. The asbestos dust is such that nearly all passes a 14 mesh testing sieve, with about 75% by weight retained on a 200 mesh testing sieve and about 25% by weight passing a 200 mesh testing sieve. In preparing the special coating material, the asphalt is heated to a heat-liquefied condition and the asbestos dust and limestone dust are thoroughly mixed. The mixing preferably is carried out at a temperature of about 375° to 475°F. It is important that the bitumen and finely divided filler be thoroughly mixed and intimately commingled so that the composition of the coating with respect to the distribution of the filler is uniform throughout. The composition should be applied to the felt base sheet while in a thoroughly mixed condition so that the resulting layer is substantially uniform in consistency and in thickness. The coating is applied to the felt base sheet at the rate of about 45 pounds per 100 square feet. Before the coating cools to atmospheric temperature, a conventional surfacing material such as slate granules can be applied to the surface for weather exposure, and these granules are partially embedded in the coating composition. The granular surfacing material is preferably noncombustible and, while employed, is not relied upon to impart fire resistance to the roofing.

On the opposite side of the roofing material there may be applied a thin coating of bituminous coating material, e.g., an asphalt having a softening point of about 200° to 240°F., so as to weigh about 5 pounds per 100 square feet, and finely divided dusting or anti-stick material such as talc or mica dust or similar material may be applied and partially embedded in the back coating.

The roofing thus prepared may be packaged in sheet form or, alternatively, may be cut into sections suitable for shingles or tile-like units. The roofing weighs about 95 pounds per 100 square feet. Of this weight, about 30% by weight is in the form of mineral granules adherent to one side and finely divided dusting material adhered to the other side, the balance of the roofing being the felt, the bituminous impregnating material and the bituminous coating material. The roofing material as described can be made up into suitable preferred roofing materials and into suitable roofing structures.

Bituminous Shingles Coated with Thermosetting Acrylic Resin

C.L. Nielsen (U.S. Patent 3,276,906; October 4, 1966; assigned to Shell Oil Company) describes a process for preparing a highly flame-retardant bituminous roofing material which comprises applying a thin coating or overlay of an aqueous solution, suspension or dispersion of a thermosetting cross-linked resin, particularly an acrylic resin, to a bituminous-felt roofing material and then allowing the solvent, which is essentially water, to evaporate.

Prepared bituminous roofing is generally manufactured in large quantities by known processes which ordinarily comprise saturating a sheet material, such as felt, with a waterproofing saturant and then applying over the surface a coating layer of a bituminous material such as asphalt, pitch or tar, and preferably blown asphalt, after which a mineral surfacing in comminuted or granular form is partially embedded in the coating layer or layers.

In general, the felt is first thoroughly impregnated with a bituminous saturant by one of many procedures such as by passing the preformed web through a bath of the molten saturant. This saturated fabric is then coated with a bituminous coating composition by any of the usual methods. Before the roofing is cooled the conventional surfacing materials such as slate and sand granules are applied to the surface and partially embedded. The roofing material is then cooled by any of the means generally employed including water quenching. The roofing material thus fabricated is either cut up into shingle units or strip shingles, and in many instances is also prepared in the form of rolls of substantial length. These prepared singles generally weigh between about 90 and 300 pounds per 100 sq. ft., although weights outside such limits may be encountered. Of this weight about 25 to 40% is in the form of the mineral granules embedded in the bituminous coating material.

The mineral surfacing overlying the coating layer of the prepared roofing material so fabricated serves to protect to some extent the coating layer from the deteriorating action of heat, cold, rain, sun and fungus attack and functions, to a minor extent, to render the shingle fire-resistant, as well as to provide a decorative finish. Although the mineral surfacing embedded in the coating layer affords some protection, it frequently happens that the coating layer does not weather evenly and is deleteriously affected after exposure to the elements for a greater or lesser period of time. This granule loss thus not only depreciates the decorative effect of such shingles, but also renders the shingles less resistant to outdoor exposure. Further, and more importantly, whatever slight flame resistance the shingles possessed originally is quickly lost with the loss of granules. It has been found that these shortcomings can be overcome by applying a thin overlay of an aqueous solution of a thermosetting acrylic resin to previously prepared bituminous shingles.

The thermosetting cross-linked resins which remain suitable for use in the process include: polyurethane resins, particularly the polyurethane varnishes; polyesters; polyepoxides, including the polyethers, polyesters and other polyepoxy condensates; and acrylic resins. A detailed description of the preparation of suitable water-soluble thermosetting acrylic resins can be found in U.S. Patent 2,906,724. Preferred water-soluble thermosetting acrylic resins are those marketed by Rohm & Haas under the trade name of "Acrysol", particularly Acrysol 2003 and Acrysol 2004. Acrysol 2003 is especially preferred due to the generally decreased amount required and the generally greater fire resistance imparted to the shingles. Acrysol 2003 has a solids content of about 36% and the solvent portion (64%) contains 90% water and 10% tertiary-butanol, a viscosity of from about 3,000 to 6,000 cp. and an approximate pH of 9.1.

In general, the thermosetting resin is applied in amounts greater than about 3 grams per square foot of roofing surface. Although this amount will vary somewhat depending upon the particular thermosetting resin utilized, a generally suitable range was found to be between about 5 and 10 grams per square foot of roofing surface. The resin is applied and the water is allowed to evaporate leaving a cross-linked surface coating. If the thermosetting resin is applied in the manufacture of the roofing materials, it may be desirable to apply heat to the coating by baking at 200° to 300°F. from 15 to 30 minutes to facilitate evaporation and curing.

In an experiment to determine the relative temperatures on an asphalt shingle, an untreated asphalt shingle was placed on an inclined plane having a slope of 5 inches per foot and a Bunsen burner flame applied to the lower end of the shingle. At the edge of the flame, the surface temperature ranged from about 600° to 700°F. About one-eighth of an inch from the flame, the surface temperature was about 450°F. and about one inch ahead of the flame the surface temperature was approximately 250°F. Between this point and the flame, the asphalt was very fluid and ran into the fire. This fire-sustaining property is of course undesirable.

It was further discovered that if a thermosetting film was applied to the asphalt shingle, the coating had sufficient film strength to prevent this observed flame spread and thus rendered the asphalt shingle flame resistant, or at least flame retardant. Based upon the burning phenomena observed when the untreated shingle was exposed to a gas flame, a temperature of 600°F. was selected as a temperature at which the film strength should be determined. In other words, if a film had sufficient strength to hold the granules together at 600°F., the flame couldn't spread since the temperature drops off to less than 600°F. at a distance of one granule diameter from the flame. The process is illustrated by the following examples.

Example 1: This example describes the technique of determining the minimum coverage by the film strength and illustrates the flame retardant properties of a representative thermosetting resin. Film specimens 2 1/4 inches long, 7/8 inch wide and from 0.003 to 0.012 inch thick were prepared. Aluminum specimen clamps were attached to the ends of the specimen and the specimen suspended in an electric furnace at 600°F. with various weights attached to the lower end. The film strength of the particular film was then determined as pounds per square inch, static load.

One-half of an asphalt shingle specimen 8" x 12" was coated with various thermosetting materials. The other half (4" x 12") was untreated and used for control. This prepared shingle specimen was then supported on a wire mesh screen (1" mesh, approximately 3/32" diameter wire) at a slope of 5" per foot. A Fischer or Bunsen burner flame was then applied to the lower end of the asphalt specimen. The amount of thermosetting resin which provided flame-retardant properties was noted.

From the static loading tests, coverages and performances of the materials in the above-noted burning test, it was found that the film must have a strength at 600°F. of approximately 0.02 pound/lineal inch or greater. The film thickness required in order to achieve this 0.02 pound/lineal inch film strength can be readily calculated from the strength at 600°F. under static loading. This will then give the minimum coverage of the particular thermosetting resin.

Using the above-described technique and methods, one-half of a Pabco, 230 lb., three-tab asphalt shingle was treated with Acrysol 2003 (a thermosetting acrylic resin in water solution, containing 36% solids, 10% t-butanol in the aqueous solvent, an average viscosity of approximately 5,000 cp., and sold by Rohm & Haas Company) so that a coverage of 6.0 grams of the polymer per square foot of shingle was obtained. The treated portion of the shingle exhibited excellent flame retardation whereas the untreated portion continued to burn when the initiating flame was removed. The tensile strength of the polymer at 6.0 grams per square foot and at 600°F. was in excess of 8.6 psi. Thus, a thin overlay of the acrylic polymer applied to the surface of the asphalt shingle renders the shingle fire retardant by preventing the asphalt flow and extinguishing the flame. In the untreated shingle, the asphalt flows down into the flame and continues to supply fuel to the flame. In addition to the fire resistance properties, the weathering life of the treated portion of the shingle was considerably improved due to greater granule retention.

Example 2: The procedure in Example 1 was substantially repeated where the thermosetting cross-linked resin employed was Acrysol 2004 (an acrylic resin in water solution having a solids content of 36% containing about 5% butyl Cellosolve in the aqueous solvent, a viscosity of about 8,000 cp., and manufactured by Rohm & Haas). Excellent flame retardant properties were obtained when the coverage was approximately 10.0 grams per square foot.

Hydrocarbon Dispersion Containing Borax and Diammonium Phosphate

O.T. Hodnefield (U.S. Patent 3,201,265; August 17, 1965) has found that a heat-resistant, flame-retarding composition may be prepared in which a fire-retarding mineral filler is uniformly colloidally dispersed in a hydrocarbon dispersion medium. Notwithstanding the combustible nature of the hydrocarbon, the composition is capable of withstanding a temperature of 2000°F. for over one hour without permitting a flame to continue spontaneously. The composition is a concentrate which may be dispersed in a larger quantity of organic material, such as asphalt and hence shipped to a consumer who supplies asphalt or other material procured locally. This concentrate is prepared in a series of peptizing and blending operations in which the components of the hydrocarbon dispersion medium are mixed and fire retardant mineral and chemicals are added. The peptizing and blending may be carried out at room temperature.

Example 1: The quantities of the following ingredients are based on preparing 100 gallons of the fireproofing composition. A mixture of 5 1/2 ounces of an emulsifying wetting agent [a sodium octyl phosphate having the formula $Na_5R_5(P_3O_{10})_2$ where R is the ethyl hexyl group] and 7 ounces of lecithin is blended into one gallon of cool oil with continuous stirring until thoroughly emulsified, at which point one and one-half gallons of perchloroethylene are blended in with constant stirring. Mix 11.2 pounds of monoethanolamine and 2.4 pounds of oleic acid and add the blended emulsion with constant stirring. Add one and three-quarters of an ounce of an antifoaming agent. With continued stirring, add a separately blended mixture of a pound of copper oxide and 3.2 pounds of hexahydrobenzoic acid. Successively add 1.5 pounds of sodium fluoride or any alkali fluoride and 445 pounds of borax, and finally add 200 pounds of 300 mesh diatomaceous earth. The resulting formulation in solid state was found very effective as a fire-retardant composition when dried and ground to a fine powder and incorporated into convenient carrier materials like asphalt for roofing felts and treating fiber glass for thermal insulation in walls or attics.

Example 2: The following materials are mixed in accordance with the above procedure:

Sodium octyl phosphate having the formula $Na_5R_5(P_3O_{10})_2$ in which R is the ethyl hexyl group	2.5 lbs.
Anti-foaming agent	0.32 oz.
Perchloroethylene	32.0 lbs.
Formaldehyde	5.5 lbs.
Urea	5.5 lbs.
Oleic acid	5.5 lbs.
Monoethanolamine	5.5 lbs.
Tricresyl phosphate	3.5 lbs.
Copper napthenate, 8%	47.0 lbs.
Peanut oil	8.0 lbs.
Pearl oil	16.0 lbs.
Sodium fluoride	2.5 lbs.
Borax (anhydrous)	609.5 lbs.
Diammonium phosphate	240.0 lbs.
Boric acid	252.0 lbs.
Antimony trioxide	157.0 lbs.
Diatomaceous earth	738.0 lbs.
Mineral rubber, 2% of the above components	

The resulting material is an excellent flame retardant.

Weathering Sheet Containing Glass Fibers

J.A. Tomlinson, J.A. Berg and J.G. Hayes (U.S. Patent 3,332,830; July 25, 1967; assigned to Owens-Corning Fiberglas Corporation) describe the manufacture of a weathering sheet made of asphalt reinforced by glass fibers. The process provides a construction of two spaced webs of continuous glass fibers imbedded in a matrix with the core portion interposed between the webs. Mineral granules of conventional type deposited over the weathering surface of the product function as a protection against the actinic rays of the sun and provide a more rugged surface to withstand the rigors of installation and subsequent need for load-bearing characteristics.

The continuous glass fibers in the construction may be distributed in the spaced webs as individual continuous fibers or in twisted or untwisted strands to impart concentrations of strength in the product where desired. In the latter sense, improved nail-holding properties can be imparted by concentration of bundles of chopped glass fibers in the zone where nails are passed through to secure the weathering product to the surface to be protected. Concentrations of glass fibers under the nailheads improve tremendously the tensile and tear strength of the product and, accordingly, its tear-resistance and nail-holding properties.

In a broader sense, the product can be likened to an I-beam structure with webs of glass fibers spaced apart across the thickness of the shingles, siding or roll roofing products. The glass fibers have a high tensile strength and high modulus practically independent of temperature conditions, thereby overcoming a limitation in stiffness of the asphalt itself with variations in temperature. Additionally, the spaced web construction with the glass fibers close to the outer surfaces of the product provides a retaining network for asphalt at the surface and imparts a greater dimensional stability to the surface material than is otherwise possible with the flowable surface of asphalt. The stability of the asphalt surface is increased by the presence of the fibers to an extent such that the need for the relatively expensive granules is greatly reduced thus permitting in many instances actual elimination of such granules. This occurs particularly with the structure since the product incorporates a weathering grade asphalt throughout, thus eliminating the need for the extra weathering barrier afforded conventional products incorporating paper felt bases by the otherwise thin layer of weathering grade asphalt. Where the usual surface granules of crushed mineral material are used, however, the close presence of the network of glass fibers provides a base or nesting zone for at least a partial nesting within the web of the topmost layer of glass fibers, thereby providing additional protection to prolong the life of the product surface.

Since the spaced mats or webs of glass fibers can be relatively thin and still provide the additional strength and dimensional stability to the asphalt matrix, the webs can be extremely porous and readily adapted to allowing saturation or passage of the asphalt. Accordingly, a higher viscosity asphalt such as a weathering grade asphalt can be utilized throughout the thickness of the product so constructed.

In addition to the above, another feature of the process lies in the fire safeness imparted to the product by its special construction. Conventional asphalt constructions for roofing materials, upon being subjected to the heat of a flame extending over a period of a few minutes, result in a charring of the surface of the asphalt, which charred surface of itself is highly fire resistant, but due to the heating and swelling of the substrate below the charred surface, cracks usually occur in the charred surface with a consequent exposure and flow of fresh asphalt for continued spread of flame over the product surface. The described construction, however, in providing an inorganic noncombustible network of glass fibers immediately under the top surface of the product, provides a base for the charred surface which during spread of the flames prevents opening of the charred matter. Accordingly, the fire resistance of the charred surface, during exposure to flame, becomes progressively more noncombustible and soon results in reduction of the tendency toward sustenance of the flame. Accordingly, if the product surface is the sole base for the flame, it soon causes burn-out, in comparison to continued spread and sustenance of the flame in conventional asphalt products.

The shingle 10 of Figures 4.1a and 4.1b is made of a weathering grade asphalt material 12, physically stabilized by having a stabilizer such as black slate to improve weatherability and dimensional stability under erosive weathering forces. The asphalt, it has been found, if of the proper type, can contribute considerably to the fire safety of the product. In this respect, an asphalt is used which is produced by a catalytic air blowing process and retains a residual or dormant catalytic reaction arising by reason of incompletion of the reaction during processing. In other words, the asphalt is an air-blown petroleum residue to which a catalyst has been added. The catalyst is added during processing of the asphalt to increase its oxidation rate and accordingly reduce its blowing cycle period. Because of the high viscosity characteristics of asphalt, however, completion of the catalyzed reaction is ordinarily incomplete. The presence of such an asphalt in the product increases the fire safety of the product in that when it is subjected to the heat of flames, a gassing and a crusting of the asphalt surface is promoted when the dormant catalytic reaction is triggered.

An asphalt which has proven successful in providing this fire safety is one catalytically reacted with a chloride-type catalyst such as ferric chloride. Phosphorus pentoxide is another example of an asphalt catalyst which provides an asphalt adapted to the process.

Webs or mats 11 and 13 of glass fibers are spaced apart in parallel relation across the thickness of the construction a distance selected for the degree of stiffness desired in the product. As illustrated, the webs are located near the bottommost and topmost surfaces of the product and thus produce somewhat of a sandwich construction, but with the asphalt 12 enclosing both webs in the construction while granules 14 of glazed crushed stone or other suitable mineral material are set in the asphalt overlying the topmost layer or web of glass fibers.

When the overall product dimension is relatively thin, it has been found desirable that the glass fiber webs 11 and 13 be spaced as far apart as possible in order to impart the maximum stiffness possible with the given components incorporated in the shingle. The usual asphalt, of itself, lacks the dimensional stability and stiffness desired in shingles to resist the constant force of uniform winds or the erratic action of gusts of wind blowing against them. Additionally, by so locating the webs in the product, the topmost web immediately under the exposed surface of the product offers a more stable asphalt surface in which the granules 14 can be solidly fixed, improving the capabilities of the construction to hold the granules in place under various weather conditions, as well as assuring a longer life for surface asphalt against the damaging action of the sun's rays and of water.

The topmost glass fiber mat 11 furthermore acts somewhat as a retainer for the underlying asphalt and consequently is in a sense a fire barrier in cases where the product is subjected to high temperatures such as during a fire occurring in the vicinity of the surface to be protected. In this way, the uppermost web of glass fiber 11, besides acting as reinforcement and as a stabilizing means for support of the granules 14, also advantageously stabilizes the asphalt at the surface against motion during emergency conditions experienced in use. Additionally, since the matrix 12 is of weathering grade and extends throughout the thickness of the construction, life of the product against erosion is greatly increased. The bottommost glass fiber mat or web 13 acts in conjunction with the topmost mat 11 to impart strength to the product against its being ripped from the nails used to secure the product to the surface to be protected.

Figure 4.1c illustrates the type of mat adaptable to incorporation in the construction of Figure 4.1a. The glass fiber mat 11 includes glass fibers in three different arrangements, namely, glass fibers in the form of strands or yarns aligned in relatively straight parallel relation extending along the length of the mat, as well as randomly distributed continuous strands or yarns each of a plurality of continuous glass fibers extending over the major dimension of the product.

FIGURE 4.1: ASPHALTIC WEATHERING SHEET CONTAINING GLASS FIBERS

a.

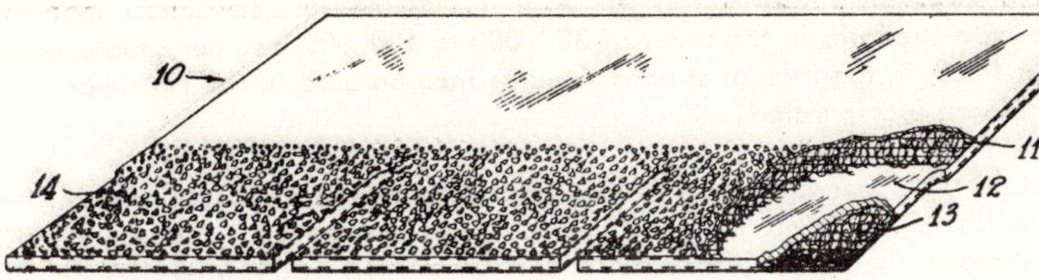

b.

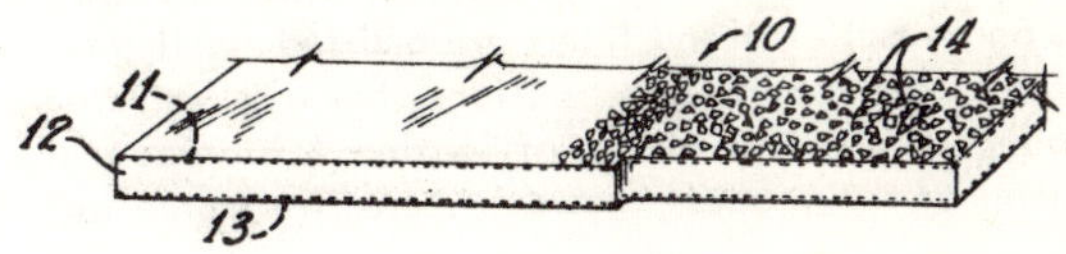

Figure 4.1a and Figure 4.1b – Partly Broken Shingle Reinforced with Glass Fibers

c.

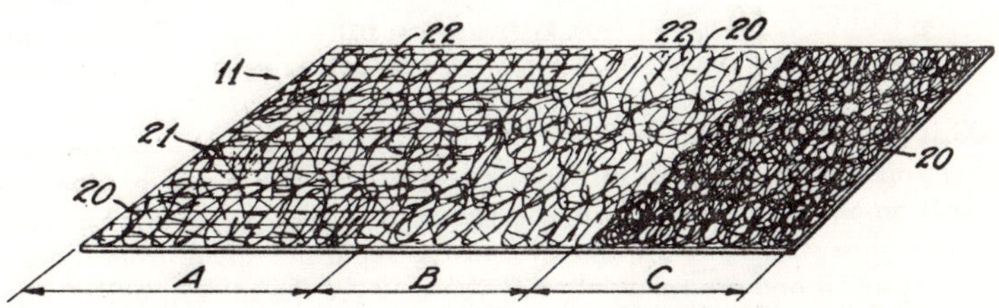

Illustration of Type of Mat or Web Glass Fibers

d.

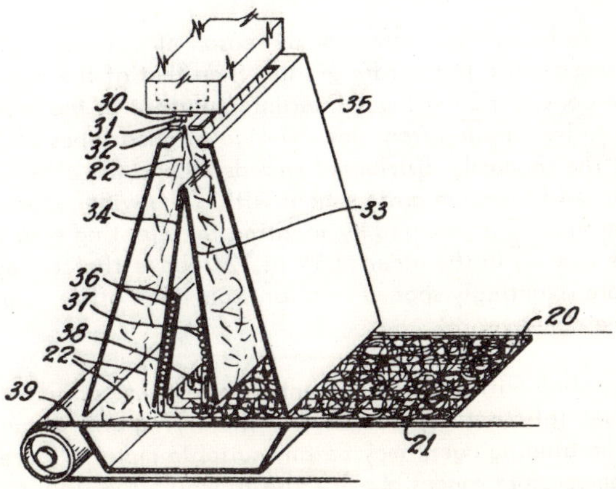

Apparatus for Producing Mat

Source: J.A. Tomlinson, J.A. Berg and J.G. Hayes; U.S. Patent 3,332,830; July 25, 1967

Additionally, discontinuous individual glass fibers are randomly distributed throughout the major dimension of the product. For tensile strength in the roofing construction in which the mat is incorporated, the continuous strands of glass fibers are most desirable whether oriented in random or parallel manner. In both instances, the strands are either twisted or untwisted, but include a bundle of continuous parallel glass fibers running for the full length of the strand. It is well established that individual glass fibers have shown tensile strengths up to 1,000,000 lbs. per square inch in laboratory tests and are commercially available with strengths in the order of 300,000 to 400,000 lbs. per square inch. Accordingly, bundles of these individual glass fibers in the form of strands or yarns incorporated in the mat impart a high reinforcing strength to matrices in which they are incorporated.

Section A of Figure 4.1c illustrates the appearance of a mat incorporating the parallel and random continuous glass strands, as well as the discontinuous glass fibers utilized to produce the product of Figure 4.1a. Regardless of the form in which these mineral fibers are incorporated in the asphalt, they act as a filler as well as a reinforcing stabilizer against thermal and mechanical forces to which the asphalt might be subjected. Section B of Figure 4.1c illustrates the combination of continuous glass fibers in the form of strands and discontinuous glass fibers with the parallel strands omitted. Section C illustrates still another section and type of mat which can be utilized successfully in the construction in which both the discontinuous and parallel strand fibers are omitted. All three sections of Figure 4.1c represent types of mats which can be incorporated in the roofing constructions, but as illustrated and as described in relation to the process by which the product is produced, the type of mat of Section A incorporating all three forms is preferred. The basic functioning elements of the mats of this process, however, are illustrated in Section C of Figure 4.1c.

Figure 4.1d illustrates the method and apparatus by which the mat of the composite type including each of the fibrous elements shown in Section A of Figure 4.1c is produced. In this arrangement, a feeder or bushing 30 is associated with a source of molten glass, such as the forehearth of a melting tank, and streams of the glass are flowed from tips 31 of the feeder and the turbulent forces of a set of blowers 32 act on opposite sides of the streams to disrupt them into discontinuous fibers 22. The discontinuous fibers 22 move downwardly under the influence of gravitational forces in addition to the forces of air flowing to the lower pressure zone established below the underlying foraminous conveyor 39 on which the fibers are deposited. The fibers are guided downwardly to the conveyor within a hood 35 surrounding the zone below the blowers 32. A separating partition 33 located below the blowers 32 dissects the space under the blowers so that the fibers 22 are divided into two moving masses or groups of fibers, namely a front group and a back group, which on subsequent deposition on the conveyor 39 form the top and bottom portions, respectively, of the final mat product.

The separator or partition 33 is a narrow elongated tent-like construction with its apex located in the upper portion of the hood in the fiber-forming zone. The separator provides a space over the conveyor from which the continuous glass strands are supplied for deposition on the conveyor between the bottom and top layer of discontinuous glass fibers. The randomly distributed continuous glass strands 20 are blown into the separator space from the creel mounted packages (not shown) disposed exteriorly of the hood and are randomly distributed over the underlayer of discontinuous fibers 22, whereupon the mass is moved forwardly by the conveyor 39 for receipt of the overlayer of intermingling discontinuous fibers at the forward part of the separator zone. In the underlying space within the separator the parallel strands 21 are also supplied for deposition on the conveyor 39 from tubes 36 leading from packages of the continuous strand (not shown) on a creel disposed laterally at the exterior of the hood 35. These strands are initially introduced into the mat-forming process by being blown against the conveyor 39 and the blowing action is halted and continued movement is maintained due to frictional adherence of the strand to the moving conveyor which draws the strands through the tubes 36 from their source located laterally of the hood but not shown herein.

Thus, the parallel strands 21 are laid on the conveyor by reason of movement of the conveyor while the randomly distributed strands 20 are blown thereon at a linear rate greater than that of the conveyor 39. To assure that the strands 21 are properly deposited on the conveyor 39 without interentanglement in the process of transfer from their source to the conveyor, they are arranged to be supplied from side-by-side aligned tubes 36 located in a separate zone 34 in the back portion of the separator 33. The randomly distributed strands 20, on the other hand, are ejected from the blowing tubes 38 which effect their distribution in random arrangement within the wider space of the separator 33. A wide distribution of the randomly distributed strands 20 is effected by locating the ejecting ends of the tubes 38 some distance above the conveyor, which for example may be in the order of 9" to 12" above the conveyor, while the parallel strands 21 which are desirably deposited in more exactingly spaced relation from each other, are supplied from the open ends of tubes 36 located within 1/8" above the conveyor 39.

The entire mass of fibers is supplied with a binding agent from a source adjacent to the hood 35 (not shown) and is then conveyed through a curing oven (also not shown) to effect the drying and/or curing of the binding agent for full integration of the fibers into the mat. The binding agent may be any suitable material, preferably a thermosetting agent, having fiber-holding properties in temperature ranges above those to which the final product is likely to be subjected in use. In this respect, phenolic binders, comprising 12 to 25% of the mat, have proven successful in having a curing temperature in the range of 300° to 400°F. and the capability of maintaining the interbonded relation for appreciable periods at temperatures in the order of 400°F. Accordingly, the integral relation of the fibers as a retaining member in asphalt or other bitumen materials is highly effective for practically all temperatures to which the product might be subjected in normal usage. The modulus of elasticity of glass fibers is in the order of from 10 to 12 million, and accordingly, it is highly effective in increasing the overall stability of the bitumen construction which has a modulus of elasticity a considerable extent lower than the glass fibers.

By way of example, mats which have performed successfully for the purposes of this process have been produced with a thickness of about 10 mils including discontinuous fibers having an average diameter of 60 hundred-thousandths of an inch and a length of from approximately 3 inches to a few feet, while the continuous fibers in bundles or strand-like form comprising about 30% of the mat by weight have a nominal diameter of 30 hundred-thousandths of an inch. One type of strand commercially available for such purposes contains 204 filaments twisted approximately one turn per inch. The strand, however, can be either twisted or untwisted, and in the case of either, a dispersion or separation between the filaments within the bundle can be provided to impart a finer texture to the mat and an intermingling interengagement to effect a bridging of the fiber gaps that might occur across the expanse of the mat. In this respect, strands such as described in U.S. Patent 2,736,676 can be incorporated in the mat as well as that shown and described in U.S. Patent 2,875,503. The mat can be made fully of such fiber-dispersed strands without need for discontinuous fibers if desired.

Details of the apparatus and process by which roofing material is produced are also given in the patent. Although described in relation to production of shingles, the process and apparatus is readily adapted to production of siding and roll roofing products as well, including asphalt and other suitable thermoplastic weathering materials. As an example of improved fire safety obtainable with structure of the process, Underwriters' Laboratory tests reveal that the flame spread rate of the shingle incorporating the catalytically reacted asphalt and chopped glass fibers at the exposed edge can be reduced to 3 feet or less in 10 minutes, much within the maximum of 6 foot spread in 10 minutes for class A rated roofing shingles.

Intumescent Coated Roofing Granules

A process described by C.E. Hinds (U.S. Patent 3,365,322; January 23, 1968; assigned to Bird & Son, Inc.) involves the use between a sheet of asphalt-impregnated roofing felt and a top layer of decorative granules of an interlayer of intumescing-coated mineral granules positioned between two layers of an asphalt coating composition comprising asphalt and fine mineral matter such as slate or rock dust. The upper layer of asphalt coating composition, i.e., the layer nearest the top layer preferably also contains between about 5 and 20% by weight of the asphalt coating composition of asbestos fiber of a class equivalent to Group 7 grade. Group 7 grade refers to a type so classified according to a screen analysis as shown in Canadian Dept. of Mines Bulletin #707 (1931) and U.S. Bureau of Mines Bulletin #403 — Asbestos (1937). The other layer of asphalt coating composition also may contain such asbestos fiber in the same amount.

In the preparation of the fire-retardant roofing, referring to Figure 4.2c, a sheet of a roofing felt, e.g., a conventional organic fiber roofing felt 2, as for example one having a thickness in the range of 0.055 to 0.070 inch and weighing about 50 to 60 pounds per 480 sq. ft. of dry felt, is impregnated with asphalt at 4 to about 175 to 185% of its dry weight. To the top face of the impregnated sheet there is applied a first or lower layer of asphalt coating composition 6 comprising asphalt and mineral reinforcing matter including asbestos fiber of the class and in the amounts above described. Although it has been found that by use of the intumescing-coated granules in the interlayer it is not essential to have any asbestos fiber in this lower layer in order to retain freedom from undesirable flow of the ignited coating, it is somewhat preferable to have about 5 to 20% by weight of the asphalt coating composition of asbestos fiber, so that the coke or char mass will possess good cohesive strength and not break up into smaller pieces. The amount of asphalt coating composition so applied and which has a softening point about 250° to 285°F. will preferably be about 20 pounds per 100 sq. ft. or may vary, for example, between about 8 and 30 pounds per 100 sq. ft.

The back face of the felt 2 optionally may also be coated with this same coating composition, or with an asphalt coating composition which contains no asbestos fiber, and then lightly coated with talc or soapstone. After the coating 6 has been smoothed to the desired thickness, it is surfaced with a layer 8 of the intumescing-coated granules consisting of mineral granules 10 having coatings 12 of a material which intumesces when heated. After pressing the resulting composite at 14 to embed the granules, any excess granules are removed and a second or upper layer 16 of asphalt coating composition similar in composition and amount to the lower layer 6 is applied and smoothed to the desired thickness. It is highly desirable that the asbestos fiber be included in this upper layer, as is indicated at 18. A top layer 20 of protective and decorative mineral granules is then applied in conventional manner and amount and the sheet cooled and cut into strips or shingles.

When the roofing of the process is exposed to the heat of a flame, as for example in the tests prescribed by the Underwriters' Laboratories for determination of Class A qualification, the coating 12 on the granules 10 of the interlayer 8 intumesces and, increasing in volume several fold, swells into a porous mass which tends to prevent flowing and running of the asphalt under the heat of the flame. The granule coating 12 first intumesces to a porous mass which, along with the asphalt and granules 10, will upon continuous heating fuse or sinter into a crust or coked mass that tends to insulate the roof deck from the flame. The intumescence of the coating takes place over all surfaces of the granules in the interlayer 8 and there is an increase in volume of the composite coatings and granule layers which on formation of the porous coked

FIGURE 4.2: ASPHALT COMPOSITION CONTAINING INTUMESCENT COATED GRANULES

a.

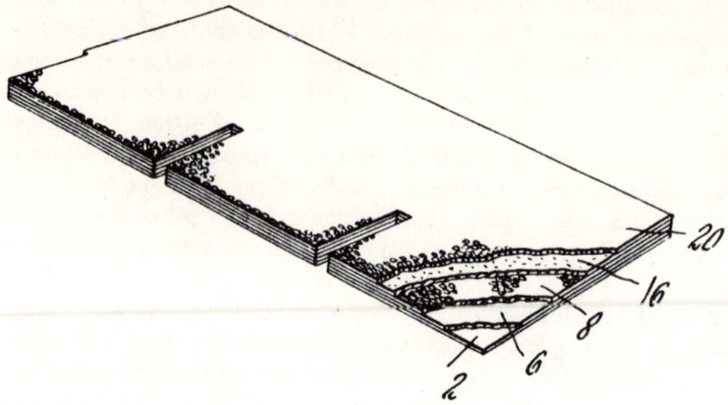

b.

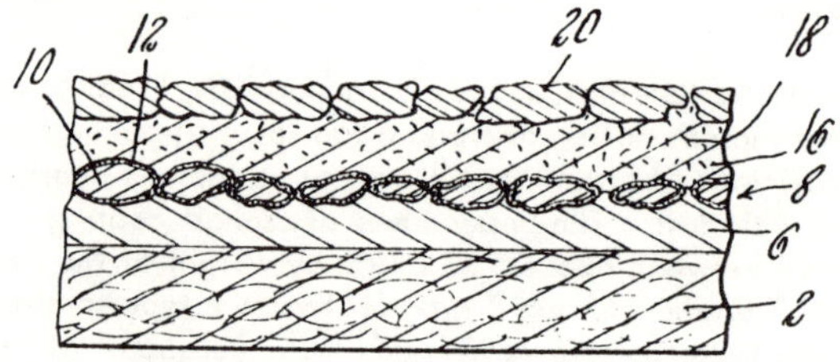

Figure 4.2a and Figure 4.2b - Detail of Roofing Shingle

c.

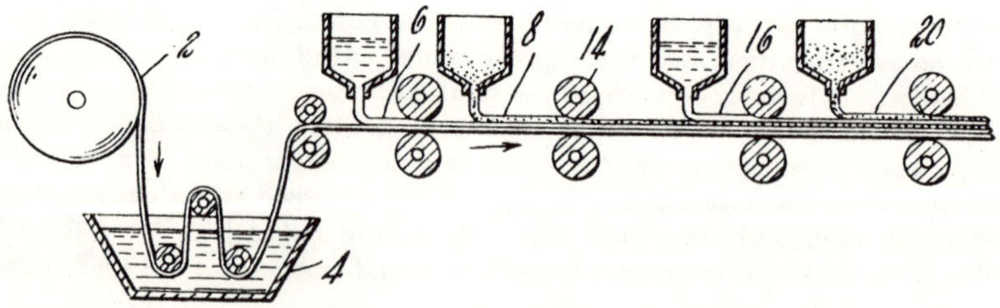

Process for Producing Shingle

Source: C.E. Hinds; U.S. Patent 3,365,322; January 23, 1968

mass increases the vertical distance between the roof deck and the flame, e.g., of a burning brand resting on the roofing surface. As further heating occurs, there takes place a shrinkage of the coating on the granules and at least a partial sintering of the intumesced coating which combines with the other materials in the char. The granules of the interlayer, individually coated with intumescing material, produce upon heating an absorbent intumesced mass which prevents sliding and easy ignition of the asphalt coating. The solid glass-like material formed on increased heating fuses or sinters with the granules and coked asphalt to form a hard, tough insulating crust which acts as a fire wall to protect the roof deck from ignition temperatures. The presence of the granules is thus highly advantageous. This intumescing granule coating, which is low in cost, requires no manufacturing equipment or operations beyond well-known normal methods for coating of roofing granules.

A preferred typical formula for an intumescing coating 12 to be applied to a mineral granule base 10 is as follows. Parts are by weight unless otherwise stated.

Base granules (#11 grading)	100
Borax (5 mol)	4
G brand sodium silicate powder	4
N brand sodium silicate solution	4
Water	2

G brand sodium silicate is a powdered sodium silicate, supplied by Philadelphia Quartz Co., having a Na_2O to SiO_2 ratio of 1:3.22. N brand sodium silicate is a sodium silicate solution, supplied by Philadelphia Quartz Co., containing 37.6% solids having a Na_2O to SiO_2 ratio of 1:3.22. The granules are coated as follows. The borax and G brand sodium silicate are added to the granules in a suitable mixing drum and the mixture tumbled for about 2 minutes to mix the dry ingredients. The water is added and the mixing continued for an added 4 minutes. The N brand sodium silicate is then added and the mixing continued for another 8 minutes. The coated granules are then dried with mild agitation at about 220° to 250°F. They are then cooled to about 100°F. and are preferably treated with slate treating oil (a very light mineral oil) containing a moisture repellent at about 10 pounds per ton of granules. The moisture repellent prevents absorption of moisture by the coating and sticking in storage. The slate treating oil aids in the embedding of the granules in asphalt.

The borax may be used in the form of sodium tetraborate pentahydrate (5 mol); sodium tetraborate decahydrate (10 mol); or sodium borate concentrates (Rasorite) passing 10 mesh to passing 200 mesh. The soluble sodium silicate solution in liquid form may have a ratio of sodium oxide to silica from 1:1.6 to 1:3.75. The soluble silicate in powder form may be any which passes 20 mesh to 200 mesh and has a sodium oxide to silica ratio of from 2:1 to 1:3.22.

Broadly, in the manufacture of the intumescing-coated granules, and depending of course upon the screen grading of the base granule used, there may be used, for each 100 parts by weight of the mineral granule base, from 1 to 50 parts by weight of borax, with the other ingredients proportioned as in the above formula. Preferred compositions are those variations of the above formula falling within the limits of 1 to 7 parts by weight of borax; 0 to 6 parts by weight of G brand sodium silicate; 3 to 10 parts by weight of N brand sodium silicate, and 1 to 5 parts by weight of water.

Polyester and Alkyd Resin Additives

P.S. Bettoli and J.J. Klimas (U.S. Patent 3,096,196; July 2, 1963; assigned to The Ruberoid Company) have found that by incorporating from about 1 to 25% by weight of a polyester (or alkyd) resin in the coating composition, the percentage of resin being based on the weight of the asphalt, it is possible to impart a pronounced fire resistance to the coating composition without materially affecting its rheological properties or altering the commercial methods by which it is applied to a saturated organic felt base to form a roofing material. Roofing material or shingles manufactured by coating a saturated web base with this polyester (or alkyd) containing asphaltic composition demonstrate markedly improved fire resistance to the point where products can be manufactured by conventional methods that will meet every test imposed by the Underwriters' Laboratories for class A and class B designation of prepared roofing.

The improved fire-retardant asphaltic coating compositions of the process comprise a blown petroleum asphalt having a softening point, when measured by the ball and ring method, in the range between 180° and 250°F., from about 1 to 25% by weight of a polyester or alkyd resin, and a finely pulverized mineral stabilizer uniformly dispersed throughout the organic components of the coating composition, the percentage of the polyester resin being based on the weight of the blown petroleum asphalt contained in the coating composition.

The tables on the following page list the chemical compositions and physical properties of eight different polyester and alkyd resins which were employed in preparing the asphaltic coating compositions which were used in carrying out the examples.

Chemical Compositions of Polyester and Alkyd Resins Used in Asphaltic Coating Compositions

Composition (Mol percent)	Resin number							
	1	2	3	4	5	6	[a] 7	[a] 8
Cis-4-cyclohexene-1,2-dicarboxylic acid anhydride	50							
Phthalic anhydride		43					38–42	37–41
Maleic anhydride			47.4					
Sebacic acid				50				
Adipic acid					50	50		
Soya oil acids							30–34	29–33
Glycerin							23–27	12–14
Pentaerythritol								11–13
Propylene glycol				50	50	25		
Diethylene glycol		43				25		
Triethylene glycol	50		26.3					
Methyl glucoside		14	26.3					

[a] Composition given in percent by weight.

Physical Properties of Polyester and Alkyd Resins Used in Asphaltic Coating Compositions

Property	Resin number							
	1	2	3	4	5	6	7	8
Physical state (25° C.)	Liquid	Solid	Solid	Liquid	Liquid	Liquid	Solid	Solid
Viscosity at 25° C. (centipoises):								
Undiluted	1,000,000			[a] 1,680	[a] 2,800	[a] 60		
In cellosolve (3:1 resin to cellosolve)	1,700–4,000	1,700–4,000	1,800–3,500					
In xylol (55%)							[b] Z_2–Z_5	[b] M–O
Specific gravity (25° C./15° C.)	1.15–1.35			1.06	1.15	1.10		
Refractive index [n_D^{23}]				1.471	1.474	1.467		
Acid number	[c] 20	15–25	15–25	1.3	1.6	2.3	6–12	6–10
Saponification number (mg. KOH/g.)				450	585	535		

[a] Viscosity in poises at 25° C.
[b] Gardner-Holdt viscosity scale.
[c] Maximum.

The table below lists eighteen different formulations of asphaltic coating compositions which were employed in the following examples. Each of the asphaltic coating compositions was prepared by mixing the particular blown petroleum asphalt together with one of the polyester or alkyd resins listed above in a mixer, and then uniformly dispersing the mineral stabilizer or mineral stabilizers throughout the two organic components of the composition.

Formulations of Fire-Retardant Asphaltic Coating Compositions

Composition (percent by weight)	Formula Number																	
	1	2	3	4	5	6	7	8	9	10	11	12	13	14	15	16	17	18
Blown asphalt 215/225 S.P.	41.9	50	44.4													41.7	42	41.9
Blown asphalt 180/190 S.P.				40	45													
Blown asphalt 210/220 S.P.						45.5	45.5											
Blown asphalt 230/240 S.P.								42.6	45	445	42.2							
Blown asphalt 240/250 S.P.												41.7	43	43	42.8			
Pulverized stonedust	51.1	35	47.7	50	50	45.5	45.5	48.5	50	50	48	47.6	49	49	48.7	47.6	49.5	49.9
Pulverized mica	5.1	5	7.5			4.5	4.5	6.9			6.9	6.8	7	7.25				
Diatomaceous earth															7	6.8	7	6
Resin No. 1	1.9	10	0.4	10	5													
Resin No. 2						4.5												
Resin No. 3							4.5											
Resin No. 4								2	5									
Resin No. 5										5	2.9							
Resin No. 6												3.9	1					
Resin No. 7														0.75	1.5	3.9		
Resin No. 8																	1.5	2.2

In each of the examples of the process, which are summarized in the table on the following page, the particular bituminous coating composition was applied to a saturated organic felt base to form a roofing material or shingle which was then subjected to a laboratory flame exposure test to determine its fire-retardant properties. The standards established for this laboratory flame exposure test, which is intended to measure the fire-retardant characteristics of asphalt-type roof coverings, particularly with regard to flame spread, combustion, and surface sliding, are designed to correlate with the the results obtained with the test procedure set forth by the Underwriters' Laboratories, Inc. as being requisite for designation of a roofing material as meeting class A standards.

Laboratory Flame Exposure Tests

Bituminous coating composition	Test results
Formula No. 1	Excellent slide and flame resistance.
Formula No. 2	Do.
Formula No. 3	Do.
Formula No. 4	Do.
Formula No. 5	Do.
Formula No. 6	All flame extinguished 1.5 minutes after turning off gas flame; excellent slide and flame resistance.
Formula No. 7	Excellent slide and flame resistance; all flame extinguished 2 minutes after gas flame turned off.
Formula No. 8	Passed laboratory flame exposure test with imperceptible to ½" slide after a 3 minute test period.
Formula No. 9	Passed laboratory flame exposure test with trace to ½" slide after a 3 minute test period.
Formula No. 10	Excellent slide and flame resistance after 3 minute test period.
Formula No. 11	Do.
Formula No. 12	Excellent flame resistance and self-coking properties.
Formula No. 13	Do.
Formula No. 14	Successfully withstands laboratory flame exposure test.
Formula No. 15	Do.
Formula No. 16	Do.
Formula No. 17	Successfully withstands 3 to 4 minutes in laboratory flame exposure test with a maximum slide of up to ½" of surface coating; excellent flame resistance.
Formula No. 18	Do.

In each of the examples of the process, asphaltic coating compositions containing a blown petroleum asphalt, a finely pulverized mineral stabilizer and a particular polyester or alkyd resin were used successfully to prepare fire-retardant roofing materials or shingles. Similar advantages are also attained using other polyester and alkyd resins. The common characteristics of these resins which make them suitable are that they are sufficiently heat stable to withstand brief periods at elevated temperatures without decomposing or vaporizing, that they are resistant to rapid gelation when incorporated in the asphalt and heated to a temperature in the general neighborhood of 400°F. for periods of five minutes, more or less, (or whatever length of time is required for preparing the mineral-filled asphalt-resin mix) so that they may be dispersed in the asphalt vehicle without thermosetting, and that they are water-insoluble or water-insensitive so that the coating product is able to retain its fire-retardant properties on prolonged exterior exposure.

Chlorinated Paraffin with Soda Ash

A process described by L.A. Bierly (U.S. Patent 2,667,425; January 26, 1954; assigned to Presque Isle Laboratories and Manufacturing, Inc.) provides a self-extinguishing asphaltic composition. An example of the composition is as follows:

Dried, powdered sludge	20 %
70% chlorinated paraffin	2 1/2 %
Asphalt or tar	Balance

These percentages are susceptible of wide variation. The sludge may vary from 5 to 50%, and the chlorinated paraffin may vary up to 40%. High percentages of chlorinated paraffin make the asphalt brittle. The presence of inert mineral fillers reduces the quantity of asphalt and therefore requires less chlorinated paraffin. The consistency of the composition may be varied by the use of solvents or thinners. If the composition is to be heated before use, e.g., for hot dipping, the chlorinated paraffin and sludge are added to molten asphalt or tar. This results in a solid composition which becomes fluid when heated above the melting point of the asphalt. Afterglow may be prevented by additions of materials such as antimony oxide and zinc borate.

At the temperatures used in hot dipping, which may be 300° to 450°F., the chlorinated paraffin breaks down, forming hydrochloric acid. It is therefore necessary that the asphalt contains a stabilizer which, at these temperatures, will either prevent decomposition of the chlorinated paraffin or which will combine with the hydrochloric acid. The sludge is that resulting from the production of caustic soda which has the following chemical analysis:

$CaCO_3$	60.0 %
$Ca(OH)_2$	29.3 %
$CaSO_4$	2.21 %
$CaCl_2$	1.16 %
$Mg(OH)_2$	1.9 %
$Fe_2O_3Al_2O_3$	1.1 %
SiO_2	1.9 %
NaCl	.064 %

It is believed that the active ingredients in this sludge which impart the fire-resisting properties are derived from the hydrated lime [$Ca(OH)_2$ and $Mg(OH)_2$]. The hydrated lime is a material which is substantially increased in volume (from 1.7 to 3.3 times) by the water of composition added during hydration. The water of composition of the calcium hydroxide is rapidly released at the temperature of 1076°F.

The water of composition of magnesium hydroxide is rapidly released at a temperature of 662°F. Both of these temperatures are below flame temperatures used in the Underwriters' tests. The portion of the asphalt coating in contact with the flame breaks down the hydrated calcium and magnesium oxides. The release of the water of composition produces a foaming action on the heated or burning asphalt, resulting in a cellular structure which serves as a heat insulator for the layers of asphalt composition not in contact with the flame. The cellular or porous structure is further increased by the expulsion of the water of composition caused by decomposition at flame temperatures of the molecules of hydrated oxides of calcium and magnesium. Both of these oxides have the property of substantially increasing in physical volume upon hydration so that upon dehydration a porous structure results.

As a guide in formulating asphalt shingle coating compositions, class A fire resistance will be imparted to asphalt in the following porportions: 1 part hydrated lime to 9 parts asphalt; 1 part sludge to 4 parts asphalt; 1 part glass or rock wool to 4 parts asphalt. The glass or rock wool also serves to mechanically anchor the coating on inclined surfaces. The sludge or hydrated lime on inclined surfaces should be supplemented by some fibrous or other material which will prevent sliding of the coating. 5% fiber is usually enough.

In U.S. Patent 2,718,479, September 20, 1955, L.A. Bierly describes the use of the above self-extinguishing asphalt coatings in a number of applications.

Halogenated Cyclopentadiene Based Additives

A process described by N.D. Blair and C.T. Bean, Jr. (U.S. Patent 3,151,992; October 6, 1964; assigned to Hooker Chemical Corporation) provides fire-retardant bituminous compositions which contain a compound selected from the group consisting of $C_{10}Cl_{12}$, $C_{10}Cl_4F_8$, $C_{10}Cl_{10}O$ and $C_{10}H_2O_3SCl_{12}$. The flame-retardant characteristics of these bituminous compositions is further improved by incorporating antimony compounds. The preparation of $C_{10}Cl_{12}$, a white, crystalline solid, is described in U.S. Patent 2,724,730. The compound, a dimer of hexachlorocyclopentadiene, is prepared by reacting hexachlorocyclopentadiene in the presence of aluminum chloride. $C_{10}Cl_4F_8$ is a colorless, crystalline solid having a melting point of about 74° to 75°C. The preparation of the compound is described in U.S. Patent 2,951,097. The compound is prepared by coupling 1,2,3,3-tetrachlorotetrafluorocyclopentene with itself in the presence of copper, while heating at the reflux temperature of the reaction mixture.

The preparation of $C_{10}Cl_{10}O$ is described in U.S. Patent 2,616,928. Hexachlorocyclopentadiene and sulfur trioxide are mixed and digested for a short period to condense two molecules of hexachlorocyclopentadiene and to form a liquid reaction product of hexachlorocyclopentadiene and sulfur trioxide. The reaction product is then hydrolyzed with an alkaline aqueous medium, after which the mixture is neutralized with sulfuric acid, whereupon the crystalline product precipitates. The reaction mixture is cooled to insure relatively complete precipitation of the product which is then separated from the mixture by filtration or centrifugation.

$C_{10}Cl_{10}O$ is also prepared by treating $C_{10}H_2O_3SCl_{12}$ with alcoholic alkali, followed by acidification. $C_{10}H_2O_3SCl_{12}$ is prepared by reacting hexachlorocyclopentadiene with chlorosulfonic acid at a temperature in the range of 30° to 170°C. as described in U.S. Patent 2,516,404.

The following examples illustrate the process. In these examples, tests were performed to determine the relative flammability of coated felt strips. All tests were carried out in a sheet metal test shield having inside dimensions 12" in width by 12" in depth by 30" in height and open at the top. The shield had a viewing window in one side, of sufficient size and in such a position that the entire length of the specimen under test could be observed. This viewing side was hinged so that the shield could be readily opened and closed to facilitate the mounting and ignition of the test specimen. Test specimens were clamped vertically, edgewise, using a spring-type paper clamp. The clamp was then attached rigidly to the shield and centered. A laboratory burner was adjusted to provide a flame of about 1" in total height and applied to the end of the specimen until it ignited but not longer than 15 seconds. If the specimen ignited on one attempt and continued to burn, the result was judged to be "burning" by this test. If the specimen did not ignite or was self-extinguishing upon the first attempt, additional fifteen second flame applications were made until the specimen ignited. The number of ignitions were noted. A specimen which did not ignite in one attempt was judged to be "nonburning" by this test. All tests were run in duplicate in a hood with the exhaust fan turned off.

Example 1: An untreated asphalt coating was prepared on a saturated asphalt film strip by pouring a heated asphalt coating material on the top surface of the felt. A heated iron was applied to give a uniform armor thickness of 1/16 of an inch. The specimen was then cut to give a final coated strip of 2" by 12" and used in the aforementioned flame test. The flame test showed that the test strip burned readily with excessive dripping of the asphalt coating.

Example 2: The procedure of Example 1 was repeated using a "stabilized" coating asphalt containing one part of calcium carbonate per part of asphalt. The flame test likewise showed the strip to be flammable and produced excessive dripping. In Examples 3 through 10, various quantities of $C_{10}Cl_{12}$ (hexachlorocyclopentadiene dimer) were incorporated in the stabilized and unstabilized asphalts used in Examples 1 and 2 by heating the asphalt to 400°F. and adding the $C_{10}Cl_{12}$ with stirring. The testing procedure used in Example 1 was followed. As indicated in the following table, the flame test data on the test strips showed that the asphalt coatings containing up to 5% chlorine were flammable and resulted in excessive dripping. The coating compositions containing 10% chlorine and greater, were nonburning, and dripping of the coating was reduced as the chlorine content increased.

Example	$C_{10}Cl_{12}$, Wt. Percent	Cl_2, Wt. Percent	Number of Ignitions	Flammability	Remarks
3	3.2	2.5	1	Burning	Excessive dripping
4	6.4	5.0	1	Burning	Excessive dripping
5	12.8	10.0	1 - 2	Nonburning	Excessive dripping
6	25.6	20.0	4	Nonburning	Moderate dripping
7	38.5	30.0	4	Nonburning	Slight dripping
8	12.8	10.0	1 - 2	Nonburning	Excessive dripping, first test was self-extinguishing
9	25.6	20.0	1 - 4	Nonburning	Slight dripping, first three tests were self-extinguishing
10	38.5	30.0	3 - 5	Nonburning	Very slight dripping, intumescent

Note: Unstabilized asphalt used in Examples 3 to 7; stabilized asphalt in Examples 8 to 10.

In Examples 11 through 17, antimony oxide was used in the asphalt compositions. The test strips were prepared in accordance with Example 1 and were tested as before except as noted below.

Example	$C_{10}Cl_{12}$, Wt. Percent	Sb_2O_3, Wt. Percent	Chlorinated Paraffin, Wt. Percent	Cl_2, Wt. Percent	Flammability	Remarks
[a] 11	None	5.0	--	None	Burning	Excessive dripping
12	12.8	5.0	--	10.0	Nonburning	Slight dripping
13	25.6	5.0	--	20.0	Nonburning	Slight dripping
14	12.8	5.0	--	10.0	Nonburning	Moderate dripping, intumescent
15	25.6	5.0	--	20.0	Nonburning	Very slight dripping, intumescent
[a] 16	None	5.0	[b] 30.0	21.0	Could not be tested.	Decomposed upon heating to 200°C.
[a] 17	None	5.0	[c] 30.0	12.0	Could not be tested.	Coating slipped off felt at room temperature.

[a] In these tests only one ignition attempt was made; the test samples were 3/4" x 6".
[b] "Chlorowax 70" (trademark of Diamond Alkali Company).
[c] "Chlorowax 40" (trademark of Diamond Alkali Company).
Note: Unstabilized asphalt used in Examples 11, 12, 13, 16 and 17; stabilized asphalt in Examples 14 and 15.

The flame test data in the table indicates that the compositions containing $C_{10}Cl_{12}$ were nonburning, dripping of the coating was minimized, and an intumescent effect was observed when antimony oxide was included in the coating composition. Addition of antimony trioxide effectively reduced the amount of chlorine necessary to render these coatings nonburning. At both the 10 and 20% chlorine levels with antimony oxide, the stabilized asphalt coating was self-extinguishing, even though the combustible substrate burned. By comparison it was found that the use of 30% of chlorinated paraffin wax was ineffective in that the coatings were so unstable that the flammability test could not be run.

In related work, N.D. Blair and C.T. Bean, Jr. (U.S. Patent 3,347,685; October 17, 1967; assigned to Hooker Chemical Corporation) describe compositions comprised of a bituminous material and a halogenated cyclopentadiene compound selected from the group consisting of: (a) bis(dichloroheptafluorocyclopentane); (b) bis(pentachlorocyclopentadienyl); (c) perchlorofulvalene; (d) dodecachlorodihydrotrindene; (e) the reaction product of bis(pentachlorocyclopentadienyl) with ethylene glycol; (f) compounds having the formula:

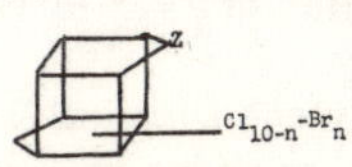

wherein Z is selected from the group consisting of Cl_2, and $(OH)[O]^-[NHT_3]^+$;

(g)

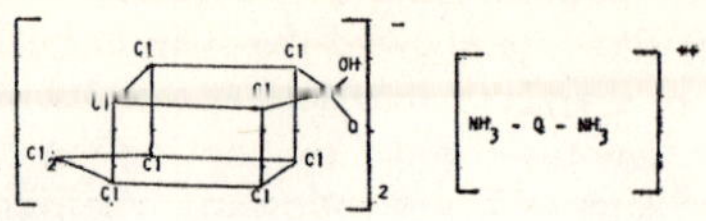

(h)

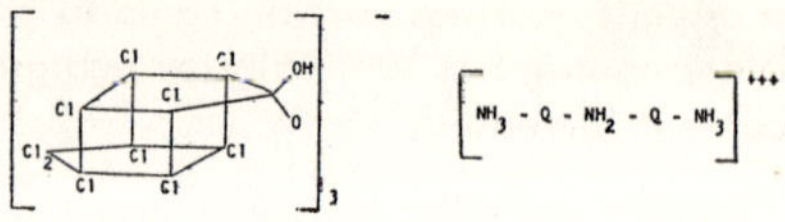

(i)

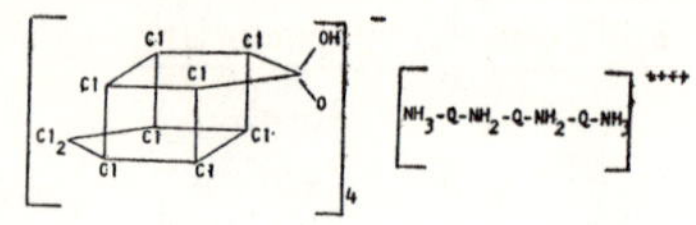

and (j) bis(trichlorophenyl) perchlorocyclopentadiene. In these formulas each T is independently selected from the group consisting of hydrogen, alkyl and alkylmonohydroxy; Q is an alkyl having 2 to 6 carbon atoms; and n is from 0 to 2 and is 1 to 2 when Z is Cl_2. $C_{10}Cl_{10}$ is prepared by the dechlorination of hexachlorocyclopentadiene. A typical method of preparation is described in U.S. Patent 2,732,409. This compound melts about 124°C. The following examples illustrate the preparation performance of the fire retardants.

Example 1: A one liter, 3-necked flask equipped with a dropping funnel, a mechanical stirrer, a reflux condenser with drying tube, and a gas inlet tube was flame dried and filled with dry nitrogen, anhydrous ferric chloride (47.25 g.) was weighed into the flask and 500 cc of distilled anhydrous tetrahydrofuran added rapidly in order to prevent excessive exotherm during the dissolution of the solid. Powdered iron metal (18.0 g.) was then added, and the mixture was stirred at reflux under a slow stream of nitrogen for 1 1/2 hours. A solution of 118.8 g. of bis(pentachlorocyclopentadienyl) in 250 cc of anhydrous tetrahydrofuran was added to the stirred mixture over a 5 minute period while still hot and the resulting dark purple reaction mixture was allowed to stand at room temperature for 16 hours.

The tetrahydrofuran was then removed by vacuum distillation and 500 cc of 1 : 1 mixture of concentrated hydrochloric acid and distilled water were added to the dark, partially solidified, residue. After a few minutes of stirring, 250 cc of benzene was added and stirring was continued for 15 to 20 minutes. The black insoluble solid was removed by suction filtration, the cake was washed with 800 cc of hot benzene and the organic layer was separated from the filtrate and was then extracted with dilute hydrochloric acid until the acid washes were almost colorless. The organic layer was then concentrated to dryness under vacuum to leave a violet crystalline residue. Recrystallization of the crystalline residue from 430 cc of hexane yielded 13.9 g. of $C_{10}Cl_8$ as violet rhombic crystals. A further recrystallization from hexane gave product with the following physical constants: Calculated for $C_{10}Cl_8$: C, 29.75; Cl, 70.25. Found: C, 29.85; Cl, 69.10. NW — 431 ebullioscopic in benzene; theory, 404. UV Spectrum — max. 389 mμ log E 4.61. UV Spectrum in Hexane — max. 603 mμ log E 2.43. MP — decomposes at 200°C. IR Spectrum — 6.51 μ, 7.95μ, 8.05μ, 8.61 μ, 13.03 μ and 14.21 μ.

Example 2: Preparation of $C_{10}Cl_{10}Br_2$ — A mixture of perchloroethylene, 280 g., aluminum chloride, 14 g., and bromine, 160 g., was stirred at room temperature for one hour, after which 273 g. of C-56 were added and the temperature was gradually raised to 112°C. over a 2-hour period. The temperature was maintained at 112° to 120°C. for 4 1/2 hours. The reaction mixture was diluted with 480 g. of perchloroethylene and 15 g. of water were added to decompose the catalyst. The reaction mixture was heated to reflux and filtered hot to remove the catalyst hydrolysis products. The filtrate was cooled to +5°C., after which the product was crystallized, filtered, dried free of perchloroethylene, washed with 150 ml. of acetone and again dried. Product weight: 225 grams. Analysis for bromine by x-ray: Percent bromine, 25.6. Theoretical for $C_{10}Cl_{10}Br_2$: Percent bromine, 24.0. A portion of the product was recrystallized from perchloroethylene and analyzed for bromine. Found: Percent bromine, 25.2.

Example 3: Preparation of $C_{10}Cl_{10}OH$ Triethanolamine Salt — To a flask containing 365 parts of water were added 203.5 parts of $C_{10}Cl_{10}O$ with a water content of 0.96%. The mixture was stirred for 8 hours at room temperature, filtered and then air dried until a water content of approximately 9.27% was obtained. 1,081.8 parts of the hydrate were dissolved in acetone and were brought to reflux temperature (approx. 62°C.). While this temperature was maintained, 298.4 parts of triethanolamine were gradually added over a 3-hour period and the refluxing was continued for an additional 8 hours. Refluxing was discontinued, the solution was gradually cooled to room temperature, and the triethanolamine salt was allowed to precipitate at the bottom of the vessel. The salt was then filtered out by conventional means. It was of light tan color, melted at 141° to 144°C. and was obtained in a yield of 57.5%.

Example 4: Preparation of $C_{10}Cl_{10}OH$ Ethylene Diamine Derivative — 300 parts of $C_{10}Cl_{10}O$ were dissolved in 244 parts of methanol. Then 38 parts of water were added with stirring. Thereafter, 100 parts of ethylene diamine were added with stirring. A light precipitate was formed as the temperature was raised to about 45°C. and the mixture was boiled for about 10 minutes. After cooling to room temperature, the white precipitate was recovered by filtering. The precipitate was washed with successive portions of methanol and was air dried at room temperature.

Example 5: Preparation of Base Material — Asphalt (60.5 parts) was heated to 200°C. until melted. The molten asphalt was then removed from the source of heat and 32.5 parts of mineral spirits were added with stirring. Stirring was continued and 7.05 parts of finely divided asbestos fiber (Quebec Asbestos Standard Grade 7R) were added. The mixture was then cooled to room temperature and was used as base material for the flame-retardant compositions of the following examples. When it is desirable to prepare compositions with other solvents, such as trichloroethylene, the above procedure is used except that the desired solvent is substituted for a like amount of mineral spirits of Example 5.

Example 6: Fire-Retardant Compositions — To 100 parts of the base material of Example 5 were added three parts (5% based on asphalt content) of bis(dichloroheptafluorocyclopentane), referred to as $C_{10}Cl_4F_{14}$, and three parts (5% based on asphalt content) of antimony oxide with stirring until uniform. The halogen content of this composition was 2.9%. A coating was prepared on 3" x 12" saturated asphalt felts (a composition shingle) by applying the above composition with a knife at a coverage rate of 2 gallons for each 100 square feet (or a film of about 1/16 of an inch in thickness). The strips were allowed to dry until substantially all of the mineral spirits was removed.

Testing was performed to determine the relative flammability of the coating by supporting the strip at an angle of 30° from the vertical, in a draft-free hood, and applying a flame from a compressed propane portable blow torch. The torch flame was adjusted to give a 1 1/2" inner blue cone, was applied perpendicular to the face of the strip for a period of 15 seconds and was then removed. The self-extinguishing time was recorded. A second 15-second flame application was made immediately after the first flame was extinguished and the self-extinguishing time for the second application was recorded. If the specimen was self-extinguishing or did not burn upon the second ignition, it was judged to be self-extinguishing by this test. Intumescence was measured at the highest point of rise.

The composition of this process was self-extinguishing in 2.0 seconds after the first ignition and 11.0 seconds after the second ignition. An intumescence of 0.19 inch was observed. The composition was considered to be nonburning. Additional compositions containing different amounts of $C_{10}Cl_4F_{14}$ were prepared and tested. Results are summarized below.

Ex.	Fire Retardant	Parts of Retardant	Parts Sb_2O_3	Self-Extg. Time (sec.)		Intumescence (in.)	Flammability
				1st Appln.	2d Appln.		
7	$C_{10}Cl_4F_{14}$	6.1	3.0	4.0	15.7	0.27	Non-burning.
8	$C_{10}Cl_4F_{14}$	12.1	3.0	2.3	7.5	0.27	Do.
9	$C_{10}Cl_4F_{14}$	18.2	3.0	5.0	6.7	0.19	Do.

Example 10: To 100 parts of the bituminous composition of Example 5 were added 3.0 parts of $C_{10}Cl_{10}OH$-triethanolamine salt of Example 3. The resulting composition was tested in the same manner as used in Example 3. The composition was nonburning and self-extinguishing in 2.3 seconds after the first 15-second flame application and 8 1/4 seconds after the second application of flame. Additional compositions containing higher amounts of the amine salt of Example 3 were prepared and tested as in Example 6. Results are summarized below.

Ex.	Fire Retardant	Parts of Retardant	Parts Sb_2O_3	Self-Extg. Time (sec.)		Intumescence (in.)	Flammability
				1st Appln.	2d Appln.		
11	Ex. 4	6.1	3.0	2.3	5.0	0.23	Non-burning.
12	Ex. 4	12.1	3.0	1.3	4.0	0.25	Do.
13	Ex. 4	18.2	3.0	2.7	6.2	0.29	Do.

Example 14: To 100 parts of the bituminous composition of Example 5 were added 18.2 parts of perchlorofulvalene (Example 1) and 3.0 parts of antimony oxide. The resulting composition was tested in the same manner as Example 6. The composition was nonburning and self-extinguishing in 2.3 seconds after the first 15-second flame application and 3.8 seconds after the second application. An intumescence of 0.19 inch was observed.

In a continuation of this work, N.D. Blair and C.T.Bean, Jr. (U.S. Patent 3,347,807; October 17, 1967; assigned to Hooker Chemical Corporation) describe a fire-retardant bituminous composition comprised of: (a) a bituminous material; and (b) a Diels-Alder adduct of a polyunsaturated aliphatic compound and a polyhalogenated cyclopentadiene having the formula as shown on the following page.

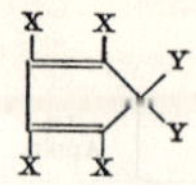

where X is selected from the group consisting of fluorine, chlorine and bromine and Y is selected from the group consisting of fluorine, chlorine, bromine and an alkoxy radical; and the Diels-Alder adduct contains at least about 0.6 mol of polyhalogenated cyclopentadiene per mol or polyunsaturated aliphatic compound.

It is within the scope of this process to polymerize the polyunsaturated aliphatic compound to a high molecular weight by conventional rubber polymerization techniques so that the resulting polymer is properly described by the term "rubber". These rubbers contain carbon to carbon unsaturation which can be as much as slightly less than 1 mol of carbon to carbon unsaturation per mol of the polyunsaturated aliphatic compound starting material. Generally these polymers have molecular weight from about 100,000 to 300,000, although they can have molecular weight as low as about 50,000 or as high as about 400,000. Illustrative of polyhalogenated cyclopentadienes suitable for use in this process are the following:

Hexachlorocyclopentadiene
5,5-dimethoxytetrachlorocyclopentadiene
Hexabromocyclopentadiene

Satisfactory polyunsaturated aliphatic compounds for use in preparing the Diels-Alder adduct include both linear and cycloaliphatic compounds, as exemplified by the following:

1,3-butadiene
1,5-hexadiene
1,7-octadiene
1,11-dodecadiene

The preparation of 2,2',3,3',4,4',5,5',7,7,7',7'-dodecachloro-1,1,1',1',6,6'-hexahydro-2,2',5,5',-methano biphenyl is described in U.S. Patent 2,606,910. For convenience this compound will be referred to by its empirical formula $C_{14}H_6Cl_{12}$. The adduct is prepared by condensing two mols of hexachlorocyclopentadiene with 1 mol of 1,3-butadiene at a temperature below 200°C. and then recrystallizing the reaction product from isopropyl alcohol. The compound melts at about 232°C.

The preparation of bis[1,2,3,4,7,7-hexachlorobicyclo(2.2.1)hept-2-ene-5-yl]butane may be effected by condensing 2 mols of hexachlorocyclopentadiene with 1 mol of 1,7-octadiene. The condensate is prepared at a temperature between 40° and 200°C. and the excess hexachlorocyclopentadiene is removed by vacuum distillation. The condensate is recrystallized from benzene to obtain a product that melts at 230° to 233°C. For convenience this compound will be referred to by its empirical formula $C_{18}H_{14}Cl_{12}$. The following examples illustrate the process.

Example 1: Preparation of Polymeric Fire Retardant — 50 parts of "Cis-4" polybutadiene rubber having a cis-1,4 content of 95% and 320 parts of hexachlorocyclopentadiene that had been treated with magnesium oxide for removal of impurities and 172 parts xylene commercial grade, were introduced into a reactor. The reactor contents were heated for 3 hours at 90° to 100°C. with agitation to dissolve the rubber, and then for 20 hours at 150°C. The resulting polymer solution was poured into methanol, redissolved in toluene and then precipitated from methanol. The resulting light polymer had a chlorine content of 57.9% and an intrinsic viscosity of 0.4504 dl./g. at 30°C.

Example 2: Preparation of Base Material — This is described in U.S. Patent 3,347,685.

Example 3: Fire Retardant Compositions — To 100 parts of the base material were added three parts (5% based on asphalt content) of $C_{14}H_6Cl_{12}$ and three parts (5% based on asphalt content) of antimony oxide with stirring until uniform. The halogen content of this composition was 2.9%. A coating was prepared on 3" x 12" saturated asphalt felt (used as an intermediate in the manufacture of composition shingles) by applying the above composition with a knife at a coverage rate of 2 gallons for each 100 square feet (a film of about 1/16 of an inch). The strips were allowed to dry until substantially all of the mineral spirits were removed.

The composition of this process was self-extinguishing in 1.5 seconds after the first ignition and 4.0 seconds after the second ignition. An intumescence of 0.25 inch was observed. The composition was considered to be nonburning.

In a manner similar to Example 3, other compositions were prepared and tested. Results are summarized in the table on the following page.

Example	Fire Retardant	Parts of Retardant	Parts Sb_2O_3	Self-Extg. Time (sec.) 1st Appln.	Self-Extg. Time (sec.) 2nd Appln.	Intumescence (inc.)	Flammability
4	$C_{14}H_6Cl_{12}$	6.0	3.0	1.0	2.0	.37	Non-burning.
5	$C_{14}H_6Cl_{12}$	12.1	3.0	1.0	1.0	.56	Do.
6	$C_{14}H_6Cl_{12}$	18.2	3.0	1.0	1.0	.75	Do.
7	$C_{14}H_6Cl_{12}$	1.8	----------	6.0	26.5	.19	Do.
8	$C_{18}H_{14}Cl_{12}$	18.2	3.0	2.2	8.1	.44	Do.
9	----------	----------	3.0	7.0	burns	none	Burning.

C.S. Hardo and R.R. Hindersinn (U.S. Patent 3,433,654; March 18, 1969; assigned to Hooker Chemical Corporation) have found that bituminous compositions can be rendered fire retardant by the addition of a Diels-Alder adduct of a non-conjugated unsaturated fatty ester and a polyhalogenated cyclopentadiene having the formula:

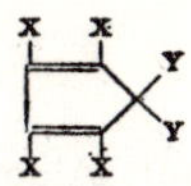

where X is selected from the group consisting of fluorine, chlorine and bromine, and Y is selected from the group consisting of fluorine, chlorine, bromine and an alkoxy radical, the Diels-Alder adduct containing about 10 to 50 weight percent halogen.

The preferred esters are the glyceride esters of fatty acids and particularly the fatty esters generally known as drying and semidrying oils, such as soya, linseed, safflower, peanut, corn, cod, anchovy, and the like, vegetable and marine oils having iodine values of about 60 or more.

Polyhalogenated cyclopentadienes suitable for use in this process include: hexachlorocyclopentadiene, 5,5-dimethoxytetrachlorocyclopentadiene, hexabromocyclopentadiene, 5,5-dibromotetrachlorocyclopentadiene, 5,5-difluorotetrachlorocyclopentadiene and 5,5-diethoxytetrachlorocyclopentadiene. The preferred polyhalogenated cyclopentadiene is hexachlorocyclopentadiene.

The adduct is made by reacting 0.1 to 1.1 equivalent polyhalogenated cyclopentadiene per equivalent of ester. With safflower oil, which has an iodine value of about 145, which corresponds to about 5 positions of ethylenic unsaturation per ester molecule, theoretically, 5 mols of polyhalogenated cyclopentadiene can be adducted per mol of ester. However, it is preferred to adduct 0.5 to 1.0 equivalent polyhalogenated cyclopentadiene per equivalent of ester to form an adduct containing 10 to 50% by weight of halogen. The following examples illustrate the process.

Example 1: Preparation of Safflower Oil Adduct Fire Retardant — Hexachlorocyclopentadiene (246 parts) and 175 parts of safflower oil were charged to a reaction vessel equipped with a stirrer, condenser, thermometer, heating mantle and nitrogen inlet and outlet. The charge was slowly brought to a temperature of 180°C. under a nitrogen atmosphere and maintained at that temperature for 24 hours. The reaction mixture was then vacuum stripped at 180°C. and an absolute pressure of less than one millimeter of mercury, to yield a product which was a dark brown semisolid at room temperature. The product analyzed 44 to 45% chlorine, had a density at 24°C. of 1.35 grams per cubic centimer and a Brookfield viscosity of 12.5 poises at 100°C.

Example 2: Preparation of Base Material — 60.5 parts of asphalt were heated to 200°C. The molten asphalt was then removed from the source of heat and 32.5 parts of mineral spirits were added with stirring. Stirring was continued and 7.0 parts of finely divided asbestos fiber (Quebec Asbestos Standard Grade 7R) were added. The mixture was then cooled to room temperature at which temperature it was a free flowing liquid. This composition was used as a base bituminous material for the flame-retardant composition of the following examples.

Example 3: Fire-Retardant Compositions — To 100 parts of the base material of Example 2 were added 6.1 parts (10% based on asphalt content) of the safflower oil adduct of Example 1 and three parts (5% based on asphalt content) of antimony oxide. The safflower oil adduct was heated to about 80°C. and then added to the base material which was at room temperature. The mixture was stirred until uniform.

A coating was applied to 3" x 12" pieces of felt previously saturated with asphalt known as composition shingle, by applying the prepared asphalt composition with a knife at a coverage rate of two gallons for each 100 square feet. This coverage provided a dried film of about 1/16 of an inch. The felt with an applied asphalt coating was then laminated by placing one piece over the other to form two ply lay-ups. The lay-ups were allowed to dry for 3 days at room temperature, then at 50°C. for 24 hours, thus removing substantially all of the solvent. The prepared composition of this example was self-extinguishing in 2.2 seconds after the first ignition and 5.0 seconds after the second ignition. An intumescence of 0.31 inch was observed.

Examples 4 to 10: In a manner similar to Example 3, other compositions in which 6.1 parts of the adduct of Example 1 were used were prepared and tested. Results are summarized in the table below.

Example	Metal compound	Parts metal compound	Self-extinguishing time (seconds)		Intumescence (inch)	Flammability
			1st application	2nd application		
4	Antimony oxide	1.7	6.9	11.1	.22	Self-extinguishing.
5	Arsenic oxide	5.0	3.0	6.5	.22	Do.
6	Antimony sulfide	5.0	2.2	4.6	.29	Do.
7	Lead arsenate	5.0	3.9	10.8	.22	Do.
8	None		4.9	8.7	0.10	Do.
9 [1]	Antimony oxide	5.0	7.0	Burns	None	Burns.
10 [1]	None		Burns	Burns	None	Do.

[1] No fire retardant adduct in these compositions.

Similar results are obtained when the fire retardant is a Diels-Alder adduct of 5,5-dimethoxytetrachlorocyclopentadiene, hexabromocyclopentadiene, 5,5-dibromotetrachlorocyclopentadiene, 5,5-difluorotetrachlorocyclopentadiene or 5,5-diethoxytetrachlorocyclopentadiene, rather than hexachlorocyclopentadiene.

Low Penetration Asphalt and Polyhalogenated Polyphenyls

R.E. Koons (U.S. Patent 3,342,614; September 19, 1967; assigned to Monsanto Company) has found that an asphaltic composition comprised of as high as 97 and as low as 80% by weight of a particular low penetration asphalt and 3 to 20 weight percent of a fire-retardant chemical possesses fire-retardant properties substantially superior to those of known fire-retardant bituminous compositions. The particular low penetration asphalt which is the basic component of the fire-retardant composition is one having the following properties: a penetration of 3 to 35 at 77°F. and a softening point of 130° to 300°F. Asphalts having these properties may be obtained by the vacuum reduction, solvent treatment, air blowing or a combination of these of the residium from the "topping" of a crude petroleum. This residium material is generally known as a topped crude oil and is that which remains after the atmospheric distillation of crude petroleum to remove gas, gasoline, naphthas, kerosene, distillates or diesel oil and gas oil.

The most preferred asphalts are those which have been obtained by: (1) the solvent extraction of a topped crude oil with normally gaseous hydrocarbons such as butane and/or propane to obtain an asphalt of less than 20 penetration at 77°F.; or (2) the solvent extraction or vacuum reduction of such residium to obtain an asphalt of less than 70 penetration at 77°F. with subsequent air blowing to less than 35 penetration. The asphalts obtained by the first method above have a penetration of 0 to 20 at 77°F., and a softening point of 150° to 250°F., while those obtained by the second method have a penetration of 0 to 35 at 77°F. and a softening point of 130° to 250°F.

The fire-retardant chemicals which are a part of the composition should be stable up to temperatures of 250°F. or higher. Further, these chemicals must be compatible with the asphalt and not exude from the composition on heating or under the exposure of service for which the composition is intended. Also, the fire-retardant chemicals must have a low volatility to prevent their evaporation from the composition at the elevated temperatures generated by approaching or nearby fires as well as under more normal service conditions. Preferred fire-retardant chemicals are halogen-containing organic compounds such as chlorinated paraffins and polychlorinated polyphenyls and also certain organic phosphorus-containing compounds.

Examples of halogenated organic compounds are such compounds as Aroclor 1248, 1254 and 1260, marketed by Monsanto Chemical Company, which are polychlorinated polyphenyls, Clorowax 40 and 70, marketed by Diamond Alkali Company, hexachlorodiphenyl oxide and hexabromodiphenyl oxide. Examples of fire-retardant organic phosphates which may be utilized in the coating compositions are tricresylphosphate, triphenylphosphate, cresyldiphenylphosphate, tri-o-chlorophenylphosphate and tris(2,3-dibromopropyl)phosphate.

Several asphaltic mastics were prepared. Fire tests on these mastics are given in the table on the following page. The asphaltic mastics were spread 3/8 of an inch thick on calcium silicate blocks and cured in an oven at 140°F. for one week to evaporate all volatile solvents from the blend. The resulting panels were then subjected to a fire test by exposure to a standard flame. After 200 seconds exposure to the flame, the percent of the panel burning was recorded and the flame removed.

The time between the removal of the flame and complete self-extinguishment of the burning asphalt was also recorded. The percent of the asphalt panel burning after 200 seconds exposure to a standard flame and the rate at which the burning asphalt self-extinguishes are indicative of the fire resistance of the asphalt composition.

	Composition of the Asphaltic Fire-Retardant Mixture	Percent of Panel Burning at 200 Seconds	Seconds Until Self-Extinguishing Occurred
A	A vacuum reduced asphalt of 90 penetration at 77°F.	40	300+
B	A vacuum reduced asphalt of 90 penetration at 77°F. containing 10 weight percent of Aroclor 1262. *	31	58
C	A vacuum reduced asphalt of 90 penetration at 77°F. which was air blown to 18 penetration at 77°F. containing 17 weight percent of Aroclor 1254. *	14	15
D	The asphalt of C containing 17 weight percent of tricresyl phosphate.	14	12
E	Propane extracted residue of 39 penetration at 77°F. which was air blown to 1 penetration at 77°F. containing 14 weight percent tricresyl phosphate.	8	1
F	Propane extracted residue of 3 penetration at 77°F. containing 8 weight percent of tricresyl phosphate.	3	1
G	Propane extracted residue of 33 penetration at 77°F. which was air blown to 1 penetration at 77°F. containing 13.5 percent of tricresyl phosphate.	1	0

* The Aroclor compounds are polychlorinated polyphenyls.

Compositions C, D, E, F and G above represent compositions prepared in accordance with the process. The E, F and G compositions represent the preferred compositions. A comparison of compositions B and F dramatically illustrates the superiority of these asphaltic compositions over a more generally known composition in regard to their fire-retardant properties. There is in the F composition greater than a 900% decrease in the percent of the panel burning after 200 seconds, and a 5700% decrease in the time necessary for self-extinguishment. This improvement with the asphalt of the process is obtained while using 2% less of a fire-retardant chemical than the other asphalt. The following examples illustrate the asphaltic compositions of the process.

Example 1: A fire-retardant mastic with low permeability to water vapor:

Components	Percent by Weight
35 penetration propane extracted asphalt, air blown at 500°F. to 1 penetration at 77°F.	43.6
Tricresyl phosphate	5.9
Petroleum naphtha, 300° to 360°F. boiling range	26.6
Slate flour	9.3
7T asbestos	14.6
	100.0

This mastic was sprayed on a prepared surface and allowed to cure free of the petroleum solvent, and it was found to have the following properties:

(1) It gave excellent protection to steel in the prevention of corrosion.
(2) It was very durable with respect to outside weathering conditions.
(3) The permeability to water vapor was found to be very low (0.002 perm.-inch)
(4) It had excellent resistance to abrasion.
(5) In a comparative fire test, as described previously, it had excellent fire-retardant properties.

Example 2: A fire-retardant mastic with high permeability to water vapor:

Components	Percent by Weight
35 penetration propane extracted asphalt, air blown at 500°F. to 1 penetration at 77°F.	25.0
Tricresyl phosphate	3.4
Aromatic naphtha, 300° to 360°F. boiling range	27.9
Perchloroethylene	27.9
Expanded vermiculite (Zonolite No. 4)	5.3
7M asbestos	10.5
	100.0

This fire-retardant mastic was found to be suitable for application over damp insulation material. It was also found to be self-extinguishing in the uncured state. On curing, it was found to possess the same desirable properties listed under Example 1 on the previous page. The major difference in the cured film was the permeability. A permeability of 0.027 perm.-inch allowed the escape of water vapor from the underlaying insulation materials without the formation of blisters, and yet prevented water on the outside surface from penetrating the film in the liquid form.

THERMAL INSULATION

Use of Phenol-Aldehyde Resins

K.N. Edwards (U.S. Patent 3,053,714; September 11, 1962; assigned to Wood Conversion Company) has found that an asphalt coating containing certain heat-hardenable phenol-aldehyde resins will intumesce upon exposure to heat of a flame. The combination of a heat-hardenable phenol-aldehyde resin with asphalt will form a coating which, upon exposure to the heat of a flame, will swell, bubble, or puff to produce a firm, resinous, expanded film which helps to shield the substrate from the fire and also to retain the asphalt in place.

The following examples illustrate the combination of coating asphalt with resoles and other desirable ingredients, such as microbiocidal agents, plasticizers, wetting agents, and flame-retardant ingredients which evolve inert, incombustible gases upon the application of the heat of a flame. The ingredients are added to each other and heated to a suitable temperature, with stirring to obtain complete and uniform mixture. The resoles are preferably added last, and the entire mixing is preferably kept at as low a temperature as possible. A suitable temperature is usually from about 210° to 230°F. In each example the coating was applied in the form of a hotmelt to one side of 43 pound kraft paper. The coating was permitted to cool and then was tested for intumescence by applying the flame from a gas burner directly in contact with the coating. The flammability of the coated paper was determined in accordance with the test described in ASTM designation D777-46. The char length of this test is reported for each example.

Example 1:

Ingredients	Percent Amount
Asphalt	40
Resole (methylol substituted phenol ethers)	40
Pentachlorophenol	20

The resole is a mixture of alkyl ethers of phenol having from 1 to 3 methylol groups linked singly at the ortho and para positions on the benzene ring. The pentachlorophenol functions to inhibit the growth of deleterious microorganisms and also advantageously functions as a fire-retardant agent in the evolution of inert gases upon the application of heat. The coating swelled into a resultant foamed mass and quickly thermoset into a heat barrier upon contact with the flame. The char length was 2.92".

Example 2:

Ingredients	Percent Amount
Asphalt	30
Resole (methylol substituted phenol ethers)	42
Chlorinated hydrocarbon resin	12
Pentachlorophenol	12
Plasticizing hydrocarbon oil (spindle oil)	3.5
Wetting agent	0.5

The resole and the pentachlorophenol were the same as in Example 1. The chlorinated hydrocarbon resin exhibits fire retardance in the same manner as the pentachlorophenol, but has no apparent microbiocidal properties. The wetting agent is a high molecular weight, nonionic surfactant prepared by condensing ethylene oxide with a hydrophobic base formed by condensing ethylene oxide with a hydrophobic base formed by condensation of propylene oxide with propylene glycol. The wetting agent reduces the run off characteristics of asphalt under heat. This coating produced a foamed thermoset mass upon contact with the flame. The char length was 3.63".

Example 3:

Ingredients	Percent Amount
Asphalt	28
Resole (methylol substituted phenol ethers)	40
Chlorinated hydrocarbon resin	12
Mixed cresyl phenyl phosphates	19
Wetting agent	1

The resole, the chlorinated hydrocarbon resin, and the wetting agent were the same as the previous examples. The mixed cresyl phenyl phosphates function as a flame-retardant plasticizer. The coating produced a foamed thermoset mass upon contact with the flame. The char length was 3.12".

Backing Sheet for Thermal Insulation

In a process described by M.G. Schmitt (U.S. Patent 3,050,424; August 21, 1962; assigned to Thilmany Pulp & Paper Company), a paper sheet or paper board is coated with bitumen, and while the bitumen is in a plastic condition, particles of a flameproofing agent are embedded in the bitumen in dispersed relation and in such manner that portions of the particles are not coated with the bitumen. It is important that the bitumen be dispersed so that the sheet may be combined with another sheet to make a duplex sheet or be combined with insulating material, such as rock wool or glass fibers.

It is quite important that portions of the particles not be coated with bitumen so that when flame is encountered by the sheet, the flameproofing agent may decompose or break down so as to form a gas which will more effectively flameproof the sheet. If the flameproofing agent is coated with bitumen, the agent does not as readily decompose and consequently the bitumen may melt or burn before the flameproofing agent becomes effective. It appears that the greatly improved results of the process result from the flameproofing agent being capable of being decomposed more readily than when coated with bitumen. Flameproofing agents include ammonium sulfate, chloride, phosphate, and bromide, monoammonium phosphate, etc. The following is an example of a mineral wool batt formed in accordance with this process and its general method of preparation.

A commercial flameproofed kraft paper weight approximately 50 pounds per ream was coated with approximately 10 pounds/1000 square feet of an asphalt having a melting point of approximately 170°F. and a penetration within the limits of 20 to 35 at 77°F. The surface of the asphalt coating was then substantially uniformly dusted with approximately 10 pounds/1000 square feet of ammonium sulfate as a flameproofing compound. The thus formed backing sheet was then laminated with tackifying heat and slight pressure to a mat of mineral fibers approximately 2" thick. The mineral fiber batt was tested for fire resistance in a simulated stud space formed of inorganic cementitious sheets and simulated studs, forming a space having an inside dimension of 3 5/8" x 14 1/4" x 24" which duplicates a wall space formed by surfacing sheets and 2" x 4" wood studs on 16" centers. The batt was positioned with the nonvapor barrier surface against one of the incombustible sheets, thereby forming a chimney effect between the vapor barrier backing and the other incombustible sheet. Three milliliters of nonleaded gasoline were burned in a trough 6" x 1" x 1/4" deep, placed 1" below the edge of the vapor barrier backing sheet.

When so tested, the flameproofed mineral fiber batts, formed as above, exhibited a burned area of only approximately 13%, and passed standard tests for a good flameproof rating. As a comparison, the same flameproofing compound, ammonium sulfate, was mixed into the same 170°F. melting point asphalt in amount as great as possible while retaining the necessary spreading and adhesive characteristics of the asphalt (approximately 50% by weight of the asphalt). Mineral wool batts employing such an asphalt adhesive and a vapor barrier were tested in the foregoing manner. Such batts exhibited an average burn area approximately 180% the burn area of the batt formed in accordance with the above example.

MISCELLANEOUS COATINGS

Aerogel Thickened Formulation

R.M. Hooks (U.S. Patent 2,734,827; February 14, 1956; assigned to Southwestern Petroleum Co., Inc.) describes a coating which comprises an asphaltic vehicle, a fire-retarding chemical, an aerogel, and for some purposes asbestos fibers. A typical formulation of the coating which has been effective under service conditions is as follows:

	Percent
Ammonium carbonate	5
Ammonium sulfate	10
7-D asbestos fiber	6
Silica aerogel	4
165°F. melting point asphalt	37.4
Perchloroethylene	37.6

The ingredients are blended together by melting the asphalt and stirring the perchloroethylene slowly into it. When the solution is cooled below the decomposition temperature of any of the fire retardants, such as the ammonium carbonate and ammonium sulfate suggested in the above typical formulation, the remainder of the dry ingredients are dispersed thoroughly in the solution by vigorous stirring. It is preferable to incorporate the silica aerogel into the solution first, since its finely divided particles are more readily dispersed in the alphalt-perchloroethylene solution in its more fluid state.

During drying of the vehicle after application of the coating, the ratio of the aerogel to the asphalt increases to a point where the amount of aerogel in the dry asphalt is approximately 8 to 12%. At temperatures incident to its normal use asphalt is a solid, but upon exposure to heat it becomes molten and flows if not inhibited. By virtue of the presence of 8 to 12% silica aerogel, a firm gel is obtained when the asphalt becomes molten. The asphalt is gelled at temperatures below its melting point.

The coating affords exceptional fire resistance to vertical or steeply inclined inflammable surfaces in addition to its waterproofing function. Under the influence of heat the asphalt does not become molten and hence cannot flow from the surfaces to be protected. At the same time, the fire retardant uniformly dispersed throughout the coating is held in place and capable of functioning throughout the heated area during the period of heat application. Thus the proofing against weather, water and flame remains constant throughout the surface to which the coating has been applied regardless of surrounding conditions or temperatures.

Gilsonite Formulation

A process described by O.T. Hodnefield (U.S. Patent 2,940,942; June 14, 1960) provides a flame-resistant formulation by blending a particular gilsonite mined in the Uintah Basin of Utah and Colorado with a roughly equal proportion by weight of a finely divided mineral filler, e.g., 300 mesh talc, to form a homogeneous mixture. The particular gilsonites used usually have a melting point between about 270° and 300°F. which is considerably lower than that of the usual gilsonites such as elaterite and uintahite and the like, which have relatively high melting points, e.g., 400° to 460°F. The preferred gilsonite has the following properties.

Melting point (B. & R. gly.)	270° to 275°F.
Penetration at 78°F.	Less than 1
Color of streak of powder	Brown
Luster	Bright
Fracture	Conchoidal
Specific gravity at 77°F.	1.05 to 1.10
Hardness (Mohs' scale)	2
Loss at 325°F., 7 hours	Less than 2%
Loss at 400°F., 7 hours	Less than 4%
Character of residue 325°F.	Original
Character of residue 400°F.	Original
Solubility in carbon disulfide	99%
Solubility in carbon tetrachloride	99%
Fixed carbon	10 to 15%

The following table and examples are illustrative of the formulations.

	Preferred Composition, pounds	Range
Gilsonite	100	100 pounds.
Filler (talc)	100	50 to 125 pounds.
Oil-soluble soap	6 to 8	4 to 10 pounds.
Urea-formaldehyde resin	½	⅛ to 1 pound.
Pale Oil (250° F.)	20	15 to 25 pounds.
Low-boiling thinner (e.g. paint thinner).	130	25 to 200 pounds e.g. (quite variable according to desired shipping consistency).
Copper 8-quinolinolate, Antimony oxide.	----------	

Example 1: 100 pounds of a mixture of the preferred gilsonite (Gilsogum) blended with sufficient pale oil (250°F. flash point), to reduce the melting point of the Gilsogum from 270°—275°F. to 210°—225°F. is heated to its melting point (about 220°F.) and there is added 4.5 pounds of oleic acid and 3 pounds of monoethanolamine. The mass is agitated with the awl of a gear driven pump operating at about 200 rpm and 100 pounds of 300 mesh talc is thoroughly dispersed into the mass. This mixture proved effective as a fire-retardant composition.

Example 2: Example 1 was repeated but the final mixture was allowed to cool to 200°F. and 130 pounds of mineral spirits were added with agitation at a rate of 300 rpm. Addition of the mineral spirits lowered the viscosity and improved the workability of the mixture.

Example 3: To the final mixture of Example 2 there was added 0.5 pound of urea-formaldehyde to obtain a satisfactory fire-retardant composition.

Example 4: 20 pounds of pale oil (250°F. flash point) were added to the composition of Example 3.

Example 5: A fireproofing composition was made by emulsifying the composition of Example 4 with high speed agitation with 20% by weight of a fireproofing solution made as follows: 1500 pounds of ammonium sulfate, 250 pounds of ammonium phosphate, 250 pounds of granular boric acid crystals, and 55 pounds of borax were dissolved in 250 gallons of water at 150°F. To this solution were added 11 pounds of ammonium aluminum sulfate in two gallons of 14% aqueous ammonia. There were then added 11 pounds of copper sodium alginate in six gallons of 14% aqueous ammonia and 15 ounces of benzaldehyde.

The mixture was cooled to 100°F. and had a pH of 7.3. This mixture was heated to 125°F. and there was added a dispersion made by dissolving 30 pounds of urea in 8.25 gallons of 37% aqueous formaldehyde and then adding two gallons of 65% copper naphthenate (8% metal) in benzine. The final fireproofing composition can then be sprayed on roofing felt.

Metal Tetraborate and Mineral Fillers

C.E. Wilkinson (U.S. Patent 2,939,794; June 7, 1960; assigned to Texaco Inc.) describes a coating composition comprised of 30 to 55% asphalt, 15 to 30% volatizable diluent, 5 to 15% asbestos fiber, 5 to 10% mica, 5 to 15% vermiculite, and 2 to 8% alkali metal tetraborate. The asphalt and diluent comprise about 60 to 70% of the composition. The following example, which illustrates the preferred composition, was prepared by sifting the particular preblended inorganic filler combinations into the asphalt cutback (asphalt + naphthenic distillate) heated to a temperature between 100° and 120°F. while stirring at 70 rpm in a 2 1/2 gallon motor driven dough mixer. After all the filler had been added to the cutback, stirring was continued for a period of 10 minutes.

The volatilizable diluent employed was a straight run naphthenic petroleum distillate having a boiling range between 300° and 400°F. The asphalt component employed was a vacuum steam distilled paraffinic-asphalted petroleum residuum having a penetration at 77°F. under a 100 gram load of 55 and a 126°F. ring and ball softening point. The asphalt cutback (asphalt + naphthenic distillate) had a viscosity of 113 to 115 SF/122°F.

In the following example a wooden utility pole was placed in a vertical position and secured by burying a portion in the ground. The asphaltic composition was then sprayed on the section of the utility pole to a thickness of about 1/8 of an inch beginning at a point at the ground level and extending 8 feet up the pole. The coating was allowed to cure for three weeks. Then two concentric cylinders of 3 x 3 inch mesh reinforcing wire were placed around the pole so that there was a 3" wide annular space between the pole and the inner cylinder and a 10" wide annular space between inner and outer cylinders. The cylinders extended from the ground level to a point 5 feet up the pole. Marsh hay was packed into the 10" annular space in a manner to completely fill the space. The hay was ignited on the windward side of the pole. After allowing the hay to burn for a period of 5 minutes, the outer cylinder was removed from each pole and the burning hay remaining at the base of the pole was removed. Twenty-four hours later, the coated pole was again subjected to fire in the manner just described.

Original coating composition:

Ingredients	Wt., %	Ingredients	Wt., %
Asphalt	43.5	Mica	10
Naphthenic distillate	21.5	Vermiculite (expanded)	10
Asbestos fiber	10	Borax	5

Fire test observations: The coating was slightly cracked at two places near the base but did not pull away from the pole. The coating did not ignite and retained its pliability over the entire exposed surface. There was no charring of the pole itself except for two small spots having a total area of about 4 x 6 inches.

In related work, C.E. Wilkinson (U.S. Patent 2,965,516; December 20, 1960; assigned to Texaco, Inc.) describes the addition of a fungitoxic agent, copper-8-quinolinolate to the asphalt-asbestos based formulations described above.

Creosote Formulations

B. Lowell (U.S. Patent 3,031,374; April 24, 1962) describes fire-retardant creosote based formulations. The following examples illustrate the process.

Example 1: This was prepared by mixing together solutions A-1 and B-1. The constituents of solution A-1 are listed on the following page.

	Percent by Weight of Final Product
Tricresylphosphate, 400 ml.	1.5
Copper resinate, 80 g.	1
Antimony oxide, 200 g.	2
Chlorinated rubber, 60 g.	--

The above constituents were mixed together at 280° to 350°F. until a solution, indicated by the homogeneity of mixture was obtained. Solution B-1 consisted of:

	Percent by Weight of Final Product
Toluene, 1,200 ml.	13.5
Chlorinated rubber, 380 g.	4
Titanium oxide, 80 g.	1
Calcium carbonate, 200 g.	2
Latex, 400 ml.	4.5

The B-1 constituents were mixed together at 200°F. until a solution, again indicated by the homogeneity of the mixture, was obtained. A-1 and B-1 were then cooled to 60° to 70°F. and mixed together. To this mixture there was added: 5,800 ml. of creosote (commercial grade), 67.5% by weight of final product.

The end mixture was then heated to 200°F. and agitated until it appeared homogeneous to the eye. The product may be brush applied or used to impregnate wood under pressure, and it serves as an excellent fire retardant and protective wood coating.

Example 2: The following were mixed together at 280° to 350°F. until homogeneity was obtained:

	Percent by Weight of End Product
Tricresylphosphate, 40 ml.	40
Quinine glycerophosphate, 1 g.	1
Chlorinated rubber, 20 g.	19

This solution was then allowed to cool to 60° to 70°F. and to it was added 40 ml. of creosote, 40% by weight of end product. The end mixture was heated until homogeneous at a temperature of 200°F. The product had fine fire-retardant properties.

Tar Composition

B. Lowell (U.S. Patent 2,861,012; November 18, 1958) describes a fireproof tar composition which comprises:

	Percent
Tar	46
Sodium silicate (powder)	11
Carbamide phosphoric acid	20
Ammonium sulfate	23

First, the tar is melted; then the balance of the constituents are mixed into the melted tar; and the mixing is continued until the mixture is cooled to substantially room temperature. Experimental work with this composition indicates that the ammonium sulfate materially enhances the water-repellent properties of the tar, while the specific combination of the carbamide phosphoric acid and sodium silicate imparts to the tar an unexpected degree of fire resistance.

When this composition was applied to a thin strip of wood by dipping the wood into the tar, the uncoated portion of the wood flamed briskly when ignited, but the flaming ceased abruptly when the coated portion of the wood was reached. The tar coating melted and foamed when the flame of a Meeker burner was played on it, but the tar did not ignite, nor was there any indication of burning when the burner was removed.

Perhalopentadienoic Acid and Derivatives

The use of perhalocarbonyl pentadiene compounds in bituminous coatings is described by N.D. Blair and C.T. Bean, Jr. (U.S. Patent 3,313,636; April 11, 1967; assigned to Hooker Chemical Corporation). Illustrative of the preferred perhalopentadienoic compounds are perchloropentadienoic acid, trichlorophenyl tetrachloropentadienoic acid, bis(trichlorophenyl) trichloropentadienoic acid, perhalo alkyl perchlorocoumalin, perchlorocoumalin, and the corresponding bromine analogs, perfluoro pentadienoic acid, and mixed halogen compounds, such as tribromo phenyl tetrachloropentadienoic acid.

The perchlorocoumalin utilized in this process may be prepared by any desired method and may be satisfactorily prepared by the following route. Octachlorocyclopentene is warmed with concentrated sulfuric acid and the resulting hexachloro-2-pentenone layer is separated and added to an aqueous solution of sodium hydroxide to form the sodium salt of cis-pentachloropentadienoic acid. The free acid is generated by acidification of the sodium salt with a mineral acid and the cis-pentachloropentadienoic acid, which can be isolated by filtration or extraction, and is then subjected to elevated temperature to effect the preparation of perchlorocoumalin. The following examples illustrate the process.

Example 1: Base Material — 60.5 parts of asphalt were heated to 200°C. until melted. The molten asphalt was then removed from the source of heat and 32.5 parts of mineral spirits were added with stirring. Stirring was continued and 7.0 parts of finely divided asbestos fiber (Quebec Asbestos Standard Grade 7R) were added. The mixture was then cooled to room temperature and used as base material for the flame-retardant compositions of the following examples.

Example 2: Fire-Retardant Compositions — To 100 parts of the base material of Example 1 were added 18.2 parts (30% based on asphalt content) of pentachloropentadienoic acid and three parts (5% based on asphalt content) of antimony oxide and the composition was stirred until uniform. The halogen content of this composition was 13.4%. A coating was prepared on 3" x 12" saturated asphalt felt (used as an intermediate in the manufacture of composition shingles) by applying the above composition with a knife at a coverage rate of two gallons for each 100 square feet (or a film of about 1/16 of an inch). The strips were allowed to dry until substantially all of the mineral spirits was removed.

The composition of this process was self-extinguishing in 2.1 seconds after the first ignition and in 3.1 seconds after the second ignition. An intumescence of 0.25 inch was observed. The composition was considered to be nonburning. In a manner similar to Example 2, other compositions were prepared and tested and are summarized in the table below.

Example	Fire Retardant	Parts of Retard-ant	Parts of Sb_2O_3	Self-extg. Time (sec.) 1st Appln.	Self-extg. Time (sec.) 2nd Appln.	Intumes-cence (In.)	Flammability
3	Trichlorophenyl tetrachloropentadienoic acid	18.2	3.0	3.0	6.5	0.36	Nonburning
4	Perchlorocoumalin	18.2	3.0	1.3	4.6	0.17	Nonburning
5	Bis(trichlorophenyl) trichloropentadienoic acid	18.2	3.0	2.9	16.5	0.50	Nonburning
Control				Burns	Burns	None	Burns

Similar results will be obtained when the substituent group of Examples 3 and 5 is alkyl rather than aryl, such as perbromo or perchloro methyl or ethyl substituents.

Resinous Hydrocarbon-Phosphorus Pentasulfide Reaction Products

F.I.L. Lawrence and M.J. Pohorilla (U.S. Patent 3,279,928; October 18, 1966; assigned to Kendall Refining Company) have found that certain high molecular weight viscous, resin-like petroleum hydrocarbons may be reacted with phosphorus pentasulfide to provide hydrocarbon condensation products which are excellent flame-retardant agents. The condensation products when subsequently reacted with sulfur- and/or oxygen-containing materials produce condensation products which exhibit not only flame-retardant characteristics, but also are noncorrosive.

The resin-like hydrocarbon starting materials employed in the process are characterized by an average ebullioscopic molecular weight of at least 1,000. An optimum average ebullioscopic molecular weight range is from about 1,200 to 1,700. Additionally, the starting materials are characterized by a viscosity of at least about 900 SUS at 210°F. The resin-like starting materials exhibit a degree of unsaturation and are characterized by a bromine number within the range of about 1 to 10. Resin-like hydrocarbons useful as starting materials can be obtained by conventional liquefied normally gaseous hydrocarbon precipitation processes from normal or vacuum distillation or residual stocks and analogous fractions of paraffin or mixed base crude oils. Pennsylvania crude oils are a particularly desired source for resin-like oils.

The condensation products are produced by contacting the resin-like hydrocarbon starting material while at a temperature of at least about 400° to 600°F., preferably about 425° to 575°F. with from about 2 to 15% by weight, preferably between about 7 and 12% by weight, of phosphorus pentasulfide for a time period requisite to produce an intermediate condensation product which of itself has flame-retardant characteristics. On contact of the phosphorus pentasulfide condensation products with sulfur and/or oxygen, a final condensation product is produced which is both flame retardant and substantially noncorrosive. The following examples illustrate the process.

Example 1: Test Procedure — In the test procedure employed to determine the flame retardancy of the compositions of the process and various other coating materials, test panels of the materials to be evaluated were prepared by coating steel plates, 1/4" x 2" x 4", and white pine panels, 5/8" x 5/8" x 6", by immersing the plates or panels in the heated materials at 350° to 400°F. When the materials were cut back with a solvent, the coating was effected by immersing at

77°F. The coated plates were positioned at an angle of 45° to a vertical test flame. The flame used in the test was provided by means of a bunsen burner capable of producing a 1" flame, measured to the top of the inside cone of the flame from the top of the burner. To evaluate completely the flame retardance of the material, the burning phase was conducted in three steps.

In the first step, the flame was applied to the test panel for 15 seconds and then removed. The length of time in seconds that the coating material continued to burn, if at all, after the removal of the flame was recorded. Thirty seconds after the removal from the initial flame test the flame was again applied to the same area of the test panel for an additional 30 seconds and then removed. Again the length of time the coating continued to burn after flame removal was recorded. In step three, 30 seconds after the removal of the second flame test, the flame was again applied to the same area of the test panel for an additional 60 seconds. Again the length of time the coating continued to burn after flame removal was recorded. If, after step one or step two, the material continued to burn throughout the 30 second delay period, the test was discontinued on that specific panel, and the coating was recorded as having "fired".

Preparation of Resinous Starting Materials — About 75,000 grams of a cylinder stock derived by distillation from paraffin base Pennsylvania crude oil and characterized by a boiling point in excess of about 850°F., a molecular weight of about 750, a viscosity at 210°F. of 225 SUS, an API gravity of about 24.8, and a flash point (Cleveland open cup) of about 600°F. are mixed with propane heated to a temperature of about 190°F. and then cooled to a temperature of about 65°F. The cylinder stock-propane solution is thereafter transferred into a chilling tank where the pressure is reduced to an extent requisite to volatilize sufficient propane to lower the temperature of the solution to about -20° to -50°F. Makeup propane is added during the chilling operation, such that the ratio of propane to cylinder stock is about 3 to 1 at the end of the chilling cycle. During the chilling cycle, petrolatum is precipitated from the solution. The chilled cylinder stock-propane solution containing precipitated petrolatum is transferred to a filter feed tank and hence passed through an appropriate filter to effect removal of the petrolatum from the chilled solution.

Propane is added to the filtrate in an amount sufficient to raise the propane-cylinder stock ratio to about 10 to 1 and the temperature of the solution so obtained is elevated to about 150° to 180°F. whereupon about 15,000 grams of high molecular weight viscous material are precipitated. This viscous material contained some propane. The precipitated material so obtained is then mixed at a temperature of about 130° to 135°F. with additional propane to increase the propane-oil ratio to about 20 to 1. The temperature of the resulting solution is lowered to about 100°F. and about 6,000 grams of viscous hydrocarbons are precipitated. These materials, after removal of all residual propane, are designated as heavy resins and are characterized by a molecular weight of about 1,400, a viscosity of about 4,100 SUS at 210°F., and a bromine number of 3.7.

The remaining oil-propane solution that is the extract from the last described step is heated to about 150°F. and 9,000 grams of additional viscous hydrocarbons, which are designated as light resins, are precipitated. Any residual propane in the resin is removed in a flash chamber. These light resins are characterized by a molecular weight of about 1,300, a viscosity of about 1,150 SUS at 210°F., and a bromine number of about 4.0.

Flame-Retardant Material — 750 grams of the viscous materials separated from the cylinder stock and designated as light resin, and 250 grams of bright stock extract are charged to a suitable apparatus and initially heated to about 500°F. Phosphorus pentasulfide is added to the mixture in increments totalling about 0.25% by weight of the mixture. Sulfide addition is maintained over a period of about 12 hours until a total amount of about 9% by weight of phosphorus pentasulfide is added. During the reaction period the temperature in the reaction zone is maintained at about 500°F. to produce an intermediate condensation product which is corrosive to a copper strip when tested pursuant to ASTM Procedure D-130. To eliminate the corrosiveness, the intermediate product is contacted with sulfur to effect further condensation.

Sulfur is incorporated into the reaction mixture maintained at approximately 500°F. in increments, for a period of approximately 4 hours until a total of about 12% by weight of sulfur has been added to provide a final condensation product which has a ring and ball softening point of about 205°F. and is noncorrosive to a copper strip when tested as described above. The bright stock extract referred to is a fraction of Pennsylvania crude oil removed from Pennsylvania bright stock by phenol extraction and characterized by a viscosity of about 1,210 SUS at 210°F., a viscosity of about 840,000 SUS at 100°F. and a gravity of about 10.5° API.

Example 2: To demonstrate the flame-retardant properties of the composition, the condensation product of Example 1 was tested according to the flame testing procedure set forth above. In test No. 1, the condensation product was melted and applied to a metal panel by dipping. In test No. 2, the product was cut back by dissolving it in a straight run overhead distillate from Pennsylvania crude oil characterized by a distillation range of about 300° to 370°F., Tagliabue flash point (ASTM D56-56) of about 100°F., a gravity of about 52° API, and an aniline point of about 135°F.

The results of the tests are noted in the table on the following page.

	Steel	
Test No.	1	2
Composition, Percent		
P_2S_5 condensed resins	100	55
Cut-back solvent		45
Panel Burning Time After Flame Application (sec.):		
1 (15 sec. Exp.)	0	0
2 (Next 30 sec. Exp.)	0	0
3 (Next 60 sec. Exp.)	0	0

Emulsion Formulations

A process described by W.E. Skelton and C.E. Wilkinson (U.S. Patent 3,224,890; December 21, 1965; assigned to Texaco Inc.) provides an aqueous asphaltic emulsion comprising asphalt, water, alkali metal salt of Vinsol resin, alkali metal hydroxide, asbestos fiber, pulverized oyster shell or mica, alkali metal carboxy alkyl cellulose and an ammonium phosphate. The emulsion has outstanding fire resistant-properties before, during, as well as after curing and does not present a fire, explosion or toxicity hazard when applied in enclosed surroundings.

Advantageously, the coating composition comprises between about 15 and 20 weight percent water, between 13 and 20 weight percent asbestos fiber, between 8 and 15 weight percent pulverized oyster shell or mica, between 0.1 and 1.5 weight percent alkali metal salt of Vinsol resin, between 0.05 and 0.35 weight percent of an alkali metal hydroxide, between 0.25 and 1.0 weight percent alkali metal carboxy alkyl cellulose, and between 0.3 and 3.0 weight percent phosphate selected from the group consisting of monoammonium phosphate and diammonium phosphate. The following examples illustrate the process.

Example 1: This example illustrates the preparation of the bituminous emulsion. To 14.73 lbs. of water there was added 0.23 lb. of the sodium salt of Vinsol resin, 0.04 lb. of sodium hydroxide, and 15 ppm of polymerized dimethyl siloxane (antifoamant) and the mixture was stirred at about 145°F. The aqueous mixture and 15.0 lbs. of asphalt (heated to 250°F.) were passed through a colloid mill at about 155°F. The asphalt employed was a 50-50 blend of air-blown and vacuum reduced petroleum residua having a ring and ball softening point of 128°F. and a penetration between 50 and 60 at 77°F. To the emulsion there was added an aqueous solution composed of 0.44 lb. monoammonium phosphate, 0.44 lb. sodium carboxy methyl cellulose, 0.10 lb. sodium hydroxide, and 34.27 lbs. water. To the dispersion there was added with stirring 13.0 lbs. of asbestos fiber and 8.7 lbs. of pulverized oyster shell. The filled bituminous emulsion product was of the following composition:

Ingredients	Weight Percent
Asphalt	17.25
Water	56.33
Sodium hydroxide	0.16
Saponified Vinsol resin	0.26
Monoammonium phosphate	0.50
Sodium carboxy methyl cellulose	0.50
[1] Asbestos fiber	15.00
[2] Pulverized oyster shell	10.00
Antifoamant, 5 ppm	

[1] 98 wt. percent passing No. 12 sieve; 45 wt. percent max. passing No. 40 sieve and 30 wt. percent max. passing No. 200 sieve.

[2] 100 wt. percent passing No. 40 sieve; 55 wt. percent passing No. 200 sieve.

Example 2: This example demonstrates the fire-resistant properties of the bituminous emulsion when cured and its outstanding ability to protect wooden structures against fire. The procedure employed to test the fire resistance of the bituminous emulsions was as follows. An 8" pine rod of 13/16" diameter was coated with the bituminous emulsion to be tested to a thickness of 1/8" to 1/4" and the thus coated wooden rod was mounted in a horizontal position for 18 hours at room temperature, then cured for 24 hours in an oven at 180°F. The cured coated rod was allowed to cool before testing.

The cured coated pine rod was mounted on a rack and placed in a galvanized iron enclosure to minimize drafts. A can of Sterno (canned heat) was opened and placed directly under the test specimen half way between the ends. The Sterno was ignited and the flame was allowed to burn for 2 minutes and then extinguished. If the specimen did not ignite during this period or did not continue to burn for more than 10 seconds after withdrawal of the flame, the coated rod was again subjected to the flame for a similar additional length of time. If burning continued after the second application of the flame, the length of burning up to 300 seconds was recorded. The specimen was inspected for bleeding during the test. After cooling the coating was inspected for charring and removed so the wood rod could be inspected.

Bituminous emulsions used in the process and comparative bituminous emulsions were tested in the fire test procedure and the results are reported below.

Ingredients	8502-O	8502-N	8486-B	8486-O	8436-B	8486-D	8402-G	8388-D
Asphalt	17.25	17.25	17.25	17.70	17.50	17.25	18.75	10.86
Water	56.33	54.83	55.83	56.96	56.69	56.33	54.50	60.37
Saponified Vinsol resin	0.26	0.26	0.26	0.18	0.26	0.26	0.84	0.54
Sodium hydroxide	0.16	0.16	0.16	0.16	0.05	0.16	0.16	0.05
Monoammonium phosphate	0.50	2.0	-	-	-	-	-	-
Diammonium phosphate	-	-	1.0	-	-	1.0	-	-
Disodium phosphate	-	-	-	1.0	0.5	-	-	-
Sodium tripolyphosphate	-	-	-	-	-	-	-	0.5
Sodium carboxy-methyl cellulose	0.50	0.50	0.50	0.50	-	-	0.75	0.18
Asbestos fiber [1]	15	15	15	14	15	15	15	15
Oyster shell [2]	10	10	10	10	10	10	10	-
Mica [2]	-	-	-	-	-	-	-	12.5
Fire Test—Flame duration:								
After first 2 min. application, seconds	0	0	0	0	[3] 300+	0	0	[3] 0
Second 2 min. application, seconds	0	0	0	180+	-	180	55	[3] 300+
Bleeding of coating	None	None	None	Severe	Trace	Trace	None	Trace
Charring of coating	None	Slight	Slight	Moderate	Severe	Moderate	Slight	Severe
Charring of wood	None	Trace	Trace	Severe	-	Moderate	Slight	Moderate

[1] 98 wt. percent passing No. 12 sieve and 45 wt. percent max. passing No. 40 sieve and 30 wt. percent max. passing No. 200 sieve.
[2] 100 wt. percent passing No. 40 sieve and 55 wt. percent passing No. 200 sieve.
[3] 90 sec.

As can be seen from the above, the representative bituminous emulsions of the process (8502-O, 8502-N and 8486-B) provide fire-resistant coatings which do not burn even after being exposed to a flame for a four minute period and further provide coatings which after exposure do not bleed, char only slightly, and allow only a trace of charring of the coated wood. In contrast the remaining comparative compositions which delete or employ substitutes for one or both members of the phosphate-cellulose additive combination produce coatings which are not fire resistant, which tend to bleed and char upon exposure and which allow substantial damage to the wood surface.

Bituminous Top Coat

J.B. Welch (U.S. Patent 2,733,170; January 31, 1956; assigned to Spring Packing Corporation) describes a coating composition comprising alternating series of relatively thin coatings of a fire-retardant, foam-forming paint and relatively thick coatings of an insulating bituminous composition. U.S. Patent 2,523,626 describes a fire-retardant paint which includes a protein, a foam-forming ingredient such as monoammonium phosphate, diammonium phosphate, phosphoric acid, ammonium sulfate, sulfamic acid or boric acid, and a binder, particularly a binder of a urea-formaldehyde resin, all suspended in a liquid carrier. In U.S. Patent 2,333,189 there is described a plastic bituminous insulating composition which can be applied to vertical structures in thick layers which will not crawl, slide or slip.

This process is not related to the composition either of the foam-forming, fire-retardant paint, nor to the plastic bituminous insulating material, but rather to the use of the two in such a way as to accomplish results not obtainable with either alone. It has been found that if a structure is protected with alternating series of the foam-forming, fire-retardant paint in relatively thin coatings and relatively thick coatings of the plastic bituminous insulating composition, it is possible not only to protect the underlying surface from fire, but to protect it from corrosion, to provide a vapor seal, and to protect it from the heating effects of a fire, particularly of a flash fire.

By means of the process, it is possible to protect steel for predetermined periods of time from reaching a temperature at which it will lose strength — as for example, 1100°F., even in a severe exposure to flame. Likewise, the combination of coatings may be used to protect masonry from spalling in a fire. The fire-retardant, foam-forming paint is applied in the usual thicknesses. A preferred foam-forming paint is now being marketed under the name of "Albi". The Albi Temp-Kote PC is available for wood surfaces, the Albi Temp-Kote 99 for surfaces generally, and the Albi-"R" is a water-base material. Normally, the dried thickness of the fire-retardant paint is approximately 0.0035 to 0.004 inch per coat. It is generally desired to use sufficient to provide a total thickness of from 0.007 to 0.012 inch, this normally being accomplished by the use of two to three coats. After the first coat of fire-retardant paint has been applied, the plastic bituminous material, preferably the cork-containing material described in U.S. Patent 2,333,189, is applied over the fire retardant. Normally, this is in a thickness of 1/8" to 5/8". The fire-retardant paint is then again applied, if necessary or desirable, followed by another coat of the asphaltic plastic.

The coating has particular value in the petroleum field, especially for oil storage tanks or in the vicinity of oil storage tanks where severe fires are likely. Normally, the thickness of the asphaltic material of the cork-base type will be from 1/8" to 1 1/2", the bulk and preferred range being from about 5/8" to 1 1/4". Instead of using successive coatings of the fire-retardant paint and plastic bituminous material, it has been found that a mixed coat comprising 15 to 30% of the fire-retardant, foam-forming paint and 70 to 85% of the plastic bituminous material provides a coating which has the advantages of both coatings without the disadvantages of either. Such coatings may be applied in thicknesses of 1/8" to 5/8", or even more.

INTUMESCENT COATINGS

PHOSPHORYLAMIDES

Polyphosphorylamide, Aminoplast Formulations

M.L. Nielsen (U.S. Patent 2,596,937; May 13, 1952; assigned to Monsanto Chemical Company) has found that fire-retarding compositions of great stability may be provided by the combination of an aminoplast resin and certain chemical compounds containing nitrogen and phosphorus. The nitrogen-phosphorus-containing component may be a condensation product of ammonia and phosphoryl chloride. A preferred compound obtainable from the reaction product of ammonia and phosphoryl chloride by heat treatment is the insoluble, infusible composition which has been designated as polyphosphorylamide. The following example illustrates the preparation of the polyphosphorylamide.

One hundred parts of phosphoryl chloride was dissolved in 300 parts of kerosene of commercial grade. The kerosene had a boiling range of about 150° to 260°C. Anhydrous ammonia gas was then passed into the kerosene solution at a slow rate of addition, and simultaneously therewith, the solution was cooled in order to maintain the temperature below about 80°C. At the same time, efficient agitation of the solution was maintained. The addition of ammonia was continued in this manner until 5 mols of ammonia had been reacted with the phosphoryl chloride. The reaction product was insoluble in kerosene and appeared as a suspended white powder. After completion of the ammonia addition, the reaction mixture, while remaining suspended in kerosene, was heated to a temperature of 200°C., heating was then discontinued and the mixture permitted to cool. During the heating period, some ammonia was evolved, the nitrogen-phosphorus product apparently undergoing polymerization and becoming water-insoluble.

At the end of the heating period, which may be carried out for a period of from one-fourth to three hours, preferably one to three hours, the solids were removed from the kerosene by filtration or centrifuging. The product was next washed with water in sufficient amount so as to dissolve the ammonium chloride present in the product. The washing operation may also be followed by a heating step which is designed to dry the product and also to remove traces of the kerosene or other hydrocarbon. A convenient method for removing the hydrocarbon consists in steaming the product while it is contained in a rotating centrifuge. The nitrogen-phosphorus containing product, resulting from the process, is substantially insoluble in water as well as organic solvents. It may be suspended or dispersed in water and will then exhibit a pH which is substantially neutral or alkaline in reaction. The final product is free from chlorine as a result of the heating and polymerization step, which makes it possible for the ammonium chloride to be completely removed to leave only the condensation product consisting essentially of nitrogen, oxygen, hydrogen and phosphorus.

The nitrogen-phosphorus atomic ratio is approximately 1.80:1, but may be varied within the range of 1.5: ; to 2.0:1 by variation of the amount of NH_3 initially reacted with the $POCl_3$ and this product to some degree polymerized, and then subjected to heating as described above, to complete the polymerization. The polyphosphorylamide is a white powder, insoluble in water, and having a substantially neutral or alkaline reaction. The nitrogen-phosphorus atomic ratio of the polymeric material lies in the range of 1.72:1 to 1.80:1, which relatively high range represents a product desirable in providing a composition which, in the presence of nitrogen containing resins such as aminoplasts, bulks up by releasing nitrogen compounds as gases when exposed to fire. Polyphosphorylamide per se is infusible except when exposed to a very great heat. It has been found that at high temperatures, such as that resulting from the application of a flame, the polyphosphorylamide present in the combination resinous composition of the process reacts chemically with an aminoplast resinous base to generate a nonburning gas which permeates and swells the coating film to form a porous, insulating layer.

This action, which is termed intumescence, forms an adherent, fluffy char which does not burn. Consequently, the combustible resin whether existing as a manufactured article or as a coating composition does not itself burn, since it is protected by the heat insulating, fire-retarding overlayer. The nitrogenous or aminoplast condensation products which may be resin-forming are essentially nitrogenous, resinous products such as the formaldehyde condensation products, preferably of urea or melamine. Specifically, these products may be mono- or dimethylol urea or mono-, di-, tri-, tetra-, penta-, or hexa-methylolmelamine. The following examples illustrate the process.

Example 1: A melamine-formaldehyde intermediate stage resin (Melmac) was mixed with 10% by weight, based upon the resin, of polyphosphorylamide. The product which was obtained as a granular paste resists combustion, and when applied without further solvents as a coating material to combustible surfaces such as wooden panels, showed no tendency to glow after removal of a test flame. The material was tested in the standard fire test as described in the American Paint Journal Convention Daily, pages 8-30, November 6, 1946, which test was modified by the substitution of a gas micro burner for the alcohol cup. The combination product described above showed vigorous intumescing action when subjected to fire. Under such circumstances, the intumescent material puffed up and released gases to form a porous, heat insulating char which protected the base material from the fire.

In general, the proportion of polyphosphorylamide or other reaction products of phosphoryl chloride and ammonia may be utilized in intumescing compositions in a concentration range of 5 to 50% by weight, based upon the weight of the aminoplast resin employed in the composition.

Example 2: To a solution of 25 parts of polyvinyl chloride in 220 parts of cyclohexanone, there was added 8 parts of polyphosphorylamide and 24 parts of urea-formaldehyde molding powder. This combination product was brushed on a wooden panel as a thin coating and then allowed to dry. When exposed to fire under standard test conditions, this coating intumesced to prevent the wood panel from burning. Various resinous constituents, in addition to the polyvinyl chloride illustrated above, may be utilized together with the aminoplast resin such as the urea-formaldehyde employed in this example.

Example 3: Ten parts of polyphosphorylamide employed as a fine, white powder was mixed with 30 parts of urea-formaldehyde molding powder. This mixture was then dispersed in a liquid form of a silicone resin, a hydrolytic product of organosilicon oxychloride, as described in R.L. Wakeman's Chemistry of Commercial Plastics, pages 758-761 (Reinhold Publishing Corporation, 1947).

The above formulation was found to be a noncombustible material which, in the presence of fire, intumesced to form a porous, heat insulating char. Various silicone resins, whether in liquid or solid form, may be utilized in the intumescing compositions of this process.

In related work, M.L. Nielsen and P.E. Marling (U.S. Patent 2,596,939; May 13, 1952) and M.L. Nielsen (U.S. Patent 2,642,405; June 16, 1953; both assigned to Monsanto Chemical Company) have found that an improvement may be effected in fire-retarding compositions based upon water-soluble, aminoplast resins and nitrogen-phosphorus compounds, by the inclusion of one or more compounds characterized as polyhydroxy compounds. Examples of preferred types of suitable compounds are sorbitol, mannitol and pentaerythritol. The following examples illustrate the process.

Example 1: A water-soluble condensation product, together with the water-insoluble dispersion of polyphosphorylamide, may be made from the following ingredients:

Urea-formaldehyde condensation product	25 g.
Mannitol	16 g.
Polyphosphorylamide	8 g.
Titanium dioxide	4 g.
Water	45 g.

The above composition, when applied to test panels of wood, gave a film of good covering power, free of checking, and acceptable fire-retarding qualities. When subjected to a test flame, as described in American Paint Journal Convention Daily, November 6, 1946, the coated panel charred and intumesced at the point of contact of the flame. The frothy, porous, heat insulating char which was thus formed by contact with a flame, protected the base wood surface from combustion. The above composition also shows good adhesion to metal surfaces. For such applications to metal, the resinous film which is produced provides a composition which will not flash when subjected to heat and hence will not propagate a flame, but on the contrary, will insulate the underlying material by providing a porous overlayer which does not burn, and which prevents the spread of fire.

Example 2:

Polyphosphorylamide	10 Parts
Titanium dioxide	4 Parts
Urea-formaldehyde (water-soluble condensation product)	25 Parts
Resorcinol	10 Parts
Paraformaldehyde	8 Parts

(continued)

(continued)	Pentaerythritol	10 Parts
	Water	36 Parts

The above formulation was tested and found to resist the propagation of a flame. The coating film, after drying, was a smooth, tough, adherent film, which did not check under weathering conditions.

In a continuation of the work, M.L. Nielsen, R.W. Arnold and E.C. Chapin (U.S. Patent 2,596,938; May 13, 1952; assigned to Monsanto Chemical Company) have found that fire-retarding compositions may be provided by the combination of certain water-soluble, alkyd resins, together with an aminoplast resin and a chemical compound to furnish nitrogen and phosphorus. The water-soluble, alkyd resins which have been found useful in this process function to cross-link with the aminoplast resinous material. Such water-soluble, alkyd resins are of particular utility for application of the fire-retardant composition in the form of an emulsion such as a water paint. When such composition has once been applied, it has been found that the resinous constituents rapidly cure to complete the polymerization. The final product is, therefore, a hard, weather resisting, insoluble, continuous film in which the phosphorus compound such as the polyphosphorylamide is dispersed uniformly throughout the film. The improvement realized in the coating composition by the use of water-soluble, alkyd resins is shown in the greater ease of application and in the rapid cross-linking or polymerization which takes place.

It is desirable in the formulation of fire-retarding compositions based upon the process that a two component mixture be supplied for final mixing shortly prior to the application of the coating composition. Thus, the phosphorus compound exemplified by polyphosphorylamide may be mixed with the aminoplast resin. This preliminary mixture may then be mixed at the time of use with the water-soluble, alkyd resin. The resultant mixture may then be applied to produce the fire-retarding coating composition. However, the phosphorus compound may also be introduced with the alkyd component, and such mixture may then be combined with the aminoplast resin. Such mixtures are characterized by a controlled rate of polymerization so that the material may be applied at a convenient rate and will then polymerize or cure very rapidly after application to give an insoluble, water resistant composition as a result of the cross-linking which takes place with the aminoplast resin. The following examples illustrate the process.

Example 1:

Urea-formaldehyde condensation product (water-soluble)	28 Parts
Water-soluble alkyd based upon 1:1 ratio of citric acid and glycerol	7 Parts
Polyphosphorylamide	11 Parts
Titanium dioxide	3 Parts
Water	51 Parts

The above composition, when applied to test panels of wood, gave dense coatings having a smooth finish, which was quite stable against weathering. The coating compositions also cured rapidly. When test panels, coated with the above composition, were subjected to the standard fire tests, such as those proposed in the American Paint Journal Convention Daily, pages 8-30 (November 6, 1946), it was found that the coating provided adequate protection as a fire-retarding material. In carrying out the fire-retarding test, the above method was modified by the substitution of a gas microburner for the alcohol cup. The wooden panels, when heated by the flame of the Bunsen burner, showed a freedom from flame propagation during the standard test period of three minutes. The final result, after applying a flame to the treated panel, was the production of an adherent, porous char overlayer which was thick and frothy, thereby preventing transfer of heat to the undersurface.

Example 2:

Urea-formaldehyde condensation product (water-soluble)	17 Parts
Citric acid glycerol resin (alkyd)	16 Parts
Polyphosphorylamide	13 Parts
Titanium dioxide	2 Parts
Water	52 Parts

Formulation of the above composition showed the liquid mixture to be stable for several hours, but upon application cross-linking and polymerization took place rapidly to yield a water-insoluble coating film. This coating, when applied as a test film on wood panels, easily passed the standard test as discussed above with reference to a fire-retarding effect.

Example 3:

Urea-formaldehyde condensation product (water-soluble)	25 Parts

(continued)

(continued)

Alkyd based upon 1:1 molar ratio of citric acid and ethylene glycol	7 Parts
Polyphosphorylamide	10 Parts
Mannitol (plasticizer)	5 Parts
Titanium dioxide	2 Parts
Water	51 Parts

The above composition, when applied as a coating film to various surfaces, showed extreme flexibility and resistance against weathering. The test panels of poplar wood, when subjected to the standard fire test, showed adequate resistance as a fire-retarding material.

The use of chlorinated rubber and the polyphosphorylamide is described by P.E. Marling and M.L. Nielsen (U.S. Patent 2,596,496; May 13, 1952; assigned to Monsanto Chemical Company). The following example illustrates the process.

Solution a	Chlorinated rubber (300 cp. viscosity)	50 Grams
	Xylene	150 Grams
Solution b	Solution a	20 Grams
	Tricresyl phosphate	2 Grams
Solution c	Polyphosphorylamide	5 Grams

When formulated in a coating composition and tested as a film, chlorinated rubber alone (solution a) burned vigorously, illustrating the unsuitability of chlorinated rubber alone as a fire-retarding coating material. When solutions a and b were combined to furnish a plasticized material and tested as a film, it was found that this composition also burned freely. When, however, solutions a and b were first mixed, and then combined with c, the resulting, dried film did not burn. Test panels of this same composition painted upon wood also showed that the film of the coating material did not burn and was of sufficient stability when subjected to a flame to protect the otherwise combustible base material.

Other related patents assigned to Monsanto Chemical Co. encompassing the use of phosphorus compounds and ammonia with aminoplasts and polyhydroxy compounds include: M.L. Nielsen and R.W. Arnold (U.S. Patent 2,680,077; June 1, 1954); R. Marotta (U.S. Patent 2,676,162; April 20, 1954) and I.V. Wilson and R. Marotta (U.S. Patent 2,600,455; June 17, 1952 and U.S. Patent 2,702,283; February 15, 1955).

Phosphorylamides, Chlorinated Binders

J. Truhlar and A.A. Pantsios (U.S. Patent 2,582,181; January 8, 1952; assigned to Rudolf F. Hlavaty) describe the use of the amides, imides and imidoamides of phosphoric and thiophosphoric acids, as well as the nitride of phosphoric acid. Tests have shown that these materials are excellent fire and flameproofing agents. In practice the fireproofing material of this process is the water-insoluble, chlorine free, reaction product of phosphorus oxychloride with anhydrous ammonia. This reaction product consists of a mixture of phosphoryltriamide, phosphorylimidoamide and polyphosphorylamides. The following examples illustrate the process.

Example 1:

Phosphorylamides	30	Parts
Polyvinyl chloride	3	Parts
Acetone	67	Parts
Pentachlorphenol	0.1	Part

The phosphorylamides in fine powder form (200 mesh) are suspended in the solution of polyvinyl chloride in acetone containing the pentachlorphenol which acts as a fungicide. Fabrics, fibers, batts and any other materials which are to be fireproofed are immersed in, sprayed with or otherwise treated with the suspension until they absorb enough of it to produce a net gain in weight of approximately 15%. After this they are dried. Experience has shown that this produces effective fireproofing.

Example 2:

Phosphorylamides	20	Parts
Water	45	Parts
Chlorinated naphthalene	10	Parts
Carbon tetrachloride	23	Parts
Emulsifier	2	Parts

This is an example of an oil-in-water emulsion using chlorinated napthalene as the thermally unstable binder. The solvent used makes this emulsion safe to handle. In the same manner, other thermally unstable chlorinated organic compounds suitable as binders for the phosphorylamides can also be used. These could be neoprene, pentachlorobiphenyl, chlorinated, natural or synthetic rubber and chloropolystyrene.

Example 3:

Phosphorylamides	50 Parts
Commercial polyvinyl acetate emulsion	45 Parts
Tricresylphosphate	5 Parts

The polyvinyl acetate film alone supports fire but in the presence of the phosphorylamides it does not burn. On the other hand, the total effectiveness as a fireproofing mixture is much less than that of Examples 1 and 2, and the use of binders which support fire is recommended only when other considerations prevent the use of the thermally unstable chlorinated binders. Other film forming, water-insoluble materials (or such as can be rendered water-insoluble) may also be used. Among these are polystyrene emulsion, synthetic and natural latexes, and alkyd resin emulsions.

Example 4:

Phosphorylamides	27 Parts
Chlorinated rubber	8 Parts
Alkyd resin	14 Parts
Pigment	20 Parts
Plasticizer	1 Part
Solvent	30 Parts

Example 5:

Phosphorylamides	20 Parts
Linseed oil	5 Parts
Castor oil	5 Parts
Chlorinated rubber	13 Parts
Pigments	30 Parts
Tricresylphosphate	2 Parts
Solvent	25 Parts

Example 6:

Phosphorylamides	20 Parts
Chlorinated rubber	17 Parts
Carbon black	11 Parts
Tricresylphosphate	2 Parts
Solvent	50 Parts

Examples 4, 5 and 6 illustrate the use of the phosphorylamides in typical fireproof surface coatings.

Dicyandiamide and Nitrogen-Phosphorus Complex

R.E. Ellis (U.S. Patent 3,102,821; September 3, 1963) describes a flameproofing and fire-retardant intumescing composition which is composed of two essential components. The first of these is the chemical compound dicyandiamide, and the second is a known water softening composition described in U.S. Patent 2,122,122 and sold under the trademark Victamide. This latter component is a substantially neutral composition consisting of a nitrogen-phosphorus complex, resulting from heating a reaction product of phosphorus pentoxide and anhydrous ammonia at a temperature of at least about 150°C. in the presence of excess anhydrous ammonia to produce a condensation reaction in which ammonia is lost by the heated product, this compound having an atomic ratio of nitrogen:phosphorus of less than 1 1/2:1 and typically containing no water of constitution. In physical characteristics, it is a substantially nonhygroscopic, substantially free-flowing powder soluble in water giving, for a 0.25% solution, a pH value of between 6 and 7. It contains both ammonium and nuclear nitrogen, with the ammonium nitrogen generally representing at least 60% and up to about 80% of the total nitrogen. It is referred to as the "nitrogen-phosphorus complex".

According to the process, merely mixing these two components together, that is dicyandiamide and the nitrogen-phosphorus complex, in a suitable vehicle for the same, preferably water, forms a composition which may be applied as a coating without further treatment to articles made of wood, cellulose, fiberboard and to structural metal members to effect flameproofing and fire-retardant protection. Thus, after the article to be protected is brush or dip coated with the aqueous solution of these two components, and allowed to dry at normal room temperatures, i.e., from about 15° to 35°C., the deposited film has the desired property of intumescing if the article is then exposed to high temperatures such as by contact with a flame. That is, upon such exposure to a flame the coating will undergo a chemical reaction with chemical decomposition, foaming and frothing taking place to form a porous, but solid, crusty protective coating entirely covering the coated surface of the substrate article. Such a porous crust acts to prevent access of oxygen to the substrate article, and further affords heat insulation thereto. At the same time, escape of any combustible distillation products is prevented so that protection is afforded against propagation of the fire as well as against burning and/or heat deformation of the substrate. The following examples illustrate the process.

Example 1: 25 pounds of hot water were placed in a vessel of suitable size equipped with a high speed mixer device. To this was added 7 1/2 pounds of the above described Victamide product, made in accordance with U.S. Patent 2,122,122. This mix was then agitated until complete solution had occurred. Agitation was continued and 5/8 pound of alginate was added and the mix stirred to the viscous solution point. After cooling to room temperature, five pounds of commercial dicyandiamide were added. This commercial product was fairly coarse granules of the material, but after agitation for some 35 minutes, the particles had been ground to minute size, had partially dissolved, and were otherwise well dispersed through the solution under agitation.

Example 2: The procedure of Example 1 was followed, except that at the time of addition of the alginate thickening agent, 3/4 pound of glycerine was added. It was found that this composition provided a somewhat softer, more flexible, coating.

Example 3: The procedure of Example 1 was followed using the same amounts of Victamide and dicyandiamide but with the addition of 3 pounds of titanium dioxide pigment prior to the addition of the thickening agent. At the same time there was added 1/4 pound of sodium N-octyl sulfate wetting agent. This white composition was then coated on 10 acoustical tile boards, class D, and the boards were then tested in accordance with Federal specification SS-A-118b, paragraph 4.3.3, except that the flame test was conducted in a draft free room larger than 9 by 15 feet. One brush coating was applied at the rate of 5/8 ounce of the coating composition per square foot, giving a coverage of about 249-250 square feet per gallon of composition. Prior to the flame test, the prepared panel was maintained at room temperature of 75° ± 5°F., at a relative humidity of 65% ±5% for 72 hours. The results of the test are shown in the following time-temperature table, recording the observations.

Time (minutes)	Temperature, °F.	Observations
Initial	80	
1	300	
2	450	Intumescence in center panel.
3	340	Slight darkening and increased intumescence.
4	800	Further darkening.
5	1,000	Large blisters in center.
6	1,200	No further change.
7	1,165	Small lick of blue flame in center panel.
8	1,210	Glowing of "blisters" in center panel.
9	1,250	Intumescence over the whole surface.
10	1,295	Licks of blue flame in center.
11	1,330	Licks of blue flame just over edge of center panel.
12	1,360	No further change.
13	1,360	Licks of blue flame over edge of center tile.
14	1,370	Blue flame coming from under edge of center tile, which apparently is delaminating.
15	1,400	Flames approaching frame on one side.
16	1,420	Flames approaching frame on second side.
17	1,425	Large blisters developed at frame and flames receded.
18	1,440	Flaming receded.
19	1,450	Glowing of blisters in center tile; center tile white-charred material gone.
20	1,460	No further change.

In this test, there was an afterflame for only 7 seconds when the test flame was extinguished, and there was no glowing after the extinguishing of the test flame. Throughout the testing the central panel remained intact, that is, no ashes or particles fell therefrom. While some smoke was given off during the test, it was not in excessive amounts; and intumescence was observed throughout the test. Flame approached the angle frame holding the acoustical tile boards 16 minutes after ignition of the test flame, but receded when intumescence of the surface increased. The flaming came from some delamination of the edge of the center tile.

Melamine-Formaldehyde, Dicyandiamide, Polyphosphorylamide

J.M. Venable (U.S. Patent 2,956,037; October 11, 1960; assigned to Vimasco Corporation) describes a coating which comprises a synthetic base resin or combination of such resins, a plasticizer, an ionic wetting agent, a spumific, an intumescing aid, and a carbonific. The synthetic base resin performs the dual functions of serving as a binder and combining with the spumific, the intumescing aid, and the carbonific to form a carbon structure on being subjected to burning action, such as flame impingement. The base resin preferably comprises a material selected from the group consisting of a butylated-amine-formaldehyde resin and an epichlorohydrin condensate, such as an epoxy resin. Butylated melamine formaldehyde and butylated urea formaldehyde are examples of preferred specific butylated amine formaldehydes. Specific examples of preferred epoxy resins are epichlorohydrin condensates including diethylene glycol tartrate polyester and ethanolamine phthalate polyester.

The plasticizer comprises a suitable phosphate, such as an organic phosphate ester, e.g., tris-B-chloroethyl phosphate (TCEP), tricresylphosphate, tributylphosphate, cresylphenylphosphate, tri-butoxyethylphosphate and tris-2 ethylhexylphosphate. These and related phosphates are preferred for the reason that they not only serve as effective plasticizers for the base resin but also contribute to the fire-retardant properties of the coating composition due to the fact that they are incombustible.

A preferred and recommended spumific is polyphosphorylamide (PPA) which aids in the swelling of the coating when the same is subjected to elevated temperatures and effects blowing up and/or forming the tiny cells of the expanding coating. Other spumifics of similar action but greater water solubility may be used, such as mono-ammonium phosphate,, di-ammonium phosphate for ammonium dihydrogen phosphate. The intumescing aid in the composition is a source of generating gas on being decomposed by heat. It coacts with the spumific and the carbonific to (1) increase cell puff, (2) obtain cell uniformity and (3) assist the spumific in accomplishing its function in the coating. Dicyandiamide is recommended as the material of the intumescing aid for best results. As the carbonific, it is preferred to use a pentaerythritol, of which pentaerythritol, dpentaerythritol and tripentaerythritol are specific examples. Tripentaerythritol is preferred and recommended as it is the least water-soluble.

Preferred coating compositions within the purview of this process contain parts by weight of principal ingredients as indicated in the following example:

Example 1:

Ingredient	Parts by Weight
Synthetic base resin	43
Plasticizer	2
Ionic wetting agent	1
Spumific	33
Intumescing aid	12
Carbonific	9
Total	100

One preferred commercial coating composition of this process is denoted by the following example:

Example 2:

Ingredient	Parts by weight	Approx. percent by weight
Butylated melamine formaldehyde	388	36.0
Tris-B-chloroethyl phosphate	20	2.0
Ionic wetting agent	8	.7
Polyphosphorylamide	300	27.8
Dicyandiamide	108	10.0
Tripentaerythritol	84	7.8
Diatomaceous earth material	50	4.6
Hydrated Silica	22	2.0
Mica	36	3.3
Non-ionic wetting agent	2	.2
Titanium dioxide	60	5.6
Total	1,078	100.00

The proportions of ingredients of Example 2 may be varied, as indicated in the following example:

Example 3:

Ingredient	Approx. % by Weight
Butylated melamine formaldehyde	28.0 to 50.0
Tris-B-chloroethyl phosphate	1.0 to 3.0
Ionic wetting agent	0.5 to 1.8
Polyphosphorylamide	18.0 to 32.0
Dicyandiamide	6.0 to 13.0
Tripentaerythritol	2.5 to 6.0
Diatomaceous earth material	1.0 to 3.0
Hydrated silica	1.0 to 3.5
Mica	3.0 to 10.0
Non-ionic wetting agent	0.1 to 0.7
Titanium dioxide	1.0 to 15.0

OTHER PHOSPHORUS-NITROGEN COMPOUNDS

Melamine-Formaldehyde Phosphate Ether Polyol Resins

In a process described by R.E. Sempert and R.R. Harris (U.S. Patent 3,440,201; April 22, 1969; assigned to The Sherwin-Williams Company) water-insoluble resinous condensation products are produced which are the products of the reaction of a water-soluble phosphate ether polyol and a water-soluble melamine-formaldehyde or alkoxy methyl-melamine-formaldehyde. These products are prepared by reacting a water-soluble phosphate ether polyol and a water-soluble melamine-formaldehyde or alkoxy melamine-formaldehyde in an organic solvent or solvents for the reactants at temperatures sufficiently high to remove or boil off from the reaction mixture the by-products of the reaction, which are either water or an alcohol, depending upon whether melamine-formaldehyde or an alkoxy methyl-melamine-formaldehyde is employed as a starting material. In most cases, the temperature of the reaction should be at least 200°F. (93°C.). In general, satisfactory results are obtained by carrying out the reaction in a temperature range from 210°F. (98.9°C.) to 260°F. (115.5°C.).

It will be noted that water-soluble melamine-formaldehyde or alkoxy melamine-formaldehyde has six reactive sites occupied by hydroxy groups or alkoxy groups. In other words, one of the starting materials used in the process has a functionality of 6. The phosphate ether polyol can have a functionality from 2 to 5. The extent of the reaction can be determined by the quantity of water or alcohol eliminated during the condensation as compared with the quantity of water or alcohol which can be theoretically eliminated. In the process, the reaction should be carried far enough so that 35% to 80% of the functional groups have reacted. The resins which are specially useful in forming coating compositions which can be applied to a substrate and cured to produce a clear coating that will intumesce are the water-insoluble, xylene soluble products of the reaction of a water-soluble phosphate ether polyol containing 4 to 15% by weight of phosphorus, having a hydroxyl number of 400 to 500, and a functionality of 2 to 5, and water-soluble melamine-formaldehyde or alkoxy melamine-formaldehyde containing 30 to 45% by weight melamine and 40 to 65% by weight formaldehyde, alkoxy groups, if present, constituting the remainder.

The optimum range for varnish compositions is 30 to 70% by weight phosphate ether polyol and 70 to 30% by weight melamine-formaldehyde resin. Intumescent varnish compositions of the process should preferably contain 1.2 to 10.5% by weight phosphorus, 9 to 31.5% by weight melamine and 12 to 49% by weight formaldehyde, based on the resin weight. Examples of suitable phosphate ether polyols are those having the following structural formalae:

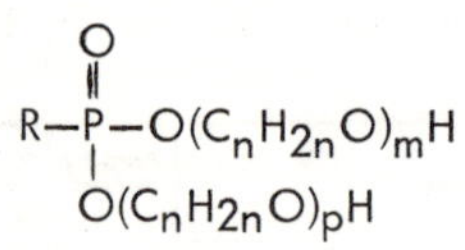

where R is aliphatic hydrocarbon (e.g., methyl, ethyl, propyl, isopropyl, butyl, isobutyl, amyl, isoamyl and higher homologues), n is 2, 3 or 4, and m and p are at least 1, the total of m and p preferably being 2 to 12. The radical R can be phenyl, tolyl, cyclohexyl or the like, and can contain halogen substituents (e.g., chlorine and bromine), or hydroxyl substituents (e.g., hydroxy ethyl, hydroxy propyl, or hydroxy butyl). A typical commercial product in this class is di(polyoxypropylene) butyl phosphate (Fyrol 19) made by reacting butyl phosphate with 1,2-propyleneoxide. The following examples illustrate the process.

Example 1: This example illustrates a preferred procedure for preparing a clear coating varnish containing the resins of the process. In accordance with this procedure, 420 parts of phosphate ether polyol having a phosphorus content of 5.5%, a hydroxyl number of 450 and a functionality of 4 (Polyol 204-S), is mixed with 290 parts of xylene (Xylene 5°C.) and the mixture is charged into a two liter three-necked flask equipped with agitation, thermometer, condenser and water separation trap. The mixture is heated to 260°F. (115.5°C.). A second mixture is prepared by mixing together 560 parts of hexamethoxy melamine-formaldehyde (Cymel 300) and 290 parts of xylene. A third mixture is prepared by mixing 18.8 parts of p-toluene sulfonic acid and 18.8 parts methyl alcohol.

The second and third mixtures are combined. 70 parts of the combined mixture is added over a period of 45 minutes to 1 hour to the first mixture, keeping the temperature at 260°F. while removing all distillate. The temperature is held at 260°F. for 10 to 15 minutes, the heat is turned off and the remainder of the combined second and third mixture is added to the reaction mixture over a period of 1 1/2 to 2 hours, while allowing the temperature to drop to 210°F. After this addition is complete, the temperature is held at 210°F. (98.9°C.) until a viscosity of T-U (Gardner-Holdt) is obtained while continuing to remove all distillate that collects in the separation trap. The product is then cooled to room temperature and is ready for use.

This product has excellent varnish properties and when applied to a wooden wall panel and cured at temperatures of 120°F. and above, gives excellent coatings which have very good intumescence. The thickness of the coating can be varied but is preferably a minimum of 3.5 mils (0.0035"). The coatings cure at about 120°F. in about 8 hours; they cure in 1/2 hour at 140°F., so that they can be handled, and in 3/4 hour at 140°F. to a hard coating. At 225°F. they cure to a hard coating in 5 to 10 minutes. The resultant coating is water white in color and is color retentive. In tests that were made on these coatings the hot portion of a Bunsen burner flame did not penetrate beyond the first ply of the wood in 5 minutes. The coatings were moisture resistant and there was no general effect except for slight softening after 18 hours at 90°F. at 100% relative humidity. The cured coatings passed the ASTMD 1211-60 cold check cycle test. These coatings had exceptional mar resistance and were not dulled when rubbed with glass wool. They also have excellent impact resistance. The gloss was excellent, being 98 to 100 at 60° on a photovolt gloss meter. The chemical resistance to all normal household chemicals was also excellent. The thickness of the coating can be varied and two or more coatings can be applied. In the tests described above the thickness was approximately 6 to 8 mils.

Example 2: This example is intended to illustrate the effect of varying the proportions of the melamine-formaldehyde resin and the phosphate ether polyol. The example also illustrates the results obtained by varying the extent of the reaction. The procedure employed was generally similar to that used in Example 1 except that the proportions of the reactants were varied within a range of 0.94:1 melamine-formaldehyde resin to phosphate ether polyol to 1.57:1 melamine-formaldehyde resin to phosphate ether polyol and the extent of the reaction was varied based on the percent of hydroxyl reacted from 32 to 63%. The results are shown in the following table.

Resin No.	A	B	C	D	E	F
Cymel 300/Polyol 204S	1.18	0.94	1.05	1.05	1.48	1.57
Percent OH reacted	32	35	50	57	59	63
Cure	Slow	Good	Good	Good	Good	(1)
Moisture resistance	Poor	Poor	Improved	Improved	Improved	Good
Intumescence	Excell.	Excell.	Excell.	Excell.	Excell.	Excell.

[1] Almost air dry.

It will be seen from the table that the weight ratio of the hexamethoxy-melamine-formaldehyde (Cymel 300 or Cymel 301) to phosphate ether polyol (Polyol 204S) has been varied from about 0.9 to about 1.6 and the extent of the reaction has also been increased. In all cases, when the composition was applied to a wood panel and cured, the intumescence of the resultant coating was excellent. The rate of curing improved with an increase in the extent of reaction and also with an increase in the amount of the melamine-formaldehyde resin. In a like manner the moisture resistance improved.

Water-Resistant Carbonific Polyurethanes and Spumific Polyaminotriazine Phosphates

D.A. Yeadon, E.T. Rayner, G.B. Verburg, L.L. Hooper, Jr., F.G. Dollear and H.P. Dupuy (U.S. Patent 3,396,129; August 6, 1968; assigned to U.S. Secretary of Agriculture) describe processes for the synthesis of (1) water-resistant carbonific polyurethanes, (2) water-resistant, nonhygroscopic spumific, fire-retardant polyaminotriazine phosphates, (3) water-resistant carbonific polycarbonates, and (4) water-resistant, intumescing, fire-retardant poly(urea-urethane-phosphate-phosphonates), and the utilization of these products in paint formulations to obtain exterior and interior, intumescing, fire-retardant paints.

The following examples illustrate the process. In these formulations, where reference is made to a "polyamide resin", the material used was a commercially available resin sold under the trademark "Versamid 930", a polyamide of ethylene-diamine and dimerized linoleic acid, as disclosed in Technical Bulletin 11-B-1 (1960) of the General Mills Chemical Division and produced in accordance with U.S. Patent 2,379,413. This resin has a maximum Gardner color of 12, a softening point of 105° to 115°C., a Brookfield viscosity of 21 to 27 poises at 160°C., and an amine value in the range of 1.4 to 7, as shown in Technical Bulletin 11-D-3 (1960) of the same manufacturer.

Example 1: Preparation of a water-resistant, carbonific polyurethane material — Two mols of pentaerythritol, five mols of phenyl isocyanate, one mol of tolylene diisocyanate, and 0.1% of diethylaminoethanol catalyst were mixed in a metal beaker equipped with a stirrer, thermometer and heating mantle. Heating the stirred slurry slowly to about 60°C. initiated an exothermic reaction of the isocyanates with the polyol, which then caused the temperature to increase spontaneously to about 180°C. or higher. Below 100°C. the temperature increased slowly but above 100°C. the temperature rose so rapidly that it was necessary to cool the reaction mixture. After this initial spontaneous reaction subsided the molten mixture was heated for three hours between 100° and 160°C. and then cured overnight for about twelve hours at about 150°C. in a gravity convection oven. When cooled the clear, almost colorless resin was pulverized and sieved through a 100-mesh screen to give a powdered, water-resistant carbonific polyurethane product.

Example 2: Preparation of carbonific, spumific, fire-retardant phosphorus-containing polyurethane material — (1) To two mols of pentaerythritol, one mol of melamine, and one mol of a dihydroxyalkyl dialkyl phosphate slurried in a metal beaker was added four mols of phenyl isocyanate and four mols of tolylene diisocyanate. To the stirred mixture was slowly added one mol of a dialkyl dialkylolaminoalkylphosphonate. The spontaneous, exothermic reaction which covers a temperature range about from room temperature to 180°C. was regulated by the rate of addition of the phosphonate component. After the initial reaction subsided the material was heated at about 120°C. for at least 16 hours. When cooled, the product was pulverized and sieved through a 100-mesh screen. In this reaction the dialkyl dialkylolamino-alkylphosphonate functions as a catalyst to promote the formation of the urethanes since it has a tertiary amino nitrogen.

(2) One mol of pentaerythritol and one mol of chloromethylphosphonodichloridate were stirred in a two-liter glass beaker equipped with a thermometer, stirrer, and heating mantle. An exothermic reaction occurred which gradually increased the temperature to about 60°C. in about one hour. The mixture was then heated to about 135°C. for about three hours or until about 88% of the hydrogen chloride had been evolved. The mixture was cooled to about 50°C. and diluted with about one volume of acetone. The reaction mixture was stirred, and one mol of tolylene diisocyanate was added. A spontaneous exothermic and foamy reaction occurred which increased the temperature to about 75°C. Foaming was moderated by rapid stirring which resulted in a homogeneous product. After curing at about 100°C. for about 24 hours, the pale yellow product was pulverized and sieved through a 100-mesh screen to yield a fine water-resistant carbonific, spumific, fire-retardant powder which could readily be formulated into fire-retardant coatings.

Example 3: Preparation of water-resistant, fire-retardant, spumific, polyaminotriazine phosphates — Six-tenths of a mol of 20% phosphoric acid was rapidly added to 0.6 mol of 2,4,6-triaminotriazine suspended in 350 milliliters of water by vigorous agitation in a Waring Blendor. After continued thorough mixing for ten minutes the water was removed by evaporation in a forced draft oven at 120°C. for about 16 hours. The dried triaminotriazine phosphates were then heated at 210°C. for five hours to obtain a fire-retardant, spumific product of maximum water-resistance.

Example 4: Preparation of water-resistant, carbonific polycarbonate — One mol of dipentaerythritol, two mols of ethylene glycol bis(chloroformate) and two mols of isopropyl chloroformate were mixed and reacted for about 24 hours between 80° and 120°C. in a container equipped with an air condenser until the evolution of hydrogen chloride ceased.

Example 5: Preparation of a dehydrated castor-tung oil alkyd — The dehydrated castor-tung oil alkyd "a" was prepared by the following procedure.

Alkyd a

Components	Parts
Dehydrated castor oil	50
Tung oil	50
Linseed fatty acids	100
Trimethylolethane	35
Chlorendic (1,4,5,6,7,7-hexachlorobicyclo-(2.2.1)-5-heptene-2,3-dicarboxylic) acid	43
Tolylene diisocyanate	19

The oil components and 50% of the trimethylolethane were alcoholyzed and gasproofed in the presence of 0.03% litharge at 300°C. for eight minutes. The temperature of the mixture was reduced to 280°C., then the linseed fatty acids, the remainder of the trimethylolethane, and 2% of triphenyl phosphite (as an esterification catalyst and color stabilizer) were added. Reaction was continued at about 250°C. until 80 to 90% of the theoretical amount of water of esterification was evolved. The temperature was reduced to 150°C. then the chlorendic acid and several milliliters of benzene were added. The mixture was esterified further at about 180°C. under reflux until practically all of the remaining water of esterification was removed azeotropically (acid value of about 6). The temperature was reduced to about 100°C. then the tolylene diisocyanate and 0.1% diethylethanolamine catalyst were added, and the mixture was then heated at 110° to 120°C. for about two hours. After standing overnight at room temperature, several milliliters of absolute alcohol was added to prevent excessive cross-linking upon aging.

Preparation of Fire-Retardant, Intumescent Paints: The alkyd, the carbonific and spumific components, pigments, and other paint ingredients of the formulations were milled together in preparing the various intumescent, fire-retardant paints. Mineral spirits or Hi-Flash naphtha was added to these paint formulations as needed to adjust their final viscosity. Listed below are several examples of intumescing, fire-retardant paint formulations prepared with the carbonific-spumific compositions described above.

Formulation A

Components[1]	Percent of Nonvolatiles
Titanium dioxide	2.86
Zinc borate	8.42
Magnesium silicate (Mistron)	0.42
Lead sulfate	4.80
Lead carbonate	4.80
Zinc oxide	2.70
Tris(2,3-dibromopropyl) phosphate	5.00
Water-resistant, fire-retardant spumific[2]	10.00
Water-resistant carbonific polyurethane[3]	20.00

(continued)

(continued)

Components[1]	Percent of Nonvolatiles
Alkyd "a"[4]	30.00
70% chlorinated paraffin	5.62
Polyamide resin[5]	4.58
Cobalt naphthenate solution (6% cobalt as metal)	0.10
Lead naphthenate solution (24% lead as metal)	0.40
Anti-skinning agent (National ASA)	0.15
Ultraviolet screening agent (UVINUL No. 490)	0.15
Total	100.00

[1] Mineral spirits was added to adjust final paint viscosity to about 90 Krebs units.
[2] 2,4,6-triaminotriazine phosphates.
[3] Prepared from two mols pentaerythritol, five mols phenyl isocyanate, and one mol tolylene diisocyanate.
[4] Alkyd was added as a mineral spirits solution of 65% solids content.
[5] 64 grams General Mills' Versamid No.930 was dissolved in 41 grams n-butanol by heating, then 75 grams of mineral spirits was added.

Formulation B

Components[1]	Percent of Nonvolatiles
Titanium dioxide	2.86
Zinc borate	8.42
Magnesium silicate (Mistron)	0.42
Lead sulfate	4.80
Lead carbonate	4.80
Zinc oxide	2.70
Tris(2,3-dibromopropyl) phosphate	10.00
Melamine	5.00
Water-resistant carbonific polyurethane[2]	20.00
Alkyd "a"[3]	30.00
70% chlorinated paraffin	5.60
Polyamide resin[4]	4.29
Cobalt naphthenate solution (6% cobalt as metal)	0.16
Lead naphthenate solution (24% lead as metal)	0.65
Anti-skinning agent (National ASA)	0.15
Ultraviolet screening agent (UVINUL No. 490)	0.15
Total	100.00

[1] Mineral spirits was added to adjust final paint viscosity to about 90 Krebs units.
[2] Prepared from two mols pentaerythritol, five mols phenyl isocyanate, and one mol tolylene diisocyanate.
[3] Alkyd was added as a mineral spirits solution of 65% solids content.
[4] 64 grams of General Mills' Versamid No. 930 was dissolved in 41 grams n-butanol by heating, then 75 grams of mineral spirits was added.

Formulation C

Components[1]	Percent of Nonvolatiles
Titanium dioxide	4.89
Zinc borate	14.38
Magnesium silicate (Mistron)	0.72
Lead sulfate	8.20
Lead carbonate	8.20

(continued)

(continued)

Components[1]	Percent of Nonvolatiles
Zinc oxide	4.61
Water-resistant carbonific polyurethane[2]	18.00
Alkyd "a"[3]	30.00
70% chlorinated paraffin	5.62
Polyamide resin[4]	4.58
Cobalt naphthenate solution (6% cobalt as metal)	0.10
Lead naphthenate solution (24% lead as metal)	0.40
Anti-skinning agent (National ASA)	0.15
Ultraviolet screening agent (UVINUL No. 490)	0.15
Total	100.00

[1] Mineral spirits was added to adjust final paint viscosity to about 90 Krebs units.

[2] Prepared from two mols pentaerythritol, five mols phenyl isocyanate, and one mol tolylene diisocyanate.

[3] Alkyd was added as a mineral spirits solution of 65% solids content.

[4] 64 grams of General Mills' Versamid No. 930 was dissolved in 41 grams of n-butanol by heating, then 75 grams of mineral spirits was added.

Formulation D

Components[1]	Percent of Nonvolatiles
Titanium dioxide	4.89
Zinc borate	14.38
Magnesium silicate (Mistron)	0.72
Lead sulfate	8.20
Lead carbonate	8.20
Zinc oxide	4.61
Pentaerythritol	18.00
Alkyd "a"[2]	30.00
70% chlorinated paraffin	5.62
Polyamide resin[3]	4.58
Cobalt naphthenate solution (6% cobalt as metal)	0.10
Lead naphthenate solution (24% lead as metal)	0.40
Anti-skinning agent (National ASA)	0.15
Ultraviolet screening agent (UVINUL No. 490	0.15
Total	100.00

[1] Mineral spirits was added to adjust final paint viscosity to about 90 Krebs units.

[2] Alkyd was added as a mineral spirits solution of 65% solids content.

[3] 64 grams of General Mills' Versamid No. 930 was dissolved in 41 grams of n-butanol by heating, then 75 grams of mineral spirits was added.

Preparation and Evaluation of Fire-Retardant Paint Films: Test specimens were prepared by applying two or three coats of paint (about three to four mils per coat) onto poplar heartwood panels (6 x 12 x 1/4 inches) for evaluation in the standard fire-test cabinet (ASTM D1360-58). The single-formulation and multiple-formulation paint systems were employed. In the multiple-formulation paint system, two undercoats of intumescing fire-retardant paint were applied followed by a topcoat of a nonintumescing, fire-retardant paint formulation. The multiple-paint formulation paint system was employed with paint formulations A and B. The different coats of paint were applied about 24 hours apart. The procedure listed in Federal Specification TT-P-34a was followed as closely as possible in the preparation, curing, and evaluation of these test specimens. The weight of each coat of paint and total paint applied on each panel, the weight loss resulting from the fire-retardancy test in the fire-test cabinet, before leaching, after leaching, and after weathering, are reported in the table shown on the following page.

Evaluation of two- and three-coat paint films, prepared from single- and multiple-paint formulations, for their fire retardancy before leaching, after leaching, and after weathering, according to Federal Specification TT-P-34a*]

Paint formulations	Paint on panel (gms.)				Panel weight loss after fire-retardancy evaluation (gms)		
	Coats			Total	Non-leached	Leached	Weathered [1]
	1st	2nd	3rd				
A	20.2	15.1		35.3	4.1		
	18.8	13.6		32.4		5.5	
	19.3	15.7	13.9	48.9	5.4		
	18.6	15.0	12.8	46.4		5.9	
	18.2	13.8	14.0	40.0			5.4
A	18.6	14.4	[2]11.8	44.8	6.4		
	18.0	14.3	[2]12.0	44.3		5.7	
	19.0	14.5	[2]12.0	45.5			4.6
B	20.0	14.9		34.9	5.0		
	17.6	13.5		31.1		4.1	
	17.7	13.7	14.8	46.2	4.5		
	16.6	14.2	13.0	43.8		4.0	
	19.1	13.6	14.0	46.7			4.6
B	18.6	14.7	[2]9.9	43.2	6.1		
	18.3	15.1	[2]9.8	43.2		4.3	
	19.6	13.9	[2]9.4	42.9			4.2
C	18.1	13.8		31.9	6.5		
	15.8	13.0		38.8		6.5	
	16.8	12.5	13.0	42.3	5.7		
	15.5	13.2	12.9	41.6		6.5	
D	16.2	13.8		30.0	8.7		
	16.0	14.1		30.1		10.5	
	16.0	14.4	13.2	43.6	6.8		
	16.3	13.4	14.2	43.9		9.0	

*Details of experimental procedure are given in Federal Specification TT-P-34a. These are average results of duplicates or triplicates.

[1] The test specimens were exposed to natural weathering at 45° facing South for about five months at New Orleans, Louisiana.

[2] A weather-resistant color-stable paint formulation.

Organo-Phosphorus Amides

R.R. Hindersinn and G. Witschard (U.S. Patent 3,449,161; June 10, 1969; assigned to Hooker Chemical Corporation) describe a fire-retardant intumescent paint composition comprising an organic halogenated coating vehicle and a water-insoluble organphosphorus amide of a formula selected from the group consisting of:

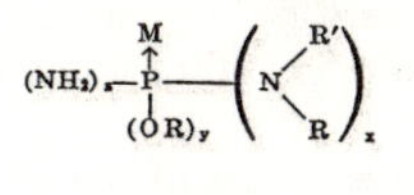

and

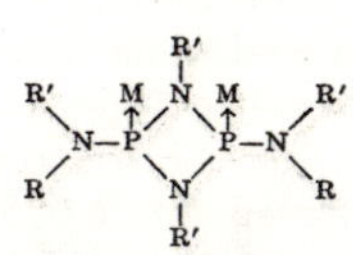

where M is selected from the group consisting of oxygen and sulfur, R is selected from the group consisting of aryl and alkyl, R' is selected from the group consisting of hydrogen, aryl and alkyl, x is 1 to 3, y is 0 to 2 and z is 0 to 1, the sum of x, y and z being equal to 3.

Compositions of the process retain their fire-retardant and intumescent properties even after prolonged leaching, scrubbing or washing with water or normal soap and water compositions. The intumescent fire-retardant compositions are comprised of a halogenated organic coating vehicle and an organo-phosphorus amide of defined structure or a nonhalogenated organic coating vehicle, an organic halide and an organo-phosphorus amide of the structure defined above. The organic coating vehicle used is preferably an oil based composition such as long, medium or short oil modified alkyd resins, acrylic resins, bodied oils, polyesters, phenolic modified oils, polyvinyl acetates, urethane modified oils and epoxidized oils.

Compounds included within the organo-phosphorus amides of this process are: Thiophosphoryltrianilide, dimeric thiophosphoric anil anilide, dimeric phosphoric anil imide, diphenylphosphoramidate, triethyltriamidothionophosphates and diphenylphosphorylamidate,

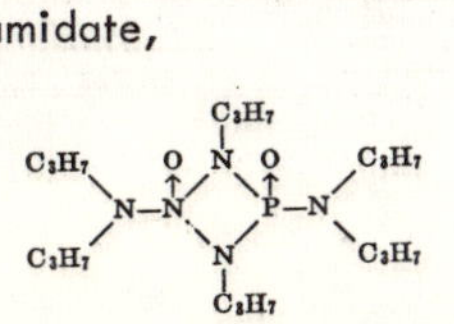

and

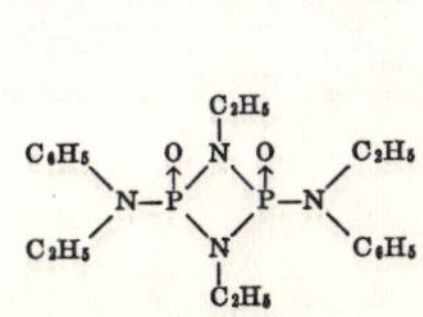

The most preferred compound is phosphoryltrianilide (PTA) having the structure:

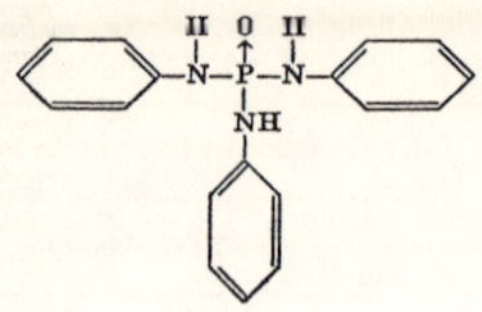

and dimeric phosphoric anil anilide (DIP)

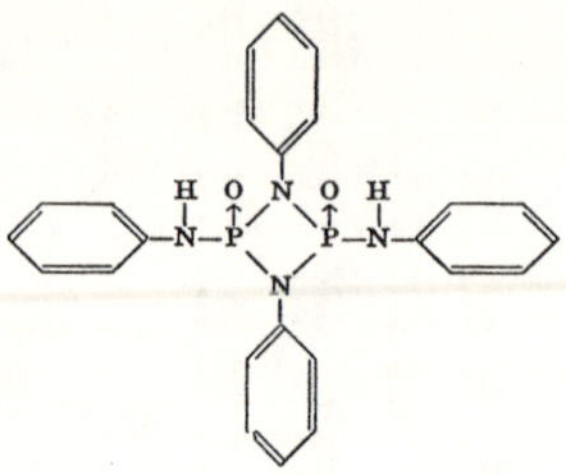

The following examples illustrate a series of five paint compositions formulated in accordance with the process compared with three commercially available fire-retardant intrumescent paints. The five paint compositions of this process are further compared with control paint formulations using the same coating vehicles but without the organo-phosphorus amides. The paints in these examples were prepared in a conventional manner using a suitable paint mill. The milling was accomplished by mixing sufficient coating vehicle with the entire pigment portion and drier system to make a paste of the correct consistency for grinding. After forming the paste, the compositions were ground and thinned to the desired viscosity by adding solvent and the remainder of the ingredients.

The paints were applied to 1/4 by 8 by 12 inch yellow poplar panels by brushing. Two coats were applied to each test panel, obtaining as uniform coverage as possible. Backs, edges and sides were also well coated with the same material. A drying period of approximately one week at room temperature was allowed between coats while an additional period of two weeks was allowed after the second coat. Prior to the burning test, the panels were placed in an oven at about 120°F. for a period of 40 hours. The flame tests were carried out on the test panels using the methods specified in American Society for Testing Materials Procedure D-1360-58, "Fire Retardancy of Paints", method C.

In addition to the flame tests, at least four replicate panels of the formulations were leached in water for 24 hours at 50°C. in the manner described in Federal Specifications TT-P-0026B. These test panels were dried and subsequently tested for fire retardancy in the same manner as the panels which were not leached. In addition to the leaching test described, paint compositions prepared in accordance with the process were also tested for serviceability in the manner described by the National Paint, Varnish and Lacquer Association. In this test, the paints were applied to ground glass plates and, after suitable aging as described for the flammability tests, the paint films were subjected to 1,000 cycles which equal 2,000 strokes of a one pound brush saturated with a 0.5% soap solution (Ivory Flakes). Any break showing in the film on the completion of the 1,000 cycles is sufficient to fail the test. All of the paint compositions of this process as illustrated in Examples 1 through 5 readily passed this test.

EXAMPLE 1.—PAINT FORMULATION A

Ingredients	Parts by Weight
Pigments:	
TiO$_2$, rutile	114.6
Titanium-calcium pigment (30–70%)	30.9
Zinc oxide	28.7
Fillers:	
Diatomaceous earth	20.1
Calcium carbonate	8.6
Halogenated Organic 1,4,7,10-dimethanocycloocta-1,2,3,4,7,8,9,10,13,13,14,14-dodecachloro-1,4,4a,5,6,6a,7,10,10a,11,12,12a-dodecahydro(1,2,5,6)dibenzene	280.2
Organo-Phosphorus Amide, Phosphorus trianilide	90.0
Coating Vehicle, Soya modified phthalic alkyl (50% nonvolatiles)	571.3
Flow Agent, Silicone resin (60% nonvolatiles)	10.0
Driers:	
Cobalt naphthenate 6%	1.4
Lead naphthenate 24%	5.7
Solvent, Mineral spirits	145.0
Anti-skinning Agent	0.15

EXAMPLE 2.—PAINT FORMULATION B

Ingredients	Parts by Weight
Pigments:	
TiO$_2$, rutile	114.6
Titanium-calcium pigment (30–70%)	30.9
Zinc oxide	28.7
Fillers:	
Diatomaceous earth	20.1
Calcium carbonate	8.6
Organic Halide, 1,4,7,10-dimethanocycloocta-1,2,3,4,7,8,9,10,13,13,14,14-dodecachloro-1,4,4a,5,6,6a,7,10,10a,11,12,12a-dodecahydro(1,2,5,6)dibenzene	90.0
Organo-Phosphorus Amide, Phosphoryl trianilide	280.2
Coating Vehicle, Soya modified phthalic alkyd (50% N.V.)	571.3
Flow Agent, Silicone resin (60% N.V.)	10.0
Driers:	
Cobalt naphthenate 6%	1.4
Lead naphthenate 24%	5.7
Solvent, Mineral spirits	146.0
Anti-skinning Agent	0.15

EXAMPLE 3.—PAINT FORMULATION C

Ingredients	Parts by Weight
Pigments:	
TiO₂, rutile	114.6
Titanium-calcium pigment (30–70%)	30.9
Zinc oxide	28.7
Fillers:	
Diatomaceous earth	20.1
Calcium carbonate	8.6
Organic Halide, Chlorinated biphenyl (68% chlorine)	185.2
Organo-Phosphorus Amide, Phosphoryl trianilide	185.2
Coating Vehicle, Soya modified phthalic alkyl (50% N.V.)	571.3
Flow Agent, Silicone resin (60% N.V.)	10.0
Driers:	
Cobalt naphthenate 6%	1.4
Lead naphthenate 24%	5.7
Solvent, Mineral spirits	108
Anti-skinning Agent	0.15

EXAMPLE 4.—PAINT FORMULATION D

Ingredients	Parts by Weight
Pigments:	
TiO₂, rutile	114.6
Titanium-calcium pigment (30–70%)	30.9
Zinc oxide	28.7
Fillers:	
Diatomaceous earth	20.1
Calcium carbonate	8.6
Organic Halide, Chlorinated paraffin (70% Cl)	185.2
Organo-Phosphorus Amide, Phosphoryl trianilide	185.0
Coating Vehicle, Soya modified phthalic alkyd (50% N.V.)	571.3
Flow Agent, Silicone resin (60% N.V.)	10.0
Driers:	
Cobalt naphthenate 6%	1.4
Lead naphthenate 24%	5.7
Solvent, Mineral spirits	154
Anti-skinning Agent	0.15

EXAMPLE 5.—PAINT FORMULATION E

Ingredients	Parts by Weight
Pigment, Titanium-calcium pigment (30–70%)	302.0
Fillers:	
Diatomaceous earth	86.0
Calcium carbonate	86.0
Organo-Phosphorus Amide, Phosphoryl trianilide	90.0
Carbonifics, Tripentaerythritol	45.0
Spumifics, Melamine	135.0
Halogenated Coating Vehicle, Chlorinated alkyd (60% N.V.) meeting Military specification MIL–R–2147 (Ships)	476.2
Driers:	
Cobalt naphthenate 6%	10.0
Lead naphthenate 24%	3.4
Solvent, Mineral spirits	122.9

CONTROL #1

Ingredients	Parts by Weight
Pigments:	
TiO₂, rutile	122.8
Titanium-calcium pigment (30–70%)	368.8
Zinc oxide	30.7
Fillers:	
Diatomaceous earth	21.5
Calcium carbonate	70.6
Coating Vehicle, Soya modified phthalic alkyd (50% N.V.)	612.2
Driers:	
Cobalt naphthenate 6%	1.54
Lead naphthenate 24%	6.14
Solvent, Mineral spirits	227

CONTROL #2

Ingredients	Parts by Weight
Pigments:	
TiO₂, rutile	122.8
Titanium-calcium pigment (30–70%)	368.8
Zinc oxide	30.7
Fillers:	
Diatomaceous earth	21.5
Calcium carbonate	70.6
Halogenated Coating Vehicle, Chlorinated alkyd (60% N.V.) meeting military specification MIL–R–2147 (Ships)	510.0
Driers:	
Cobalt napthenate 6%	1.54
Lead naphthenate 24%	6.14
Solvent, Mineral spirits	230

In Example 5 and Control #2, the chlorinated alkyd was a commercially available alkyd resin produced by the reaction of chlorendic acid, phthalic anhydride, soya acids and glycerol. The table below shows the results obtained, the figures being the average of at least four panels.

	Film Thickness, mils	ASTM D 1360-58 Cabinet Flame Test			
		Weight-Loss (grams)	Intumescence (inches)	Char Area (square inches)	Char Volume (cubic inches)
Example Number:					
Control #1	10	17.5	None	37.63	5.30
Example 1	11	7.7	0.63	10.43	0.96
Example 1 after 24 hour leach	11	6.9	0.73	9.40	0.82
Example 2	10	6.5	0.92	17.90	2.21
Example 3	10	9.9	0.61	18.65	2.53
Example 4	10	5.5	0.83	13.12	1.19
Control #2	10	12.5	None	24.81	4.86
Example 5	10	9.9	0.48	14.40	2.39
Example 5 after 24 hour leach	10	8.9	0.43	11.88	2.04
Commercial #1	9.2	10.6	None	27.31	5.99
Commercial #1 after 24 hour leach	9.2	7.6	None	24.15	5.32
Commercial #2	10	2.0	0.53	11.19	0.53
Commercial #2 after 24 hour leach	10	21.2	None	49.86	10.47
Commercial #3	14.7	3.5	0.91	3.50	0.13
Commercial #3 after 24 hour leach	14.7	24.8	None	46.02	10.00

The commercial paints illustrated above were as follows: Commercial #1 was a fire-retardant paint of the soya alkyd type containing 10% antimony oxide based on the nonvolatile content. Commercial #2 was a fire-retardant paint containing an organo-phosphate fire retardant. Commercial #3 was a latex paint containing monoammonium phosphate, dicyaniamide, pentaerythritol and starch. A comparative analysis between the control examples and the examples of this process illustrate the pronounced fire retardant effect produced by incorporating the organo-phosphorus amides into the paint composition. Comparing the commercial paint composition with the examples of the process illustrates the pronounced effect of fire retardant characteristics and intumescence properties both prior to leaching and after leaching. The commercial products tested tended to lose their fire retardance characteristics after leaching with water.

Tris(Haloalkyl)Phosphate, Nitro-Containing Polyhydric Alkanol

C.C. Clark, H.M. Schroeder and L.R. Garrison (U.S. Patent 3,375,220; March 26, 1968; assigned to Textron Inc.) describe a polyurethane fire-retardant coating composition consisting essentially of the reaction product of aromatic hydrocarbon diisocyanate and nitropolyhydric alkanol; the composition contains tris(haloalkyl)phosphate to improve the fire-retardant properties. Representative nitro-substituted polyhydric alkanols include tris(hydroxymethyl) nitromethane, 2-nitro-2-methyl 1,3-propanediol, 2-nitro-2-ethyl-1,3-propanediol, etc.

Among the polyol esters of phosphorus which are useful in the process are the phosphates and phosphonates of the formula:

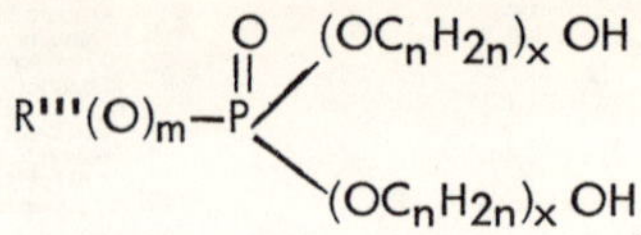

where m is 0 or 1, n is 2 to 6, especially 2 to 3, x is 1 to 10, preferably 1 to 3, and R''' is a hydrocarbon radical, for instance of 1 to 8 carbon atoms, and may be aliphatic, including cycloaliphatic, aromatic or mixed aromatic-aliphatic, and is preferably lower alkyl.

Illustrative of these materials are di(trioxypropylene)-methyl phosphonate

$$CH_3P(O)[(OCH_2\underset{\displaystyle CH_3}{\underset{|}{C}}H)_3OH]_2$$

di[oxypropylene]butyl phosphate

$$C_4H_9OP(O)[(OCH_2\underset{\displaystyle CH_3}{\underset{|}{C}}H)_xOH]_2 \quad \text{where x is 1 or 2}$$

di[trioxyethylene]hydroxymethyl phosphonate

$$HOCH_2\text{—}P(O)[(OCH_2CH_2)_3OH]_2$$

The fire-retardant and intumescent characteristics of the reaction product depend to a great extent on the use of the nitro-substituted polyhydric alkanol and the presence of a haloalkyl phosphate, preferably a tris(haloalkyl)phosphate. The halogen component of the phosphate has an atomic number of 17 or 35, that is chlorine or bromine, and preferably the alkyl is monohalo-substituted. Chlorine is the preferred halogen. The alkyl, including cycloalkyl, groups are preferably lower alkyl radicals and in general, alkyls of a lesser number of carbon atoms, e.g., 2 to 4 are preferred, as they have lower fuel and smoke potential. One or more of the alkyl radicals may be substituted with materials other than halogens and may have an interrupted carbon chain. The amount of haloalkyl phosphate in the composition is sufficient to have the desired effect and is often about 20 to 60, preferably about 25 to 50 weight percent on the basis of the diisocyanate, nitroalkanol, dissimilar polyol and the haloalkyl phosphate. If desired, the phosphate can be added to the composition before or after reaction of the diisocyanate, nitroalkanol and dissimilar polyol, but preferably the addition is after the reaction.

The fire-retardant and intumescent characteristics of the haloalkyl phosphate-containing coating composition of the process may be enhanced by including an effective amount of a lower alkylene diamine tetraacetic acid. The alkylene members may have 2 to 4 carbon atoms, and it is preferred to use ethylene diamine tetraacetic acid. In general, the coating composition will contain about 20 to 60 weight percent of the alkylene diamine tetraacetic acid, preferably the amount is about 25 to 50 weight percent, on the basis of the polymer composed of diisocyanate, nitroalkanol and dissimilar polyol, and the alkylene diamine tetraacetic acid. The coating compositions are normally applied to substrates as films of less than 10 mils thickness and can contain other additives to impart special properties such as plasticizers, etc. Also, the substrate for the coating may be specially treated materials, including flammable members impregnated with fire-resistant chemicals or coated with a sealant. The following examples illustrate the process.

Example 1: 2,4-tolylene diisocyanate (870 grams, 10 equivalents) was charged to a reaction flask provided with a reflux condenser, water trap, thermometer, glass mantel heater and stirrer. Cellosolve acetate (ethyl ether of ethylene glycol acetate) (374 grams) and 374 grams xylene were also added. Two hundred fifty two grams (5 equivalents) of tris(hydroxymethyl)nitromethane were added to the mixture in the flask while stirring. The temperature during 2 hours reached 60°C. when it was reduced to 45°C. and kept at this point overnight (about 16 hours). At the end of this time some of the tris (hydroxymethyl) nitromethane was not dissolved. Temperature was raised to 60°C. and after 7 1/2 hours at this point the percent NCO was 23.92. After another hour 1.1 grams of dibutyl tin dilaurate was added. The temperature rose in 45 seconds to 132°C. The mixture was then cooled to 45°C. Temperature was kept at 40°C. for another 16 hours when the percent NCO was 12.78. Temperature was then raised to and kept at 60°C. for another 5 hours, when the percent

NCO was 12.02, the Gardner viscosity J, the total solids 63.7 and the free TDI 9.72. This vehicle dried to a mar free condition in 3 hours on glass. One-third of the weight of the solids of this vehicle as tris(2-chloroethyl) phosphate was dissolved in it. The fire retardancy and intumescence of this vehicle were greatly improved over those of a vehicle made by a similar procedure but containing tris(hydroxymethyl)aminomethane in place of tris(hydroxymethyl)nitromethane.

Example 2: Fifty grams (1 equivalent of tris(hydroxymethyl)nitromethane, 74.6 grams (1 equivalent) of 2-nitro-2-ethyl-1,3-propanediol, 127 grams of Cellosolve acetate (ethylene glycol ethyl ether acetate) and 127 grams of xylene were charged to a reaction flask provided with a thermometer, stirrer, inlet for nitrogen, reflux condenser, water trap and electric mantel heater. Water was removed from this mixture by heating until it ceased collecting in the water trap. This required one hour with the temperature at 90°C. The temperature was then allowed to drop to 50°C. when 348 grams (4 equivalents) of tolylene diisocyanate (80% 2,4-isomer, 20% 2,6-isomer) were added along with 0.5 gram of dibutyl tin dilaurate. The temperature was maintained at about 80°C. for 65 minutes and the mixture was cooled to 27°C. At this point the percent NCO was 11.44, the Gardner viscosity V, the Gardner color 6 and total solids 65%. Tris(2-chloroethyl)phosphate was mixed with this vehicle at the rate of one part to two parts of the vehicle solids. This product showed good intumescence when burned as a coating on wood.

Example 3: Two hundred twenty-five grams of 55.5% pentachlorophenoxy glycerol ether in xylene or 0.67 equivalent, 46 grams (0.68 equivalent) of 2-nitro-2-methyl-1,3-propane diol, 53 grams of ethylene glycol ethyl ether acetate and 125 grams of xylene were charged to a flask which had a thermometer, stirrer, reflux condenser, water trap, glass mantel heater and nitrogen inlet. This mixture was azeotroped for two hours in order to remove water and dry the system. After all of the water had been removed, the batch was cooled to room temperature and 235 grams (2.7 equivalents) of tolylene diisocyanate were added. The temperature rose rapidly to 88°C. due to heat liberated by the reaction to form polyurethane. The temperature of the batch was held at about 100°C. for 50 minutes, then allowed to fall to room temperature. This vehicle analyzed 74.9 percent nonvalatile, 5.11 percent NCO, 1.03 free tolylene diisocyanate, greater than Z10 viscosity and 12 minus Gardner color.

The vehicle was diluted to 60 nonvolatile before evaluating it. One form of it included one percent N-cocomorpholine based on vehicle solids along with 1/2 part by weight of tris(2-chloroethyl)phosphate per part of vehicle solids. This vehicle dried in 2 hours on a Morest chart to a film possessing good flexibility and a Sward hardness of 46. When 1/2 part by weight of ethylene diamine tetraacetic acid per part of vehicle solids was stirred into the latter vehicle and the resultant vehicle coated on a Morest chart, it dried to a fairly flexible film in 3 hours. This film had a Sward hardness of 20. The film containing the phosphate ester showed fair fire retardance and intumescence. The film containing the phosphate ester and the ethylene diamine tetraacetic acid showed good fire retardance and intumescence.

AMINOPLASTS AND INORGANIC PHOSPHATES

Aminoplast and Ammonium Phosphate Formulations

A process described by H.A. Scholz and E.E. Saville (U.S. Patent 2,566,964; September 4, 1951; assigned to U.S. Gypsum Company) provides a paint which may be first mixed in dry powdered form, the mixture being then mixed with water immediately prior to use, and it may be applied by any suitable means such as by a brush or spray gun. The binders of the fire-retardant paint are suitable water-soluble aminoplast resins, for example, either a urea formaldehyde resin, a urea thiourea formaldehyde resin, or a melamine-aldehyde resin.

In order that the paints may produce the proper intumescent effect, they must contain certain pigment-fillers of two different kinds, one selected from one group and another selected from another group; that is to say, one or more of a pigment filler in each group must be used together with one or more from the other group. The groups are as follows:

Group 1	Group II
Barytes	Silica
Gypsum	Clay
Lithopone	Pyrophyllite
Zinc sulfide	Talc
Titanium dioxide	Mica
Titanium dioxide — barium sulfate	
Titanium dioxide — calcium sulfate	

It has also been found advantageous to add a small quantity of gum arabic to the composition to give it the desired slip, freedom from curtaining, and also to make it more easily spread by means of a brush, as ordinarily the aqueous solutions of aminoplast resins are rather short and do not brush out well. The addition of the gum arabic overcomes this defect. The following examples illustrate the process. All of the paints made in accordance with the nine examples seen in the table on the following page, are described in terms of the dry ingredients which may be formed into a powdered mixture

	Examples								
	1. %	2. %	3. %	4. %	5. %	6. %	7. %	8. %	9. %
A zinc sulfide pigment (50%) (Such as "Cryptone BA 40")	21.60	21.60	—	—	—	—	—	23.7	23.5
A lithopone (such as "Albalith #362")	—	—	23.10	—	—	—	—	—	—
Barytes (Barium Sulfate)	—	—	—	24.25	—	—	—	—	—
Zinc sulfide (98%) (such as Cryptone ZS 180")	—	—	—	—	22.50	—	—	—	—
Ground calcium sulfate dihydrate	—	—	—	—	—	14.80	—	—	—
A titanium oxide pigment (such as "Ti Pure R 110")	—	—	—	—	—	—	22.50	—	—
Silica powdered so as to pass a 325 mesh screen	—	—	—	—	—	—	—	—	15.0
Talc	—	—	15.00	—	—	—	—	—	—
Ground pyrophyllite (such as "Pyrax B")	5.00	5.00	—	5.45	5.62	6.40	5.62	5.5	—
Clay	8.35	8.35	—	9.21	9.51	10.78	9.51	9.3	—
Mica	4.20	4.20	4.50	4.46	4.60	5.23	4.60	4.5	4.5
Monoammonium phosphate	50.00	30.00	27.40	26.19	27.80	31.60	27.80	27.2	27.2
Diammonium phosphate	—	20.00	5.50	5.45	5.62	6.40	5.62	5.5	5.5
Gum arabic	.65	.65	2.30	2.28	2.35	2.67	2.35	2.3	2.3
A urea formaldehyde resin (such as "Uformite 500")	10.20	10.20	22.20	22.71	22.00	22.12	22.00	—	—
A melamine formaldehyde resin (such as "Melmac 421")	—	—	—	—	—	—	—	22.0	22.0

that is then mixed with water to a spreading consistency and then applied in any suitable manner to the article or surface to be painted. Such powdered mixture should, however, be used promptly, as it has been found that there is a tendency for the gradual insolubilization of the aminoplast resins as the result of the action of the ammonium phosphates. It is therefore advisable to utilize such powdered mixtures within a day or so after they have been mixed. However, dry paint mixtures can be made up by keeping the aminoplast resins and the ammonium phosphates apart. Thus all of the dry ingredients, with the exception of the aminoplast resins, may be mixed into a powdered composition, while the aminoplast resin, also in powdered or at least granular form, may be packed in a separate container or package, with directions to the user to either blend the aminoplast resin with the rest of the ingredients, and then to stir the resulting powder into a suitable quantity of water, or else to dissolve the resin in the water and then stir in the other powdered ingredients.

The behavior of the paint upon application to a given surface, such, for example, as insulation board, is as follows: As the paint dries, the aminoplast resin will become insoluble in water and will permanently hold the pigments in place. When a flame is applied to the dried paint film, the gases resulting from the thermal decomposition of the aminoplast resin and the ammonium phosphates will be trapped by the viscous dough-like mass formed by the melting of the ingredients, and which have been stiffened to a heavy consistency by the mixture of fillers, so that there will be formed a thick stiff frothy sponge-like mass which is entirely uncombustible and which will preserve its insulating properties, even though exposed to a naked flame for a period of from 5 to 10 minutes.

G. Jones, W. Juda and S. Soll (U.S. Patent 2,523,626; September 26, 1950; assigned to Albi Manufacturing Co., Inc.) also describe a number of urea-formaldehyde intumescent coatings. The following examples illustrate the compositions.

Example 1:

	Parts
Paraformaldehyde	12
Monoammonium phosphate	67
Urea	15
Starch	8

In addition the following may be added if desired:

	Parts
Titanium-barium pigment	4
Boric acid	2
Tragacanth gum	1

The paraformaldehyde is kept in a separate container. The other ingredients are well mixed and ground. Before use, the paraformaldehyde is mixed with the other ingredients and 50 parts of water are added. A liquid composition of good painting qualities is thus obtained. It is allowed to stand for one half hour before use.

Example 2:

	Parts
Monoammonium phosphate	70
Urea	16
Starch	8
Tragacanth gum	0.8
Pigment	4
Water	30
(This mixture is not reactive; it is stable on storage)	
Aqueous formaldehyde[1] (37%)	32.5

[1] To be kept separately.

The aqueous formaldehyde is added before use to the other ingredients with stirring. The resulting composition will heat slowly. It is allowed to cool before it is applied. The following example is a composition containing a preformed urea-formaldehyde resin.

Example 3:

Diammonium phosphate	65 grams
Dextrin	25 grams
Urea-formaldehyde resin (Beetle No. 250V.)	32 grams
Water	120 cc.

When a preformed urea-formaldehyde resin is used, it is often advantageous to store it separately, either in aqueous solution or in water-soluble solid form.

These authors also describe similar systems using proteins, such as glue, casein and gelatine as the nonresinous carbonaceous substance in place of the carbohydrate. (U.S. Patent 2,452,054; October 28, 1948).

Urea-Formaldehyde and Ammonium Orthophosphates

G.L. Cunningham (U.S. Patent 2,679,489; May 25, 1954; assigned to Waterloo Corporation) describes a composition which comprises urea and formaldehyde together with a relatively large amount of a material, such as the ammonium or zinc compounds of orthophosphates and borates and a small amount of very finely divided inert, inorganic, solid to stabilize and strengthen the foam structure when formed. In making up a coating composition or paint the urea and formaldehyde are kept separated until they are to be used. They are then mixed in an aqueous or water medium together with the ammonium phosphate or other flame-retarding reagent.

To provide the thickness or body to the aqueous solution thus attained a thickening agent is provided which is not salted out of the solution by the presence of the large amount of ammonium phosphate or other salt. Such salting out would cause the composition to become "ropy" and unworkable as a paint. Examples of such thickening agents are carboxyl methyl cellulose, methyl cellulose and polyvinyl alcohol. These thickening agents may be used in small quantities and generally not over 5% by weight of the composition. Since the composition is thickened initially, the formation of the resin is not required to give it the requisite thickness or body and the formation of the resin may be retarded. For this purpose a small amount of hexamethylene tetramine is added to this composition. The presence of this reagent retards the formation of the resin and so slows down the reaction rate that the composition may be retained in a usuable form for upwards of fifty to sixty hours. Temperature conditions, therefore, do not have any substantial or major influence on the thickening. The paint may be used shortly after the elements of the composition are mixed and remain usuable for two days or more regardless of weather conditions.

After the composition has been applied to a surface it will be set (due to the evaporation of water from the coating), the setting taking about two or three hours depending upon the temperature and humidity. Although the coatings at this stage will not be very hard and tough a chemical reaction sets in so that within twelve to twenty-four hours after the composition has been applied the coating will be hard and tough. The composition when applied as a coating has good leveling properties and after being completely set and hard has elastic and tough characteristics rather than being brittle. Examples of the composition are as follows.

Example 1: 37.5 parts monoammonium dihydrogen phosphate, 5 parts titanium dioxide (water dispersion type), and 10 parts urea are mixed together and ground to 100 mesh, or better to 200 to 325 mesh. (The use of fine powdered mixtures increases the rate at which the orthophosphate goes into solution and enhances the smoothness of the resulting coating.) To this mixture is added an aqueous solution made of the following ingredients:

10 parts 40% commercial formaldehyde solution, 20 parts water, 4 parts hexamethylene tetramine and 0.23 part carboxy methyl cellulose, high viscosity type. The major part of the ammonium dihydrogen orthophosphate will quickly go into solution. With stirring at intervals during the first few minutes after being mixed the mixture is a very smooth, creamy slurry. It can be used immediately as a coating material to provide fire-retarding properties. After being mixed for about twelve hours the paint has thickened a little, and is now very "oily" and is at the best time for use. The titantium oxide is an insoluble pigment but it will have little or no tendency to settle out since even from the beginning of the mixture the paint has had body and viscosity. The thickening agents also tend to keep the titanium dioxide in suspension. In any case a slight agitation of the paint will restore the titanium to a state of complete suspension. When this composition is applied, for example, to wood by dipping, spraying or painting, it will go on very smoothly giving a uniform surface which is hard, tough and elastic. It has good levelling qualities. It will set in about two or three hours to a firm surface which permits another coating to be added. The coating will not peel, flake or dust off in use.

Example 2: 37.5 parts monoammonium dihydrogen orthophosphate, 5 parts titanium dioxide (water dispersion type), 10 parts urea and 0.23 part carboxy methyl cellulose are mixed together and ground to 200 to 325 mesh. A separate mixture is made which is composed of 4.0 parts paraformaldehyde and 4.0 parts hexamethylene tetramine. To prepare the paint the two powder mixtures are mixed together and 30 parts of water added. This will give a smooth creamy slurry which is immediately suitable for use as a fire-retardant coating. However, it will improve slightly within the next twelve hours. The mixture will remain usuable for at least sixty hours.

Water-Soluble Melamine-Formaldehyde

P. Simon (U.S. Patent 3,021,293; February 13, 1962; assigned to The Sherwin-Williams Company) describes a fire-retardant coating composition which comprises an ammonium phosphate (either mono or diammonium phosphate or mixtures thereof may be selected) in finely divided form; a nonalkylated, water-soluble aqueously condensed, melamine-aldehyde condensate product in a dried powdered form; pentaerythritol in a dried powdered form; prime pigments including the generally used pigments of high opacity, for example, titanium dioxide, zinc sulfide, lithopone, etc., along with any inert extender. It is preferred to use inert extender pigments along with the prime pigment which are mixed together and wetted with a solution of chlorinated rubber in a volatile organic hydrocarbon solvent. Normally the chlorinated rubber useful and preferred contains more than 20% but not more than about 70% by weight of chlorine and the chlorinated rubber solids constitute from about 5% to not more than about 12% of the total liquid paint composition.

The various dry stocks and the liquid halogenated rubber solution in a volatile hydrocarbon solvent is mixed in conventional paint mixing equipment and subjected to a zone of high shear to prepare a very fine heterogeneous dispersion of the various powdered dry stocks. Dispersion may be obtained by employment of a ball mill, by means of a roller mill, a Hy-R speed mill, colloid mill or bead mill. The melamine-formaldehyde component is particularly critical as to composition. The mol ratio of melamine to aldehyde in the condensation product should fall within the range of from 1:1 to 1:4 and preferably within the range of from 1:2 to 1:3. The aldehyde-amine condensation is conducted in an aqueous mixture in the absence of alkylating alcohol. In other words, the amine-aldehyde condensation product is nonalkylated. It is most general and usual with most amine aldehyde resins intended for coating composition end use that they be alkylated. Further, the amine-aldehyde condensation is soluble in water and is insoluble in the usual organic solvents for paints.

Pentaerythritol appears to be unique in its particular ability to form the residual carbonaceous matrix essential to the entrapment of the foam which forms upon exposure of a deposited paint film to flame conditions. It has been found that pentaerythritol is specific in its usefulness and that other materials which may be a source of carbonaceous foam lack coherence when compared on an otherwise equivalent basis with pentaerythritol. Using other materials suggested in the prior art, for example, starch, the intumescence is lacking or negligible and the flame spread and charred volume of test panels is so much increased that the coating compositions fail to pass standard fire-retardant tests. To establish the quality of the fire-retardant paints of this process, the products were tested in accordance with the standard test method for fire hazard classification for building materials, ULC Subject, C-723, NFPA-255, ASTME 84-50T, a method which is also very similar to that used by the Underwriters Laboratories in the U.S.

Among the data collected as a result of the test are the "flame spread", the "fuel contribution value" and a figure known as the "classification factor". By way of comparison, the standard asbestos cement panel has standard "flame spread" of about 2 1/2 feet after a 10 minute time of burning test. The "flame spread" value is defined as the flame propagation in feet divided by the total potential length of the flame travel, which in this test was 19 1/2 feet. Utilizing a material essentially as described in the illustrative example which follows, the "flame spread" of the paint product of the process was of the order of 1 foot or the "flame spread" value of the order of 5.1 feet. In a test of ten common fire-retardant paint products, "flame spread" under the same testing procedures were compared. The lowest "flame spread" value recorded in a test of any of the ten products was of the order of 15 to 20. All others of the products tested were greater than 20 and some paint materials which were submitted as classifiable as fire-retardant paints had values of the order of 50 to 70. In determining the factor called "fuel contribution value", time is plotted as an abscissa and temperature as an ordinate, measuring the temperature developed at the end of the combustion chamber. The "fuel contribution value" using the standard asbestos board in the test box is of the order of 1880, while an unpainted red oak panel used as standard is of the order of 6920. The products of this process under exactly the same test conditions were found to have a "fuel contribution value" of the order of 2035. A somewhat arbitrary "classification factor" is used by the Underwriters Laboratories which reflects the nonburning quality of panels subjected to flame tests. The standard asbestos cement panel has a "classification factor" of zero. The products of this process in two separate tests were found to have a "classification factor" of 3.2. In prior art materials subjected to a similar rating, the lowest "classification factor" based on a similar test procedure was of the order of 15. In a test for scrubbability, a glass panel was coated with two coats of the paint product of the subsequent example, the first coat of 0.0015 and the second coat at 0.003 inch in film thickness. After three weeks of drying, the paint was subjected to 5,000 cycles using the Gardner straight line washability machine Model 105A. 1% Ivory soap was used as the test solution. Only a slight polishing was noted. After 500 cycles of washing, the paint film intumesced upon exposure to flame conditions, maintaining practically completely its original intumescence depth. The following examples illustrate the process.

Example 1:

	Parts
Melamine (1 mol)	126
Formalin (3 mols)	243

The pH of the mixture of these substances is adjusted to 6.3 as determined with bromthymol blue indicator (0.04%) in conjunction with a Hellige color disk, bromthymol blue D (No. 170-D). The mixture is heated up to about 80°C. over about one-half hour. The temperature is then allowed to drop to about 60°C. requiring about 10 minutes, and it is maintained at that temperature for one-half hour measured from the time that the mixture originally attained 60°C. After this time, the syrup is adjusted to a pH of about 8 to 8.5 and it is filtered to remove any insoluble impurities which may be present, producing a clear hydrophilic resin syrup. The syrup is now concentrated by heating under a vacuum of about 9 to 10 inches of mercury absolute pressure, the vacuum being maintained at such a point that the temperature of the resinous syrup will be high enough to keep it fluid, i.e., above the melting point. This temperature will vary from about 50°C. to a temperature slightly above 100°C., e.g., 105° to 115°C. as water is removed. When the resinous syrup has been substantially dehydrated, it is removed from the concentrating kettle and cooled to produce a substantially clear, solid resin which is ground to a fine white powder.

Example 2: Example 1 is repeated except that a mixture of 126 parts of melamine (1 molecular proportion) and 162 parts (2 molecular proportions) of 37% aqueous formaldehyde solution is reacted in the presence of 177 parts of added water. Due to the lower proportion of formaldehyde as compared with Example 1, the outlet gas temperature in the spray drier may be higher than that employed in Example 1, for example, 80°C. without causing melting of the powdered product. The product has an amorphous structure and possesses unlimited solubility in water at ordinary temperatures.

Example 3:

200 parts mono ammonium phosphate
80 parts melamine-formaldehyde condensate powder (of Ex. 1)
85 parts pentaerythritol
100 parts titanium dioxide pigment
25 parts asbestine pigment (inert)
500 parts 20% solids content, 125 cps. grade chlorinated rubber (about 67% chlorine content by weight), 80% volatile (xylol) high flash naphtha hydrocarbon solvent
30 parts hydrogenated rosin mono ester of glycerine
5 parts lecithin

The above components are weighed out in a pony mixer, the solids thoroughly dispersed in the liquids by coarse mixing and the contents of the mixer are passed over a three roll grinding mill of conventional design. The product is further capable of being reduced in viscosity with hydrocarbon solvents, illustratively xylene. As the hydrocarbon solvents are dissipated in the air upon application of the paint in a film, they do not materially interfere with the fire-retardant quality of the dry paint product.

Insoluble Metal Metaphosphates

J.H. Basto, M.M. Capp, C.R. Davison and G.H. Garbaden (U.S. Patent 3,037,951; June 5, 1962; assigned to Benjamin Moore & Co.) describe a fire-retardant composition which comprises a water-insoluble metal metaphosphate salt, and a water-insoluble polyol and preferably, a water-insoluble aminoplast. Compounds such as ammonium phosphate, diammonium phosphate, phosphoric acid, ammonium sulfate, sulfamic acid, ammonium sulfamate and ammonium bromide have been utilized in various fire-retardant compositions, but these materials have the disadvantage of leaching out or washing out because of their water solubility. This makes for a much reduced fire-retardancy when these compositions are subjected to the action of water. To overcome these faults this process employs the water-insoluble metal metaphosphates. Examples of such salts are insoluble potassium metaphosphate, insoluble sodium metaphosphate, zinc metaphosphate and calcium metaphosphate.

Water-insoluble polyhydric compounds may be used in the compositions of this process. The polypentaerythritols are preferred. As specific examples of polypentaerythritols may be mentioned dipentaerythritol and tripentaerythritol. The aminoplasts or amino aldehyde condensation products employed in the compositions are water-insoluble and include a wide variety of materials which are prepared from aldehydes, organic compounds containing at least one and preferably two amino groups having at least two and preferably three replaceable hydrogen atoms. The preferred aminoplasts for use in the compositions are the water-insoluble ureaformaldehyde condensation products. The molecular ratio of formaldehyde to urea may be between 1 and 2.5 mols formaldehyde to 1 mol of urea. Such condensation products are used as a dry solid. The use of the alkylated modifications of such resins is likewise contemplated, particularly the butylated modifications of conventional urea-formaldehyde and melamine formaldehyde resins. The butylated urea-formaldehydes are water-insoluble and can be used as a solution without drying, if necessary.

The coatings formed from these compositions possess the property of intumescing when exposed to an open flame or to heat and do not appreciably lose this property even when the coating is washed with water periodically or exposed to high humidity conditions over the normal life of the coating. Compositions which contain a combination of the aminoplast, water-insoluble metal metaphosphate and the polypentaerythritol provide greater intumescence than is possible with a composition containing only the metaphosphate and polypentaerythritols. The coating is highly resistant to leaching with water. For example, a substrate coated with a composition according to the process was submerged for 30 minutes in water and then dried overnight at room temperature. This coated substrate was then exposed to the direct flame of a Bunson Burner for 5 minutes, and the coating intumesced to form a puffed char of substantially the same volume as was obtained with the same coating on a substrate that was not submerged in water. The following examples illustrate the process.

Example 1:

	Parts
Insoluble postassium metaphosphate	40
Dipentaerythritol	20
Aminoplast resin powder	15
Titanium dioxide	17
Antimony oxide	10
Halogenated alkyd resin	30
Thinner	59
Cobalt drier	1

The aminoplast resin powder is prepared by reacting 1 mol of formaldehyde and 1 mol of urea. The halogenated alkyd resin is prepared from dehydrated castor oil (representing 25% of total solids), glycerine and hexachloroendomethylenetetrahydrophthalic acid in accordance with the following procedure. The glycerine and castor oil are introduced into a reaction kettle and heated to 212°F. and the acid is added. The temperature is then raised slowly, in order to avoid foaming, to 370°F., at which temperature the oil, glycerine and acid are reacted in the presence of carbon dioxide to an acid number of 5 or less. The resulting alkyd resin is reduced with mineral spirits to a non-volatile of 56%.

The thinner consists of 80% mineral spirits (aliphatic hydrocarbon solvent) and 20% aromatic solvent. The drier is cobalt naphthenate having a metal content of 6%. The paint is prepared in conventional paint making equipment. The ingredients are mixed in a paste mixer or other suitable equipment to produce a paste consistency. When a smooth resulting coating is desired, the resultant paste from the paste mixer or other equipment may be ground on a roller mill, a ball mill, high speed stone mill, or other types of mills. The above paint is applied by conventional means to an unpainted or painted surface, and, upon drying produces a decorative and serviceable film of acceptable appearance. This paint is comparable in stability to commercial paints.

Example 2:

	Parts
Insoluble sodium metaphosphate	40
Tripentaerythritol	20
Titanium dioxide	10
Antimony oxide	20
Halogenated alkyd resin	30
Thinner	47

(continued)

(continued)

	Parts
Cobalt drier	1
Lead drier	2

The halogenated alkyd resin, thinner and cobalt drier in the above example are the same as those described in connection with Example 1, while the lead drier is lead naphthenate having a metal content of 24%.

Example 3:

	Parts
Calcium metaphosphate	40
Dipentaerythritol	20
Aminoplast	20
Titanium dioxide	15
Antimony oxide	15
Halogenated alkyd resin	30
Thinner	53
Manganese drier	2

The aminoplast is prepared by reacting 2 mols of formaldehyde with 1 mol of urea. The halogenated alkyd resin and thinner of Example 3 are the same as described in connection with Example 1, while the manganese drier is manganese naphthenate having a metal content of 6%.

Urea-Formaldehyde-Acrolein Resin

A process described by W. Schulenburg (U.S. Patent 2,881,088; April 7, 1959; assigned to Alim Corporation) provides a fire protective or retardant composition essentially composed of a resin component, an organic nonresinous carbon forming component, such as a carbohydrate or proteinaceous material, and an inorganic foam forming agent. It has been found that fire protective or retardant compositions of the three component type indicated above can be substantially improved by replacing all or a part of the urea-formaldehyde resin component by a urea-formaldehyde-acrolein resin. When exposed to a flame, the presence of the urea-formaldehyde-acrolein resin causes the formation of a particularly fine-pored carbon layer which has a nonpeeling impervious surface and combustible articles coated with such composition are protected against combustion upon exposure to a flame for longer periods than articles protected with compositions in which the resin component is only a ureaformaldehyde resin.

Furthermore, the presence of the urea-formaldehyde-acrolein resin in the compositions improves the flexibility of the coatings produced and furthermore increases their resistance to water. Preferably, the resin component of the fire retardant or protective composition is composed of a mixture of 40 to 60% of urea-formaldehyde resin and 60 to 40% of urea-formaldehyde-acrolein resin. Suitable urea-formaldehyde-acrolein resins for the composition are, for example, shown in German Patents Nos. 748,842, 802,846 and 833,707. A particularly suitable urea-formaldehyde-acrolein resin is one in which the molar proportions of urea, formaldehyde and acrolein is 1:2.0 to 2.5:0.12. Preferably, the molar ratio of urea to total formaldehyde is about 1:2.6. The urea formaldehyde resins employed in the compositions preferably are those produced by condensing a source of formaldehyde in a ratio between 1 and 2.5 mols of formaldehyde per mol of urea.

As the foam forming inorganic component, the following can, for example, be employed with advantage in the compositions according to the process: mono-ammonium phosphate, diammonium phosphate, phosphoric acid, ammonium sulfate, sulfamic acid, ammonium sulfamate, ammonium bromide, sodium tungstate, sodium borate and boric acid. The third essential component of the composition which is of carbohydrate or proteinaceous nature, is such that it is converted to carbon in the composition rather than volatilized at elevated temperatures. Suitable carbohydrates and proteinaceous materials include starch, modified starches, dextrin, gum tragacanth, flour (which contains both carbohydrate and protein), solid polyhydric alcohols, such as mannitol or glucosamine, glue, casein and gelatin. The following examples illustrate the fire protecting agents according to the process.

Example 1: 200 grams of urea-formaldehyde resin and 170 grams of urea-formaldehyde-acrolein resin were ground to a particle size corresponding to a screen of 1,600 meshes per $cm.^2$ and then mixed with 484 grams of monoammonium phosphate, 20 grams of calcium phosphate, 76 grams of soluble starch and 50 grams of asbestos powder, all of the same particle size. This mixture upon dilution with water provided an excellent protective coating after painting upon the surface to be protected and drying. The urea-formaldehyde resin employed was produced as follows: 35 kg. of disodium phosphate were dissolved in 812 kg. of 30% formaldehyde (243 kg. formaldehyde) with adjustment of the pH to 8.2 and 243 kg. of urea were added to such solution. The resulting mixture was stirred for about 4 to 5 hours while maintaining a temperature up to about 40°C. and then allowed to stand over night. The resulting thick mass was then heated to about 55°C. and the water content reduced to 5 to 10% under a vacuum of 80 to 100 mm. Hg. As the water evaporates, the temperature is is gradually raised to about 120°C. The resulting semi-dried product was then put into aluminum pans which were placed in a drying oven maintained at 103°C. under a vacuum of 100 mm. Hg for 24 hours. The originally tough mass rose similarly to a cake dough and a coarsely porous frangible product resulted which was easy to remove from the pans and grind.

The urea-formaldehyde-acrolein employed was produced as follows: 17.76 kg. of urea, 4.43 liters of distilled water and 0.30 kg. of glacial acetic acid were mixed and heated to 50°C. Upon reaching this temperature, 2 kg. of acrolein were added over a period of 5 to 6 minutes at a rate that the temperature of the mixture did not exceed 58°C. After the acrolein had been added, the mixture was heated to 70°C. for 30 minutes and then cooled to 45°C. The product obtained was a yellowish thinly viscous precondensate. At the same time, 55.43 kg. of 30% formaldehyde, 4.35 kg. of hexamethylene tetramine and 15.55 liters of dilstilled water were mixed in another vessel and the resulting solution mixed with the precondensate after it had been cooled to 45°C. The resulting mixture was heated in 10 minutes to 90°C. and maintained at this temperature for 30 minutes and then cooled to 20°C. and the pH adjusted with NaOH to 7.2. The resulting urea-formaldehyde-acrolein resin solution was then dried under vacuum.

Example 2: 200 grams of urea-formaldehyde resin and 150 grams of urea-formaldehyde-acrolein produced as described in Example 1 were ground with 480 grams of monoammonium phosphate, 24 grams of boric acid, 75 grams of pea flour, 25 grams of gum tragacanth and 50 grams of asbestos flour to a particle size corresponding to a screen of 1,600 meshes per $cm.^2$. In order to compare the properties of fire protective agents according to the process containing a urea-formaldehyde-acrolein resin with a fire protective agent containing only a urea-formaldehyde resin as the resin component, the following tests were carried out. In each instance the fire protective agent was an aqueous suspension of solids composed of 67% of monoammonium phosphate, 8% starch and 25% resin. The urea-formaldehyde resin and the urea-formaldehyde-acrolein resin employed were prepared according to Example 1. In the first test the resin component was the urea-formaldehyde resin, in the second test the resin component was a mixture of equal parts of urea-formaldehyde resin and urea-formaldehyde-acrolein resin. Plywood sheets of 5 mm. thickness were coated with these mixtures to give a coating with 300 grams of solid substance per square meter. After drying, the plywood sheets were held horizontally with the coated side lowermost and each was exposed to the flame of a Bunsen burner under exactly the same conditions. The temperature on the upper side of each sheet was measured with a thermoelectric element. The tests were discontinued when the sheet in question burned through. The results obtained are given in the following table.

Time, Minutes	100% Urea-Formaldehyde-Resin, °C.	50% Urea-Formaldehyde-Resin-50% Urea-Formaldehyde-Acrolein Resin, °C.	100% Urea-Formaldehyde-Acrolein Resin, °C.
5	78	80	82
10	118	120	122
15	145	140	142
20	154	147	145
25	160	148	148
30	169	151	150
35	190	154	153
40	[1]210	154	155
45	—	159	157
50	—	162	165
55	—	164	179
60	—	169	[1]205
65	—	[1]204	—

[1] Burnt through.

HALOGEN-CONTAINING FORMULATIONS

Chloroprene Binder, Diammonium Phosphate

R.P. Stock (U.S. Patent 2,650,206; August 25, 1953; assigned to The B.F. Goodrich Company) describes a coating composition which comprises an elastomeric material which is incapable of supporting combustion, a carbonific or carbon yielding substance, a spumific or foam producing substance capable of foaming the carbonific into a spongy porous frothy carbonaceous mass when the composition is subjected to heat above a temperature critical to the stability of the composition, and a volatile liquid medium. Although any rubberlike material which does not support combustion is suitable as the elastomeric ingredient in a heat-retardant coating composition, it is preferable that the elastomeric material be a rubbery polymer of a monomeric material consisting predominantly of chloroprene (neoprene rubbers), or rubbery silicones. It is necessary that the elastomeric material which does not support combustion comprise from 20 to 60% by weight of the total weight of the coating composition to impart the desired properties to the coating composition. The following examples illustrate the compositions of the process.

Example 1:

Material	Parts by Weight
Rubbery chloroprene polymer (GR-M)	100.0
Methyl ethyl ketone	120.0
Diammonium phosphate	200.0
Dextrin	200.0
Zinc oxide	4.0
Potassium arsenate	25.0
Total	649.00

The rubbery chloroprene polymer, diammonium phosphate, dextrin, zinc oxide, potassium arsenate and methyl ethyl ketone are mixed together and the resulting composition is agitated until a homogeneous mixture is obtained.

Example 2:

Material	Parts by Weight
Rubbery chloroprene polymer (GR-M)	100.0
Solvent-naphtha	100.0
Sugar	100.0
Ammonium bromide	300.0
Zinc oxide	4.0
Para-phenylphenol	35.0
Total	639.0

The rubbery chloroprene polymer is dissolved in the solvent naphtha, the sugar, ammonium bromide, zinc oxide and para-phenylphenol are formed into a paste using solvent naphtha as a moistening agent, the paste is mixed into the naphtha containing the rubbery chloroprene polymer, and the resulting composition is agitated for about 30 minutes.

Example 3:

Material	Parts by Weight
Aqueous dispersion of rubbery chloroprene polymer (50% total solids)	200.0
Diammonium phosphate	150.0
Urea-formaldehyde resin	125.0
Dextrin	100.0
Zinc oxide	5.0
Potassium arsenate	20.0
Water	75.0
Total	675.0

The neoprene latex, diammonium phosphate, urea-formaldehyde resin, dextrin, zinc oxide and potassium arsenate are mixed into the water and the composition is agitated for about 15 minutes.

An article coated with a heat-retardant composition will withstand extreme heat without deterioration of the article itself. Because of the fire-retardant properties of the composition its use decreases the rapidity with which a fire would otherwise spread and permits the fire to be readily extinguished.

Halogen-Containing Film Former, Dicyandiamide

In a process described by B.B. Kaplan (U.S. Patent 2,984,640; May 16, 1961; assigned to Albi Manufacturing Company) the acid requirement for the formation of a suitable puffable carbonaceous mass from the carbonaceous materials used is supplied by the inclusion of an organic film-former which releases acid at elevated temperature. Since this acid requirement can thus be satisfied, the spumific can be chosen without regard to any necessity for acid-supplying requirements, thus permitting the use of a relatively water-insoluble, non-leaching organic spumific. The composition consists of a film forming, halogen-containing organic material which liberates a hydrogen halide upon exposure to elevated temperatures, a relatively low molecular weight polyhydroxy alcohol of low water-solubility as the carbonaceous material, and a water-insoluble organic spumific such as dicyandiamide or melamine pyrophosphate. The entire composition may thus be organic in nature and contain solely materials of low water-solubility. Among the halogen-containing, organic, film forming materials which may be employed are chlorinated rubber; rubber hydrochloride; polymers and copolymers of vinyl chloride, particularly those containing 5% by weight or more of combined vinyl chloride, such as polyvinyl chloride itself and copolymers of vinyl chloride with vinyl acetate or other vinyl esters; polymers and copolymers of vinylidene chloride, particularly those containing 5% or more of combined vinylidene chloride, including copolymers of vinylidene chloride, including copolymers of vinylidene chloride with vinyl chloride or with vinyl acetate or other vinyl esters; polychloroprene; and like materials which have relatively high molecular weights, are capable of forming films, are thermally unstable, and which liberate hydrogen chloride when subjected to elevated temperatures of the order of 125° to 400°C. or even higher.

The relative proportions of halogen-containing, film forming material, polyhydroxy alcohol, and water-insoluble organic spumific may vary over a wide range. In general, the ratio of film-forming material to alcohol may be from 3:1 to 1:3 by weight, although best results are obtained in the range from 1:1.5 to 1:2 by weight. The ratio of film forming material to spumific may likewise vary from 3:1 to 1:3 by weight, but is preferably within the range from 1:1 to 1:2 by weight. The following specific examples illustrate the process.

Example 1: To 100 parts by weight of a 15% solution in xylene of chlorinated rubber (67% chlorine) there were added 23 parts by weight of tripentaerythritol and 23 parts by weight of dicyandiamide. A steel panel coated with a lead oxide primer was painted (brushed) with the composition (1 gallon/100 sq. ft.) and dried at room temperature, then used as a closure for a muffle furnace heated at 1650°F. with the painted face exposed directly to the heat. A thermocouple mounted at the face of the panel remote from the furnace showed that it required 12 minutes' exposure for the temperature at the face of the panel to reach 1000°F. When an identical primer-caoted panel, free from the coating of the process was exposed under the same conditions, only 1 1/2 minutes' exposure was required for the face of the metal to reach a temperature of 1000°F. The exposed face of the treated metal exhibited a blackened, charred or carbonized appearance with the characteristic puff resulting from intumescence.

In similar fashion wood, fabric and other combustible materials coated or impregnated with the composition and allowed to dry exhibited remarkable resistance to heat and to burning. Similar results are obtained when melamine pyrophosphate is substututed for dicyandiamide.

Example 2: A composition was prepared by mixing together 100 parts by weight of a 15% by weight solution of chlorinated rubber (67% chlorine) in xylene, 9 parts of dipentaerythritol, and 30 parts of dicyandiamide. The composition was readily applied to surfaces to be protected by brushing or dipping and when dry the coating was highly effective in retarding the transmission of heat and the propagation of flame. As in the case of Example 1, melamine pyrophosphate could be substituted for dicyandiamide.

Isobutylene Copolymers with Vermiculite

W.P. Ellis and I.J. Steltz (U.S. Patent 3,090,764; May 21, 1963; assigned to Benjamin Foster Company) describe a fire-retardant mastic coating which contains as the intumescing agent a mineral material which is substantially insoluble in water so that the mastic is not rendered ineffective through exposure to water or weather. The insoluble mineral intumescing agent is one which contains combined water and which is capable of expanding or bloating to many times its original volume when heated. It is raw unexpanded vermiculite ore, the term "vermiculite" designating a group of minerals which have a platy, laminated structure resembling that of mica. It differs from mica principally by the fact that when heated it exfoliates to a remarkable extent in a direction at right angles to the laminae. The chemical composition of vermiculite much resembles that of biotitephlogopite mica which has had removal of much of its alkalis and taken up water. Analyses of commercial vermiculite show a rather wide variation in relative amounts of the chief constituents, but it would appear that the main factor which affects the degree of exfoliation on heating is combined water, which varies between about 4.3 and 13.8%.

The composition of this process consists essentially of a fire resistive mastic which does not burn or support combustion and in which water-insoluble expandable mineral ore is mixed. The fire resistive mastic component of this process may be prepared by a special procedure described in U.S. Patent 2,861,967. The mastic which is preferred is composed of several ingredients, namely (1) an isobutylene polymer, or co-polymer, and (2) a chemical condensation product which acts inter alia as a plasticizer for the polymer or co-polymer. Another ingredient, designed to contribute the fire-retardant properties of the composition is (3) a resinous fire-retardant chlorinated compound, which may be resinous chlorinated paraffin wax or resinous chlorinated biphenyl. The contained chlorine content of this ingredient will be above 50% and preferably in the range of 60 to 70%. A filler (4) which will usually be incorporated into the composition preferably consists of a mixture of fibrous and nonfibrous mineral fillers, and to facilitate application of the composition it will desirably also include a proportion of a gelling agent. Another ingredient (5) is the intumescing agent, pulverized unexpanded vermiculite ore. The vehicle includes from 20 to 40% by weight of an isobutylene polymer or co-polymer, from 20 to 60% by weight of plasticizer, and from 10 to 35% by weight of a resinous chlorinated compound, preferably chlorinated paraffin or chlorinated biphenyl. The fillers will be 10 to 50% and the intumescing agent 20 to 65% by weight of the coating composition.

A typical applied coating of the composition described and containing 60% by volume of solvent, when applied 0.25 inch thick will shrink by evaporation of the volatile solvent to an approximate 0.10 inch. When this dried coating is exposed to a fire such as that from a gas flame, the resulting intumescence reaches about 0.75 inch thickness. This swelling amounts to a volume increase of 7 1/2 times. Nominally dry, unexpanded American vermiculite when heated above 700°F. will expand 12 to 18 times in volume, so it is seen that the mastic binder has a restricting effect on the vermiculite expansion. For this reason, it is preferable to use a mineral intumescing agent which has a dry expansion of at least 10 times.

The compositions of this process are preferably prepared by dissolving the isobutylene co-polymer in the solvent, adding the chlorinated resin and the plasticizer, and then adding the filler component and the intumescing component. The fire-retardant coating composition may be applied to the work in any convenient way, for example, either by spraying or troweling. Typical formulas found to be satisfactory in all of the foregoing respects are as follows.

Example 1:

	Percent by Weight
Isobutylene-styrene copolymer	9.82
Plasticizer	16.28
Filler	16.84
Gelling agent	2.50
Intumescent agent	44.74
Resinous chlorinated compound	9.82
	100.00

Example 2:

	Percent by Weight
Isobutylene-isoprene copolymer	7.12
Plasticizer	11.76
Asbestine	10.00
Asbestos fiber	3.00
Antimony oxide	1.00
10/20 mesh vermiculite ore	60.00
Resinous chlorinated paraffin	7.12
	100.00

Example 3:

	Percent by Weight
Linear isobutylene polymer	13.34
Plasticizer	22.07
Asbestine	12.50
Pulverized mica	10.00
Asbestos fiber	1.50
6/20 mesh vermiculite ore	25.00
Gelling agent	2.26
Resinous chlorinated paraffin	13.33
	100.00

The compositions described above produce fire retardant coatings which have an optimum combination of toughness, flexibility, and hardness under most weather conditions, and which intumesce strongly when exposed to fire.

Hexachlorocyclopentadiene Condensation Products

A process described by N.D. Blair and J.A. Pawlak (U.S. Patent 3,347,684; October 17, 1967; assigned to Hooker Chemical Corporation) provides a decorative fire-retardant paint which is comprised of a volatile vehicle component, a nonvolatile vehicle component, an inert pigment and a fire-retarding pigment derived from hexachlorocyclopentadiene, which may be selected from the group consisting of (a) bis(pentachlorocyclopentadienyl), (b) 2,2',3,3',4,4',5,5',7,7,7',7'-dodecachloro-1,1',2,2',5,5',6,6'-octahydro-2,2',5,5'-methanobiphenyl, (c) compounds which conform to the following formula:

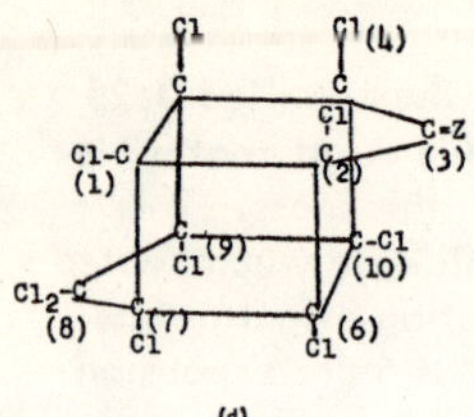

where Z = $(OH)-NRC(=Y)-R_1$ or $(OH)[O]^-\ [NH(R_4)_3]^+$

(d) 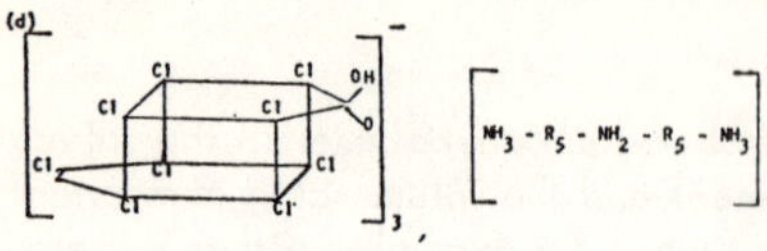

(e) 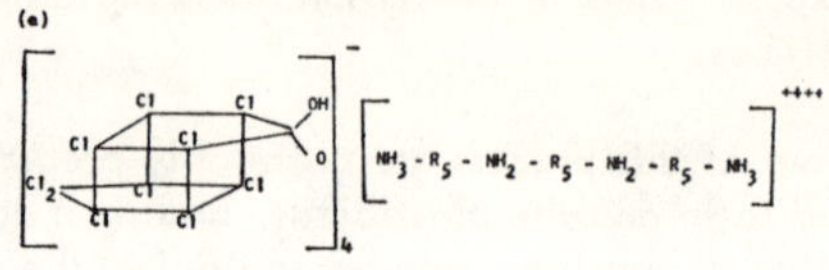

(f) 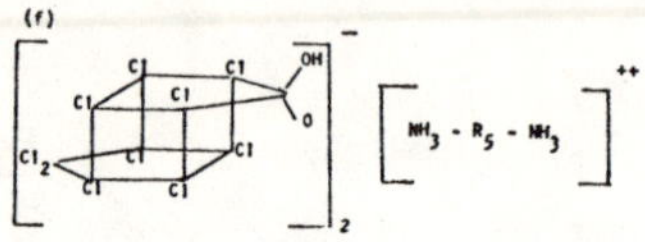

and (g) mixtures thereof. R is selected from the group consisting of hydrogen, alkyl, substituted alkyl, aryl and substituted aryl. R_1 is selected from the group consisting of hydrogen, alkyl, substituted alkyl, aryl, substituted aryl, amino, substituted amino, alkoxy, and aryl alkoxy. Y is selected from the group consisting of O and S, R_2 and R_3 are selected from the group consisting of H, alkyl and substituted alkyl.

The fire-retardant pigments of this process are condensation products of hexachlorocyclopentadiene where 2 mols of hexachlorocyclopentadiene are condensed either with one mol of a diene or with itself and are subsequently reacted with suitable reactants to yield the desired side chain structure. The former type of compound may be properly characterized as a Diels-Alder adduct product. The preparation of 2,2',3,3',4,4',5,5',7,7,7',7'-dodecachloro-1,1',2,2',5,5',6,6'-octahydro-2,2',5,5',6,6'-octahydro-2,2',5,5'-methanobiphenyl is described in U.S. Patent 2,606,910. For convenience this compound will be referred to by its empirical formula ($C_{14}H_6Cl_{12}$). $C_{14}H_6Cl_{12}$ is prepared by condensing two mols of hexachlorocyclopentadiene with one mol of 1,3-butadiene at a temperature below 200°C. and then recrystallizing the reaction product from isopropyl alcohol. This compound melts at about 232°C. The compound bis(pentachlorocyclopentadienyl) (short title $C_{10}Cl_{10}$) is prepared by the dechlorination of hexachlorocyclopentadiene. Methods for the preparation of $C_{10}Cl_{10}$ are described in U.S. Patent 2,732,409.

Since the balance of the fire-retardant pigments of this process may be considered as derivatives of the previously described pentacyclo structure, nomenclature difficulties may be avoided by referring to this parent structure (see below) as $C_{10}Cl_{10}$:

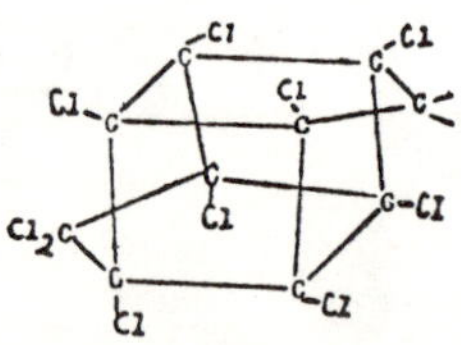

$C_{10}H_2O_3SCl_{12}$ is prepared by reacting hexachlorocyclopentadiene with chlorosulfonic acid at a temperature within the range of 30° to 170°C. as described in U.S. Patent 2,516,404. The exact structure of this compound has not been stated in the literature. The reaction product is characterized by a melting point of about 146°C. The ketone known as decachlorooctahydro-1,3,4-metheno-2H-cyclobuta [cd]-pentalene-2-one (short title $C_{10}Cl_{10}O$), whose preparation is described in U.S. Patent 2,616,928 (Re. 24,435) and 2,616,825 (Re. 24,749), may be hydrolyzed with water to yield the monohydrate where the Z term is $(OH)_2$, to form the dihydrate where the Z term is $(OH)_2 \cdot H_2O$ or the polyhydrate where the Z term is $(OH)_2 \cdot xH_2O$. The amide fire-retarding pigments of this process where the Z term of the generic formula is:

$$(OH)-NRC(=Y)-R_1$$

are prepared by heating the above described $C_{10}H_2O_3SCl_{12}$ in an appropriate solvent with at least one molar equivalent of an amide or thioamide of the structure $NHRC(=Y)R_1$.

The amides as defined before, include the following compounds:

$C_{10}Cl_{10}(OH)-NH-CHO$

$C_{10}Cl_{10}(OH)-NH-COCH_3$

$C_{10}Cl_{10}(OH)-NH-COCH_2CH_3$

$C_{10}Cl_{10}(OH)-NH-COCH_2Cl$

$C_{10}Cl_{10}(OH)-NH-COC_{17}H_{35}$

$C_{10}Cl_{10}(OH)-NH-COCHOHCH_3$

$C_{10}Cl_{10}(OH)-NH-COCH_2CH_2CH_2CH_2CONH[C_{10}Cl_{10}(OH)]$

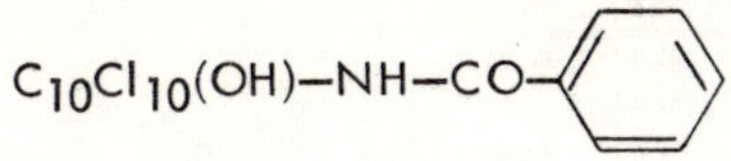

$C_{10}Cl_{10}(OH)-NHC(=S)-CH_3$

$C_{10}Cl_{10}(OH)-NHC(=S)-NH_2$

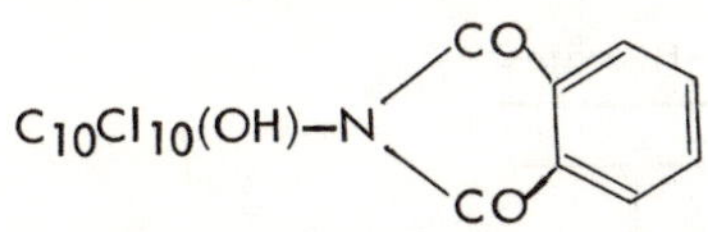

Amine salts of the type exemplified by the following formulas

$$[C_{10}Cl_{10}(OH)O]^{-}[NH(R_4)_3]^{+}$$
$$[C_{10}Cl_{10}(OH)O]_2^{-}[NH_3-R_5-NH_3]^{++}$$
$$[C_{10}Cl_{10}(OH)O]_3^{-}[NH_3-R_5-NH_2-R_3-NH_3]^{+++}$$

and

$$[C_{10}Cl_{10}(OH)O]_4^{-}[NH_3-R_5-NH_2-R_5-NH_2-R_5-NH_3]^{++++}$$

are suitable fire-retarding pigments in the paints of this process. These amine salts are readily prepared by reacting, e.g., by reflux, the $C_{10}Cl_{10}$ hydrates with the desired amine in a suitable solvent for a period of time at moderate temperatures. Suitable amines include both aromatic and aliphatic amines whether primary, secondary or tertiary.

The fire-retardant pigments of this process offer advantages over fire-retarding agents employed in paint by the prior art, namely, insensitivity to water and second these materials are true substitutes for previously used pigments in paints and are not merely additional additives to be overcome by the tinting power of the balance of the paint pigments. While in some cases it may be desirable to substitute the fire-retarding pigments of this process, for the entire pigment volume of the paint under consideration, generally only from about 20 to about 80% of the pigment will be substituted by retardant pigments of this process. Especially useful and decorative paints are prepared when the fire-retarding pigments of this process are substituted for the so-called filler or low covering power pigments of the paint. Very effective decorative fire-retarding paints are prepared when from about 30 to about 70% of the total pigment is substituted, emphasis being placed on the substitution of the fire-retarding pigments for the filler pigment of the paint so that the tint strength remains unchanged.

The paints of this process may be prepared in a conventional manner on suitable jar and paint mills by mixing sufficient vehicle with the entire pigment portion and dryer system to make a paste of the correct consistency for grinding, after which the desired dispersion or grind is then thinned down with the remainder of the vehicle and the formulation. The paints were applied to 1/4 x 8 x 12 inch yellow poplar panels by brushing. Two coats were applied to each test panel, for a total coverage of approximately 125 sq. ft. per gallon, obtaining as unifom coverage as possible. Backs, edges, and sides were also well coated with the same material. A drying period of approximately 1 week at room temperature was allowed between coats while an additional period of two weeks was allowed after the second coat. Prior to the burning test, the panels were placed in an oven at about 120°F. for a period of 40 hours. Other panels were aged exposed to sun and weather for 1 year after painting and were then tested. The flame tests were carried out on the test panels using the methods specified in American Society for Testing Materials Procedure D-1360-58, "Fire Retardancy of Paints", Method C. The following examples illustrate the process. The use of varying amounts of $C_{14}H_6Cl_{12}$ in an exterior linseed oil paint is illustrated in the table below, comprising Examples 1 to 4.

Fire Retardant Paints with $C_{14}H_6Cl_{12}$

Example	1	2	3	4
Pigment:				
Titanium Dioxide	200	200	200	200
Zinc Oxide, Leaded, 35%	401	330	260	220
Magnesium Silicate	251	180	110	70
$C_{14}H_6Cl_{12}$	--------	142	282	362
Vehicle:				
Heat Bodied Linseed Oil (Z-2 Viscosity)	142.4	172.6	201.6	218.1
Raw Linseed Oil	230.3	320.5	374.4	405.2
Mineral Spirits	153.8	143.8	130	81.3
24% Lead Naphthenate, 24%	6.2	8.2	9.6	10.4
6% Manganese Naphthenate, 6%	1.7	2.5	2.9	3.1
Total	1,386.4	1,529.6	1,570.5	1,570.1
Pigment Volume Concentration	32.0	32.0	32.0	32.0
$C_{14}H_6Cl_{12}$, percent (volume) of pigment	--------	32.1	54.6	64.7
Viscosity, Brookfield, poises, #4 spindle, 60 r.p.m.	12.3	13.0	11.5	13.2
Flame Test:				
Average Weight Loss, gram	12.9	7.7	5.9	4.7
Average Char Area, square inches	20.7	14.8	14.5	18.9
Average Char Volume, cubic inches	3.25	1.97	1.13	0.57

Exterior linseed oil paints containing $C_{10}Cl_{10}O$-hydrates are illustrated in the table below.

Fire-Retardant Paints with $C_{10}Cl_{10}O$-Hydrates

Example	5	6	7	8
Pigment:				
Titanium Dioxide	200	200	200	200
Zinc Oxide, Leaded, 35%	260	220	180	330
Magnesium Silicate	110	70	30	180
$C_{10}Cl_{10}O$ Hydrate (See Ex. 18)	282	362	442	142
Vehicle:				
Heat Bodied Linseed Oil (Z-2 viscosity)	203.2	220	237.2	173.4
Raw Linseed Oil	377.5	409.3	440.6	322.1
Mineral Spirits	183.7	184.2	182.5	180
Lead Naphthenate, 24%	9.7	10.5	11.3	8.3
Manganese Naphthenate, 6%	2.9	3.2	3.4	2.5
Total	1,629.0	1,679.2	1,727.0	1,538.3
Pigment Volume Concentration	32.0	32.0	32.0	32.0
Fire Retardant, percent (volume) of pigment	54.9	65.0	73.7	32.4
Viscosity, Brookfield, poises, #4 spindle, 60 r.p.m.	13.3	13.4	13.7	--------
Flame Test:				
Average Weight Loss, gram	8.0	9.2	10.0	10.0
Average Char Area, square inches	18.4	21.8	24.4	17.5
Average Char Volume, cubic inches	2.45	3.0	3.7	2.6

The amine salts of this process are described in the table below. This table shows the testing or $C_{10}Cl_{10}O$-triethanolamine salts in a decorative exterior linseed oil paint formulation. In addition to being decorative and fire-retardant these paint formulations showed satisfactory intumescent properties.

Fire-Retardant Paints with $C_{10}Cl_{10}O$-Triethanolamine Salts

Example	9	10	11	12	13
Pigment:					
Titanium Dioxide	200	200	200	200	200
Zinc Oxide, Leaded, 35%	330	330	360	220	180
Magnesium Silicate	180	180	110	70	30
$C_{10}Cl_{10}O$-Triethanol amine salt (See Ex. 19)	142	142	282	362	442
Vehicle:					
Heat Bodied Linseed Oil (Z-2 viscosity)	176.3	175.5	197.4	212.8	228.2
Raw Linseed Oil	284.6	317.0	366.8	395.3	424.0
Mineral Spirits	93.8	102.5	100.0	175.0	177.5
Lead Naphthenate, 24%	7.7	8.1	9.4	10.1	10.9
Manganese Naphthenate, 6%	2.3	2.4	2.8	3.0	3.3
Total	1,416.7	1,457.5	1,528.4	1,648.2	1,675.9
Pigment Volume Concentration	32.0	32.2	32.0	32.0	32.0
Fire Retardant, percent (vol.) of pigment	33.5	31.2	53.6	63.8	72.7
Viscosity, Brookfield, poises, #4 spindle, 60 r.p.m.	13.2	--------	12.9	12.9	13.0
Flame Test:					
Average Weight Loss, grams	8.9	7.5	6.3	9.4	8.4
Average Char area, square in	20.1	13.3	19.7	24.4	23.1
Average Char Volume, cubic in	2.50	1.64	2.5	3.2	3.54

The previous table is illustrative of results obtainable when other amine salts of the formulas discussed are employed in place of the triethanolamine salt. The table below illustrates the use of various other $C_{10}Cl_{10}OH$ based compounds in decorative fire-retardant linseed oil paints. Intumescence was also obtained from these paints.

Fire-Retardant Paints — $C_{10}Cl_{10}O$ Adducts

Example	1	14	15	16	17
Pigment:					
Titanium Dioxide	200	200	200	200	200
Zinc Oxide, Leaded, 35%	401	330	330	330	330
Magnesium Silicate	251	180	180	180	180
$C_{10}H_2O_2SCl_{12}$	-------	142	-------	-------	-------
$C_{10}Cl_{10}O$-Ethylene Diamine (See Ex. 21)	-------	-------	142	-------	-------
$C_{10}Cl_{10}$	-------	-------	-------	142	-------
$C_{10}Cl_{10}O$-ethylene glycol (Ex. 20)	-------	-------	-------	-------	142
Vehicle:					
Heat Bodied Linseed Oil (Z-2 viscosity)	142.4	139.6	170.0	170.6	175.0
Raw Linseed Oil	230.3	225.3	274.3	271.5	282.3
Mineral Spirits	158.3	75.0	81.3	92.5	87.5
Lead Naphthenate, 24%	6.2	6.1	7.4	7.4	7.6
Manganese Naphthenate, 6%	1.7	1.8	2.2	2.2	2.3
Total	1,386.4	1,299.8	1,387.2	1,396.2	1,406.7
Pigment Volume Content	32.0	32.0	32.0	32.0	32.0
Fire Retardant, percent (volume) of pigment	-------	28.4	31.1	30.3	33.0
Viscosity, Brookfield, poises #4 spindle, 60 r.p.m.	12.3	13.0	12.3	13.0	11.5
Flame Test:					
Average Weight Loss, gram	12.9	8.1	9.8	9.5	10.5
Average Char Area, square in	20.7	21.5	25.5	28.3	24.0
Average Char Volume, cubic in	3.25	3.5	3.39	3.76	3.44

As previously noted, many of the fire-retardant paints contained water-soluble fire-retarding agents which were soon leached out. The table below illustrates the water-insolubility of the fire-retarding agents of this process. Additional poplar test panels were coated in the standard method with the fire-retardant paint containing varying amounts of $C_{14}H_6Cl_{12}$. After being completely dried in the manner of the other test panels of this process, these panels were then immersed in water at approximately 72°F. for a period of 30 days. At the end of this time, the panels were removed and conditioned for 4 days at room temperature and 40 hours at 50°C. The panels were then burned in accordance with the previously described method. In each case the linseed paint formulations had a pigment vehicle content of 32%. Example 1 contained no fire-retarding agent.

Water Leached Fire-Retardant Paint

Paint of Example 1: Leached 30 days in distilled water at approx. 72°F:

Flame test:

Average weight loss, gram	9.1
Average char area, square inch	13.8
Average char volume, cubic inch	1.76

It is seen that the paints of this process retained their fire-retardancy after prolonged periods of exposure to water.

MISCELLANEOUS

Phenol-Sulfonate-Urea Condensates

W. Juda (U.S. Patent 2,616,866; November 4, 1952; assigned to Pyrotron Development Corporation) describes fire retardants which may be prepared by directly reacting phenol with sulfonic acid, and by reacting the resulting reaction product directly with urea. This reaction may be followed, preferably, by a neutralization with ammonia or other volatile base. These urea phenol-sulfonates — and especially upon being neutralized — are stable under ordinary conditions and may be prepared, stored, mixed with dispersing liquids or fluid vehicles, in suitable condition for various applications, in various proportions, without being subject to further significant changes in their compositions or properties, before use or after.

Such compositions, in which the molecular ratio of the sulfonic acid group or groups, linked to a molal quantity of the phenol, may vary from 1/2 to not over 3, and the molecular ratio of the urea which is combined with a molal quantity of the phenol in the resultant phenol sulfonate reaction product may vary from 1/3 to not over 3, are especially characterized by manifesting the property of fire-resistance when subjected to high, i.e., flame, temperatures. They also possess the additional property of undergoing, under such continued high temperature conditions, a finely dispersed intumescence. They thus form a minutely porous structure of high heat insulating properties. In acquiring this porous, fine, vesicular structure, the pore spaces formed are presumably filled with noncombustible gases which are the by-products of the heat-induced decompostion. This presents a fire-retarding surface and also a potentially heat-insulating layer of considerable thickness over the combustible material or other article to which the composition was initially applied as a thin, though preferably continuous coating. A preferred and representative group of such compounds are: monourea phenol monosulfonate, monourea phenol disulfonate, diurea phenol disulfonate, monourea phenol trisulfonate, diurea phenol trisulfonate and triurea phenol trisulfonate. In place of phenol or hydroxy benzene, C_6H_5OH, in the preparation of such compounds, cresols and other phenols may be used. The following examples illustrate the process.

Example 1: Phenol sulfonic acid was prepared by melting 510 grams of phenol, bringing the temperature to 75°C. and adding slowly to the phenol a mixture of 516 grams of 95.5% sulfuric acid and 66 grams of water. The temperature of this reaction mixture — containing at the end of the reaction mostly ortho- and para-phenol sulfonic acid — was maintained at about 100°C. for 24 hours. Six grams of urea were added to 21.5 grams of the above phenol sulfonic acid. The urea dissolved; then, within a few minutes, the mixture warmed to 50°C. and solidified. The resulting crystalline solid salt was a water-soluble urea phenol sulfonate having marked fire-retardant properties due to the formation of a substantially incombustible carbonaceous foam in the flame.

Example 2: 768 grams of phenol were melted and heated to 90°C. 760 grams of sulfuric acid (95.5%) were then added. The mixture turned slightly pink, the temperature rose to 139°C. during the exothermic reaction in which the sulfonic acid was formed. The sulfonation may be followed by a base titration; almost one-half the initial acidity had disappeared during sulfonation at the end of 30 minutes. The sulfonic acid was cooled to 34°C., at which temperature it started to solidify. 423 grams of urea were then added, with stirring. The temperature rose to 55°C. in the exothermic formation of the urea phenol sulfonate. Upon cooling the urea phenol sulfonate solidified. A yield of about 1930 grams was obtained. 1800 cc of formalin (containing about 37% formaldehyde) were added to 1874 grams of the above urea phenol sulfonate with stirring. The sulfonate dissolved gradually and the temperature rose to 86°C. while a progressive gradual thickening occurred. The temperature was maintained between 80° and 86°C. for ten minutes and the composition was then substantially neutralized to a pH between 6 and 8 with 660 cc of aqueous ammonia (containing about 28% NH_3).

Upon cooling this mixture was extended with water, using about 75 parts by weight of the above mixture and 30 parts by weight of water. Two coats of this material were applied to cardboard and wooden panels, imparting marked fire-retardant properties, as illustrated by the following comparative test. A specimen (designated as specimen B) of double-faced corrugated paper (8" x 8") was treated on one side with the composition of this example by brushing the upper surface of the corrugated paper while held horizontally. The surface was allowed to dry for twenty-four hours at 25° to 30°C. A second application of the composition was then made in a similar manner on top of the first coating. The specimen was then allowed to dry for three days in an oven at 60°C. The dry coating of specimen B weighed then 35.2 grams; the thickness of the coating being of the order of 1/16".

A similar, but untreated, or control specimen A and specimen B were then subjected in succession to the following fire tests. Each specimen was supported horizontally (specimen B with its coated surface facing downward) 6 1/2" above the top of a Bunsen burner which was adjusted to deliver a colorless flame 10 1/2 to 11 1/2" high with a tall indistinct inner cone. The time was recorded at which the upper center part of the test panels showed browning and the time at which the upper top center part started flaming. The following table gives the results of this test:

	Specimen A	Specimen B
Time of browning at top center.	15 seconds	10 minutes, 40 seconds.
Time of flaming at top center.	20 seconds	no flame at 15 minutes (test stopped).

Specimen A was a complete loss at the end of four minutes. Specimen B, at the end of 15 minutes exposure, showed a voluminous layer of carbonaceous protective mat, on the underside, which was about two inches thick at the center and which had protected the test panel from damage except for slight browning of the center part (due to heat transfer).

Microcrystalline Wax, Hexamethylene Triamine and Ammonium Phosphate

M. Wolf (U.S. Patent 3,475,199; October 28, 1969; assigned to Tri-Wall Containers, Inc.) describes fire-retardant compositions which comprise one or more intumescent agents and a binder which contains a hydrocarbon wax as its principal and preferably its sole constituent. The hydrocarbon wax is preferably a microcrystalline wax having a melting point in the range of about 130° to about 180°F. In addition to the hydrocarbon wax binder, the compositions contain a combination of intumescent agents which includes hexamethylene tetramine and a substantially nonvolatile and heat-decomposable inorganic compound. The inorganic compound is either an ammonium salt, a phosphate, or a sulfate. It has been found that a hydrocarbon wax makes an excellent binder for fire-retardant coating compositions, in spite of its high flammability. Hydrocarbon waxes, when compounded with intumescent agents, give coating compositions which form a thicker carbonaceous crust than that formed by any previously known fire-retardant compositions when exposed to flame. As a result, the fire retardancy of the compositions of this process is greater than that of previously known fire-retardant coating compositions, in spite of the high flammability of the hydrocarbon wax.

The compositions of this process are preferably in the form of aqueous dispersions. Such dispersions are frequently markedly acidic because of the high acidity of the inorganic intumescent agent. Often the pH is as low as 2. The percentage of solids in the dispersions is usually in the neighborhood of 50%, as, for example, from about 45 to about 55%. The high acidity of the dispersion necessitates the use of an acid-stable wax, as already explained. It is preferred to prepare the compositions as aqueous dispersions because they can be more readily applied in this form to the wallboard or other structural member to be coated. In order to maintain the various water-soluble constituents in suspension, a suspending agent such as sodium carboxymethyl cellulose may be used. Other suitable suspending agents include methylcellulose, hydroxyethyl cellulose, and sodium alginate.

The minimum coating weight giving effective fire-retardant properties to kraft paper liner for paperboard is about 30 pounds of solids per 1,000 square feet of surface coated. Effective fire retardancy is obtained at this coating weight, although slightly greater coating weights can be used. Substantially greater coating weights are generally avoided because the paper liner becomes unduly stiff. The compositions impart outstanding fire-resistance to inflammable cellulosic materials when compared with previously known coating compositions. However, in common with such previously known compositions, the wet rub resistance of the coatings is fairly low. Stated another way, the compositions of this process are susceptible to water. This is because of the high amounts of water-soluble compounds and especially inorganic salts, which are components of the intumescent agent.

It has been found that the web rub resistance can be substantially improved with virtually no sacrifice in intumescence, by applying a thin overcoat of wax on top of the intumescent coating composition after it has dried. Excellent results are obtained at a coating weight of about 14 pounds per 1,000 square feet, although this weight is merely illustrative and not critical. The wax constituting this overcoat may be applied in aqueous emulsion or as a hot melt, the melt being preferable. The following examples illustrate the process. In making up the compositions, ingredients are added in the order listed below.

Example 1:

Ingredient	Parts by Weight
Water	60
5% water solution of medium viscosity sodium carboxymethyl cellulose	34.5
Hexamethylene tetramine	30
Monoammonium phosphate	90
Starch	15
Clay	30
Dicyandiamide	30
Potassium aluminum sulfate	15
"Mobilcer X"	75

"Mobilcer X" is an acid stable aqueous emulsion of a micro-crystalline hydrocarbon wax having a solids content of approximately 50%.

Example 2:

Ingredient	Parts by Weight
Water	60
5% Sodium carboxymethyl cellulose solution	34.5
Hexamethylene tetramine	30
Monoammonium phosphate	90
Clay	30
Dicyandiamide	30
Potassium aluminum sulfate	15
"Mobilcer X"	75

Example 3:

Ingredient	Parts by Weight
Water	60
5% Sodium carboxymethyl cellulose solution	34.5
Hexamethylene tetramine	30
Monoammonium phosphate	90
Starch	15

(continued)

(continued)

Ingredient	Parts by Weight
Clay	30
Potassium aluminum sulfate	15
"Mobilcer X"	75

The above ingredients in each of the Examples 1 to 3 are added to a mixing vessel in the order named. All of the ingredients except the wax and the latex are added to a vessel equipped with a propellor type agitator and mixed until smooth. The resulting mixture is then transferrer to a Waring blender and agitated for about 5 minutes. The finely dispersed mixture is transferred back to the vessel having the propellor type agitator, after which the wax emulsion, and the latex when used, are added. The dispersions prepared as described above are applied to the surface of cellulose material such as Kraft paper in one or more passes through a coating machine, the coating or additional coating being dried after each application or pass. The total weight of coating after drying should be at least 30 pounds per 1,000 square feet.

Example 4: Kraft paper liner for paperboard is coated with the composition of Example 1, which is permitted to dry. Then an overcoat layer comprising "Mobilcer C" wax in 50% aqueous emulsion is applied at a coating rate of 14 pounds of wax per 1,000 square feet of paper liner. "Mobilcer C" is an aqueous emulsion containing about 44% by weight of an acid sensitive paraffin wax having a melting point of about 130°F.

Use of Glass Fibers in Coatings

B.B. Kaplan (U.S. Patent 3,284,216; November 8, 1966; assigned to Albi Manufacturing Company, Inc.) has found that the durability and effectiveness of intumescent paints can be greatly improved by incorporating a mass of glass fibers, which are generally from one-tenth inch to as much as six inches or more in length, those from one-quarter to one inch in length being preferred as giving optimum results with optimum appearance of the painted surface. The individual fibers may be dispersed in discrete form or as a randomly oriented mass throughout the mass of the composition or yarns of about the same length composed of several individual fibers may be dispersed in the composition. A variety of commercially available glass fibers may be employed; excellent results have been obtained with those having a fiber diameter of 3 to 5 microns (Superglas 1,000 and Superglas 1,500) which are readily wetted with aqueous solutions and which do not absorb liquids. Glass fibers having larger diameters, up to 100 or 200 microns or even more, may also be used.

Glass fibers precoated with a vinyl resin such as polyvinyl chloride (Fiberglas screen yarn) in order to facilitate their dispersion in the composition and to ensure that the composition be readily sprayable in conventional spray equipment may also be used. The amount of vinyl resin coating on the fibers may range upwardly of 100% by weight of the inorganic portion. The glass fibers may be used in amounts (not including the weight of any coating on the fibers) ranging up to about 30% by weight of the remainder of the composition, including solvents, without seriously impeding the spraying characteristics of the mix. The following examples illustrate the process.

Example 1: To 191 parts by weight of a solvent mixture containing 50 parts butyl lactate, 50 parts toluol, 30 parts isopropanol, 6 parts diisobutyl ketone, and 55 parts highflash naphtha, there were added 37.5 parts by weight of nitrocellulose as a film-forming binder, 37.5 parts of a mixed plasticizer for the binder containing 22 parts of tricresyl phosphate and 17.5 parts of triethyl citrate, 59.5 parts of a chlorinated material containing 30 parts of chlorinated biphenyl (Aroclor 1254) 14.5 parts of chlorinated polypropylene, and 15 parts of polyvinyl chloride. There were also stirred into the mixture 62.5 parts of pentaerythritol as the carbonaceous ingredient, 62.5 parts of a preformed particulate urea-melamine-formaldehyde resin, and 250 parts of a spumific agent consisting of melamine pyrophosphate.

Into four parts by weight of the above composition was mixed one part of glass fibers precoated with polyvinyl chloride, the fibers being approximately 0.1 to 1.0 inch in length (Fiberglas screen yarn). The composition was applied by spraying through conventional spray gun equipment (of the type used to undercoat automobile bodies) to a steel I-beam at the rate of one pound of the composition for each five pound weight of I-beam (one-half pound per square foot of I-beam surface) and allowed to dry at room temperature for one week. The beam was then subjected to a high temperature test in accordance with ASTM procedure E-119-58 (Sec. 18-22). The temperature of the furnace increased slowly from 1000°F. at the end of the first five minutes to over 1800°F. at 100 minutes. The average temperature at three locations within the beam during this period increased from about 65° to less than 980°F. at 100 minutes.

The exposed face of the metal treated in accordance with the process exhibited a blackened, charred, or carbonized appearance with the characteristic puff which results from intumescence. Similarly, wood, fabric, and other combustible materials, as well as synthetic plastic compositions, coated or impregnated with the composition and allowed to dry exhibited resistance to heat and to burning.

Example 2: A composition was prepared in accordance with the following recipe, in which the parts are by weight, by mixing the several ingredients listed on the following page.

	Parts
Hydroxyethyl cellulose	250
Aromatic hydrocarbon solvent (Amsco D)	1,560
Isopropanol	100
High-flash naphtha	400
Tricresyl phosphate	150

To 1,300 parts by weight of this mixture there were added with stirring 200 parts by weight of titanium dioxide, 150 parts by weight of pentaerythritol, 150 parts by weight of urea-melamine-formaldehyde resin, and 600 parts by weight of melamine pyrophosphate. While substantial protection against high temperatures was found to be provided by applying this composition when applied to a steel I-beam by spraying as described in Example 1 above, superior results were obtained by mixing with 4 parts by weight of this composition one part of glass fibers precoated with polyvinyl chloride, the fibers being approximately 0.1 inch in length.

Isano Oil and Polyamide Resins — Intumescent Agents

A.O. Allen, T.M. Murray and F.P. Liberti (U.S. Patent 2,754,217; July 10, 1956; assigned to Vita-Var Corporation) have found certain materials which have many of the film-forming properties required in the nonvolatile components of paint vehicles, which are insoluble in water and which have outstanding intumescent properties when exposed to flame. One of the film forming materials of this type is isano oil, also known as ongokea oil. This oil is obtained by pressing boleko nuts, which are native to Africa. This intumescent property is not known to exist in any other film forming vegetable oil. The intumescent qualities of isano oil are remarkable as compared to other vegetable oils and film forming materials because of the violence of the chemical reaction which takes place when this oil or a film containing it is subjected to heat. This reaction releases gases which cause a swelling and thickening of the film to form an improved insulation for the underlying flammable surface by preventing the underlying flammable surface from reaching the temperature necessary for combustion. It is noted that raw isano oil is used since treating or processing it in any manner using heat has been found to destroy this intumescent property.

Another class of film forming intumescent materials suitable for use in paint vehicles comprise nitrogen-containing resinous materials which decompose to liberate gaseous nitrogen or nitrogen-containing gas upon heating. Outstanding in this class are the polyamide resins, although many of the formaldehyde resins, notably melamine, urea and triazine formaldehyde resins have been found to be very satisfactory. One example of such a resin is Polyamide Resin No. 93, obtainable from General Mills, Inc., Chemical Division. Since isano oil has film forming properties, it is possible to make a paint using it as the only nonvolatile component of the vehicle. Similarly, the polyamide resins have film forming properties which make it possible to use them as the only nonvolatile component in a paint vehicle. It has also been found that a paint commercially effective as a fire-retardant paint may be made by using isano oil and polyamide resin as the only nonvolatile vehicle components.

Paint vehicles may be prepared using a mixture of a drying oil, e.g., linseed oil, and either isano oil or a resinous material selected from the class consisting of polyamide resins and amine formaldehyde resins as the only film forming materials in the vehicle. It is preferred, however, that paint vehicles prepared in accordance with the process should contain from about 10 to about 60% isano oil and from about 5 to about 25% polyamide resin. They should also contain conventional film forming material such as vegetable drying oil, preferably linseed oil, in a proportion of about 15 to 50%. The following compositions illustrate the process.

Example 1:

Pigment and other Solid Components	Weight (Lbs.)	Percent Pigment	Percent of Total Paint
Titanium Dioxide	100.0	10.30	6.28
Zinc Borate	292.0	30.04	18.31
Blanc Fixe	148.0	15.21	9.27
Lead Sulfate	168.5	17.33	10.56
Lead Carbonate	168.5	17.33	10.56
Zinc Oxide	95.0	9.79	5.96
Total Pigment	972.0	100.00	60.94
Vehicle Components	**Weight (Lbs.)**	**Percent Vehicle**	**Percent of Total Paint**
XX Refined Linseed Oil	170.0	27.28	10.66
Raw Isano Oil	170.0	27.28	10.66
Chlorinated Paraffin 70%	70.0	11.23	4.39
Polyamide #93 Resin	64.0	10.27	4.01
Cobalt Naphthenate Solution (2% Cobalt as Metal)	5.5	88	34
Lead Naphthenate Solution (8% Lead as Metal)	24.0	3.85	1.50
Manganese Naphthenate Solution (2% Manganese as Metal)	3.5	.56	.22
Volatile Organic Solvents	116.0	18.65	7.28
Total Vehicle	623.0	100.00	39.06
Grand Totals	1,595.0		100.00

Example 2:

Pigment and other Solid Components	Weight (Lbs.)	Percent Pigment	Percent of Total Paint
Titanium Dioxide	87.0	8.35	5.46
Zinc Borate	253.0	24.29	15.89
Blanc Fixe	128.0	12.29	8.04
Lead Sulfate	146.0	14.02	9.17
Lead Carbonate	146.0	14.02	9.17
Benzene Sulfonyl Hydrazide	200.0	19.20	12.55
Zinc Oxide	81.5	7.83	5.12
Total Pigment	1,041.5	100.00	65.40

Vehicle Components	Weight (Lbs.)	Percent Vehicle	Percent of Total Paint
XX Refined Linseed Oil	147.0	26.68	9.23
Raw Isano Oil	147.0	26.68	9.23
Chlorinated Paraffin 70%	60.5	10.98	3.80
Polyamide #93 Resin	55.2	10.02	3.47
Cobalt Naphthenate Solution (2% Cobalt as Metal)	4.7	.85	.29
Lead Naphthenate Solution (8% Lead as Metal)	20.8	3.77	1.30
Manganese Naphthenate Solution (2% Manganese as Metal)	3.0	.55	.20
Volatile Organic Solvents	112.8	20.47	7.08
Total Vehicle	551.0	100.00	34.60
Grand Totals	1,592.5		100.00

Example 3:

Pigment and other Solid Components	Weight (Lbs.)	Percent Pigment	Percent of Total Paint
Titanium Dioxide	100.0	10.30	6.41
Zinc Borate	292.0	30.10	18.73
Blanc Fixe	148.0	15.22	9.47
Lead Sulfate	168.5	17.30	10.77
Lead Carbonate	168.5	17.30	10.77
Zinc Oxide	95.0	9.78	6.09
Total Pigment	972.0	100.00	62.24

Vehicle Components	Weight (Lbs.)	Percent Vehicle	Percent of Total Paint
XX Refined Linseed Oil	170.0	28.84	10.89
Raw Isano Oil	170.0	28.84	10.89
Chlorinated Paraffin 70%	70.0	11.87	4.48
Melamine Formaldehyde Resin	64.0	10.86	4.10
Cobalt Naphthenate Solution (2% Cobalt as Metal)	5.5	.93	.35
Lead Naphthenate Solution (8% Lead as Metal)	24.0	4.07	1.54
Manganese Naphthenate Solution (2% Manganese as Metal)	3.5	.59	.22
Volatile Organic Solvents	82.5	14.00	5.29
Total Vehicle	589.5	100.00	37.76
Grand Totals	1,561.5		100.00

The table on the following page sets forth test results which were obtained with each of the above examples. In each case, sample poplar panels, 6" x 12" x 1/4", were made up and painted, each with two coats of one of the paints undergoing test. After drying, the panels were given a standard burn test which was recommended by the Engineer Research and Development Laboratories, U.S. Army, Fort Belvoir, Virginia. In this test, each panel is placed in a standard cabinet known as a New York Paint and Varnish Production Club Test Cabinet, having standard size and ventilation provisions. A cup containing 5 cc of absolute alcohol is placed one inch from the face of the panel and ignited. The test continues until all burning ceases. The charred area and the loss in weight of each panel are measured. The times at which the flames go out in the fuel cup and on the panel are also noted. Although other measurements are made and other criteria are considered in evaluating the test results, three of the most important criteria are the the percentage loss in weight, the time of persistence of the flame on the panel after the alcohol burns out, and the charred area.

The same test is repeated for each paint sample, with panels which are leached in distilled water for 40 hours at 120°F. The leached panels are dried for 72 hours at 120°F. before the burn test. The results for the leached and unleached panels are given in the table. Each test figure in the table represents an average of several tests. A negative figure in the columns headed "Persistence of panel flame" in the table indicates that the panel flame went out before the alcohol flame. From the results of the tests shown in the table, it may be seen that all the examples formulated in accordance with the process produced markedly better results as to percentage weight loss than did the commercial paint. These improved results are particularly outstanding in the case of the leached panels. The leached panels also show another remarkable improvement in the "char area" measurement.

	Unleached Panels			Leached Panels		
	Percent Wt. Loss	Persistence of Panel Flame (sec.)	Char Area (sq. in.)	Percent Wt. Loss	Persistence of Panel Flame (sec.)	Char Area (sq. in.)
Example						
I	5.33	−4	13.75	5.68	0	13.9
II	7.42	0	19	7.05	1	16.92
III	7.57	−3	18.96	8.06	1	24.8
Commercial Flame Resistant Paint	8.31	−1	21.38	15.64	−4	44.28

Polyesters Prepared from Conjugated Diacetylene Diols — Intumescent Agent

E.T. Rayner, D.A. Yeadon, L.L. Hopper, Jr., H.P. Dupuy and F.G. Dollear (U.S. Patent 3,312,565; April 4, 1967; assigned to U.S. Secretary of Agriculture) have found that compounds represented by the formulas

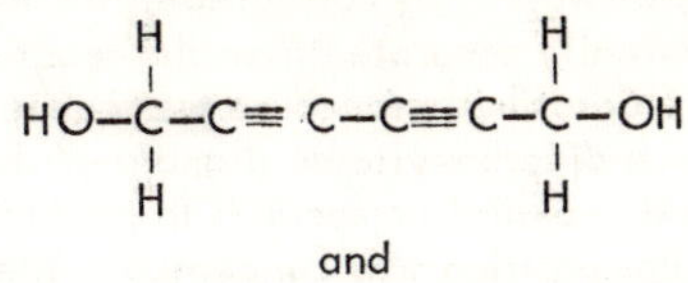

and

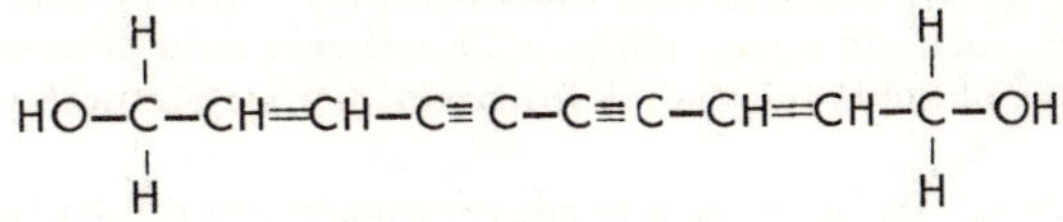

which are conjugated diacetylene diols where the hydroxyl groups are primary groups and where the hydroxyl groups are located in a position alpha to the conjugated unsaturation in the molecule and compounds which are the polyesters of these compounds with various dibasic acids are highly efficient gas producing materials when subjected to heating at decomposition temperatures. These particular types of materials will be referred to as "diyne-diol compounds" and "polyesters". The conjugated diyne-diol compounds when esterified with dibasic acids, such as oxalic, maleic, succinic, adipic, acetylenedicarboxylic or sebacic acid, to form the respective polyesters exhibit enhanced gas producing properties of the polyesters are clearly the result of a synergistic action between the dibasic acid and diyne-diol compound.

The diyne-diol compounds of this process which are useful for preparing the polyesters are made by an oxidative coupling reaction. For example, propargyl alcohol (2-propyn-1-ol) is subjected to an oxidative coupling reaction to yield 2,4-hexadiyne-1,6-diol according to the following reaction scheme:

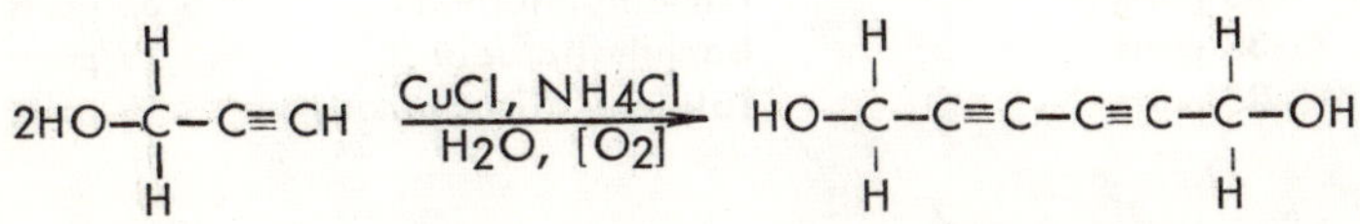

Similarly, pent-2-en-4-yn-1-ol is coupled to produce 2,8-decadiene-4,6-diyne-1,10-diol. The oxidative coupling reaction used for preparing the diyne-diols is essentially as described by Armitage et al., J. Chem. Soc. 1952, pp. 1998 to 2005. The coupling reactions were carried out at room temperature in a slightly acidic water solution using cuprous chloride as a catalyst. Oxygen at atmospheric pressure was added from a measuring burette. Termination of gas absorption was used as the indicator for reaction completion.

These diyne-diol compounds and the related polyesters when subjected to heat decompose vigorously with the evolution of heat and gases. The diyne-diols and the related polyesters were subjected to thermal decomposition tests (100° to 300°C.) in which a mercury filled gas burette was used to determine the volume of gas evolved. The volumes, after correction to standard conditions of temperature and pressure, of typical tests follow. The first compound (2,4-hexadiyne-1,6-diol) will evolve on thermal decomposition about 150 milliliters of gas, measured at standard temperature and pressure, per gram of material. The second compound (2,8-decadiene-4,6-diyne-1,10-diol) will evolve about 200 milliliters of gas, measured under standard temperature and pressure, per gram of material. The same diyne-diol compounds converted to polyesters via esterification reactions with maleic anhydride, for example, will evolve 350 milliliters of gas, measured under standard temperature and pressure, per gram of polyester material.

The above recited volumes of gas when compared indicate the increase in gas producing properties that attend the transformation of the diyne-diol compounds to polyesters. In addition, the esterification reactions render the diyne-diols water-insoluble. In contrast, a natural product, isano oil, used in the formulation of conventional intumescent paints, evolves only about 18 milliliters of gas, measured at standard temperature and pressure, per gram of material. The following examples illustrate the process.

Example 1: Preparation of 2,4-hexadiyne-1,6-diol. The oxidative coupling reaction of 2,4-hexadiyne-1,6-diol was carried out in an aqueous reaction mixture consisting of 56 grams (1 mol) of 2-propyn-1-ol (propargyl alcohol), 230 grams (4.3 mols) of NH_4Cl, 148.5 grams (1.5 mols) of CuCl and 600 milliliters of water. The aqueous solution was adjusted to pH 6.5. The reaction mixture was kept thoroughly agitated at room temperature and oxygen was added under slight pressure to the reaction mixture from a gas burette. The reaction was considered complete when the mixture ceased to absorb oxygen. The entire reaction mixture was extracted with 1,800 milliliters of ether, the extract stripped under vacuum to remove the ether and unreacted propargyl alcohol and the extracted reaction mixture finally crystallized from benzene. The yield was 51 grams (about 93%). The diyne-diol, 2,8-decadiene-4,6-diyne-1,10-diol, was prepared in the same manner using pent-2-en-4-yn-1-ol as the starting material.

Example 2: Preparation of the polyesters. 12.1 grams (0.11 mol) of 2,4-hexadiyne-1,6-diol, 0.1 mol of dibasic acid, 2 to 5% of p-toluenesulfonic acid (weight percent based on total solids) and sufficient benzene to produce a 20% solution were added to a 500 milliliter flask equipped with a stirrer, reflux condenser and a moisture trap. The reaction mixture was heated at reflux until the theoretical amount of water was collected in the moisture trap. The polyesters were insoluble in benzene solvent, thus the benzene was mechanically separated from the reaction mixture and the remaining traces of benzene were removed under vacuum. The above described procedure was carried out using oxalic acid, succinic acid, adipic acid, sebacic acid, maleic acid, acetylene-dicarboxylic acid and o-phthalic acid. Each of the dibasic acids except phthalic acid when incorporated with the diyne-diol compounds to produce a polyester imparted marked synergistic activity insofar as gas evolution upon thermal decomposition was concerned. The polyester of 2,8-decadiene-4,6-diyne-1,10-diol and oxalic acid, maleic acid, succinic acid, adipic acid, sebacic acid or acetylenedicarboxylic acid was prepared in a similar manner. The polyester of butyne-1,4-diol with oxalic or maleic acid was prepared in a similar manner.

Example 3: Preparation of tung oil alkyds. The procedure employed to prepare the various types of tung oil alkyds is essentially as described by Leo A. Goldblatt, Lucien L. Hopper, Jr. (Fellow of the Pan American Tung Research and Development League), and Eric T. Ravner, U.S. Patents 2,999,104 and 3,880,910. Several alkyds were prepared and evaluated with gas producing components. The formulations are long, long oil alkyds which are thought to be particularly essential to impart plasticity and softness, and to minimize brittleness so that the gases evolved from the gas producing components in the intumescing paint film will be confined and retained by the vehicle, the whole mass becoming the protective insulating barrier. Of these, the drying oil modified alkyds which can be employed most advantageously to prepare gas producing undercoats with the gas producing materials of this process are shown below:

(a)

Tung oil	100 parts
Linseed oil fatty acids	100 parts
Trimethylolethane	35 parts
Phthalic anhydride	33 parts
Epoxidized soybean oil (90%)	25 parts

(b)

Tung oil	100 parts
Linseed oil fatty acids	100 parts
Trimethylolethane	35 parts
Isophthalic acid	19 parts
Tolylene diisocyanate	19 parts

(c)

Tung oil	100 parts
Linseed oil fatty acids	100 parts
Trimethylolethane	35 parts
Dimerized linoleic acid	67 parts
Tolylene diisocyanate	19 parts

These alkyds, the gas producing components, pigments, and other paint ingredients can be formulated into gas producing paint formulations as illustrated below.

Formulation A

Pigment and Other Solid Components	Percent Pigment	Percent of Total Paint
Titanium Dioxide	11.87	7.48
Zinc Borate	35.00	22.06
Magnesium Silicate (Mistron #5)	1.75	1.10
Lead Sulfate	19.95	12.57
Lead Carbonate	19.95	12.57
Zinc Oxide	11.26	7.09
Polyester of 2,4-hexadiyne-1,6-diol and maleic acid	0.21	0.13
Total Pigment and Other Solids	100.00	63.00

(continued)

(continued)

Non-volatile Vehicle Components[a]	Percent Vehicle	Percent of Total Paint
Alkyd "a"[b]	35.14	13.00
Refined Linseed Oil	35.14	13.00
Chlorinated Paraffin 70%	15.48	5.73
Polyamide Resin (Versamid #930)[c]	12.98	4.80
Cobalt Naphthenate Solution (6% Cobalt as Metal)	0.13	0.05
Lead Naphthenate Solution (24% Lead as Metal)	0.59	0.22
Anti-Skinning Agent (National ASA)	0.27	0.10
Ultra Violet Screening Agent (Uvinul #490)	0.27	0.10
Total Non-volatile Vehicle	100.00	37.00
Total Paint Non-volatiles		100.00

[a] Mineral spirits was added to adjust final paint viscosity to about 90 Krebs units.
[b] Alkyd was added as a mineral spirits solution of 50-80% solids content.
[c] Sixty-four grams General Mills' Versamid #930 was dissolved in 41 grams n-butanol by heating, then 75 grams of mineral spirits was added.

Formulation B

Pigment and Other Solid Components	Percent Pigment	Percent of Total Paint
Titanium Dioxide	11.86	7.47
Zinc Borate	34.95	22.02
Magnesium Silicate (Mistron #5)	1.73	1.09
Lead Sulfate	19.92	12.55
Lead Carbonate	19.92	12.55
Zinc Oxide	11.22	7.07
Polyester of 2,4-hexadiyne-1,6-diol and Maleic Acid	0.40	0.25
Total Pigment and Other Solids	100.00	63.00

Non-volatile Vehicle Components[a]	Percent Vehicle	Percent of Total Paint
Alkyd "a"[b]	35.14	13.00
Refined Linseed Oil	35.14	13.00
Chlorinated Paraffin 70%	15.48	5.73
Polyamide Resin (Versamid #930)[c]	12.98	4.80
Cobalt Napththenate Solution (6% Cobalt as Metal)	0.13	0.05
Lead Naphthenate Solution (24% Lead as Metal)	0.59	0.22
Anti-Skinning Agent (National ASA)	0.27	0.10
Ultra Violet Screening Agent (Uvinul #490)	0.27	0.10
Total Non-volatile Vehicle	100.00	37.00
Total Paint Non-volatiles		100.00

[a] Mineral spirits was added to adjust final paint viscosity to about 90 Krebs units.
[b] Alkyd was added as a mineral spirits solution of 50-80% solids content.
[c] Sixty-four grams General Mills' Versamid #930 was dissolved in 41 grams n-butanol by heating, then 75 grams of mineral spirits was added.

Formulation C

Pigment and Other Solid Components	Percent Pigment	Percent of Total Paint
Titanium Dioxide	11.70	6.55
Zinc Borate	34.47	19.30
Magnesium Silicate (Mistron #5)	1.71	0.96
Lead Sulfate	19.64	11.00
Lead Carbonate	19.64	11.00
Zinc Oxide	11.05	6.19
Synthetic Zeolite (Union Carbide's MS-5A saturated with water)	1.79	1.00
Total Pigment and Other Solids	100.00	56.00

Non-volatile Vehicle Components[a]	Percent Vehicle	Percent of Total Paint
Alkyd "a"[b]	37.50	16.50
Refined Linseed Oil	37.50	16.50
Chlorinated Paraffin 70%	12.73	5.60
Polyamide Resin (Versamid #930)[c]	9.76	4.29
Cobalt Naphthenate Solution (6% Cobalt as Metal)	0.36	0.16
Lead Naphthenate Solution (24% Lead as Metal)	1.47	0.65
Anti-Skinning Agent (National ASA)	0.34	0.15
Ultra Violet Screening Agent (Uvinul #490)	0.34	0.15
Total Non-volatile Vehicle	100.00	44.00
Total Paint Non-volatiles		100.00

[a] Mineral spirits was added to adjust final paint viscosity to about 90 Krebs units.
[b] Alkyd was added as a mineral spirits solution of 50-80% solids content.
[c] Sixty-four grams General Mills' Versamid #930 was dissolved in 41 grams n-butanol by heating, then 75 grams of mineral spirits was added.

Typical drying oil modified alkyds which can be employed advantageously to prepare resistant topcoats for these gas producing undercoats are shown below:

(d)

Tung oil	100 parts
Linseed oil fatty acids	100 parts
Trimethylolethane	35 parts
Chlorendic acid	43 parts
Tolylene diisocyanate	19 parts

(e)

Tung oil	100 parts
Linseed oil fatty acids	100 parts
Trimethylolethane	35 parts
Chlorendic acid	43 parts
Trichlorophenyl diisocyanate	29 parts

These alkyds, pigments and other paint ingredients can be formulated into weather- and gas-resistant, topcoat paint formulations as illustrated below:

Formulation D

Pigment and Other Solid Components	% Pigment	% of Total Paint
Titanium Dioxide	11.90	7.50
Zinc Borate	35.08	22.10
Magnesium Silicate (Mistron #5)	1.75	1.10
Lead Sulfate	20.00	12.60
Lead Carbonate	20.00	12.60
Zinc Oxide	11.27	7.10
Total Pigment and Other Solids	100.00	63.00

Nonvolatile Vehicle Components[a]	% Vehicle	% of Total Paint
Alkyd "d"[b]	70.27	26.00
Chlorinated Paraffin 70%	15.49	5.73
Polyamide Resin (Versamid #930)[c]	13.46	4.98
Cobalt Naphthenate Solution (6% Cobalt as Metal)	0.05	0.02
Lead Naphthenate Solution (24% Lead as Metal)	0.19	0.07
Anti-Skinning Agent (National ASA)	0.27	0.10
Ultraviolet Screening Agent (Uvinul #490)	0.27	0.10
Total Nonvolatile Vehicle	100.00	37.00
Total Paint Nonvolatiles		100.00

[a] Mineral spirits was added to adjust final paint viscosity to about 90 Krebs units.

[b] Alkyd was added as a mineral spirits solution of 50 to 80% solids content.

[c] Sixty-four grams General Mills' Versamid #930 was dissolved in 41 grams n-butanol by heating, then 75 grams of mineral spirits was added.

Preparation of fire-retardant paint films — One or two coats (about 3 to 4 mils per coat) of the gas producing undercoat paint formulations were spread with a brush on poplar heartwood panels (6 x 12 x 1/4 inches). Then a coat of the weather- and gas-resistant topcoat paint formulation was applied. The different coats were applied about 24 hours apart. The procedure listed in Federal Specification TT-P-34a was followed as closely as possible. The total weight of paint on each panel was computed after the paint films had been cured for about 2 weeks, but the weight of the undercoat or undercoats was determined as soon as they were tack-free.

Evaluation of fire-retardant paint films — The cured paint films were evaluated according to Federal Specification TT-P-34a with a standard fire test cabinet (New York Paint and Varnish Production Club Test Cabinet). Some of the paint films were evaluated for their fire-retardency before leaching, some after leaching, and other after weathering for about 14 months. Numerous different types of intumescing fire-retardant paint formulations were developed and evaluated. The data obtained from representative examples is illustrated in the table on the following page.

Evaluation of Two and Three Coat Paint Films, Prepared from Single and Multiple Formulations for their Fire Retardancy Before Leaching, After Leaching, and After Weathering, According to Federal Specification TT-P-34a[a]

Paint Formulations		Paint on panel (gms.)				Panel weight loss after fire retardancy evaluation (gms.)		
Under coat	Topcoat	Undercoats		Topcoat	Total	Nonleached	Leached	Weathered [b]
		1st	2nd					
A	D	10.8	10.9		21.7	14.2		
		15.2		14.6	29.8	5.6		
		15.7		15.6	31.3		10.5	
		15.4		15.2	30.6			8.5
		15.5	15.5	15.2	46.2	7.8		
		15.4	15.4	15.5	46.3		9.7	
		15.7	15.5	15.9	47.1			6.8
B	D	10.6	11.0		21.6	13.5		
		15.5		15.4	30.9	10.8		
		15.8		15.1	30.9		9.7	
		15.9		16.0	31.9			9.0
		15.0	15.7	15.5	46.2	9.4		
		15.9	15.1	15.4	46.4		9.0	
		15.4	15.1	16.3	46.8			7.9
C	D	14.9	13.9		28.8	14.0		
		15.2		14.4	29.6	6.2		
		15.5		15.5	31.0		10.1	
		15.5		14.6	30.1			7.7
		15.3	14.7	14.8	44.8	6.7		
		15.5	14.9	15.1	45.5		7.8	
		15.9	15.3	14.9	46.1			6.7

[a] Details of experimental procedure are given in Federal Specification TT-P-34a. These are average results obtained from duplicates or triplicates.

[b] These test specimens were exposed to natural weathering, facing South under 45°, for about 14 months in New Orleans.

GENERAL COATING FORMULATIONS

SILICATES

Silicate and Phosphate Formulation

J.J. Rykos (U.S. Patent 2,593,400; April 22, 1952) describes the use of the following fireproofing composition:

	Percent
Sodium silicate	48 to 60
Trisodium phosphate	20 to 25
Lithopone	20 to 25
Quartz	10 to 15
Sodium hydroxide	1
Deterioration and insect inhibitor	1 to 2
Pigment	1 to 2

In preparing the fireproofing composition, substantially 1% sodium hydroxide is carefully introduced to 6% to 10% of the sodium silicate in a suitable mixer, as a conventional centrifuge mixer and mixing continued until the reaction is completed. The reaction product (a liquid) is then introduced to the remaining sodium silicate in a conventional centrifuge mixer and mixing continued for at least one hour. The finely divided quartz is then introduced to the mixture and mixing continued for substantially two hours, following which the formaldehyde and/or chloro phenol is added and mixing continued for approximately one hour. The mixture is then introduced, with the pigment, to a suitable conventional paint mixing machine and mixing continued until the pigment is evenly distributed throughout the mixture. The composition thus obtained is the fireproofing composition, minus the lithopone which is poured into cans or drums for marketing.

Tests conducted employing the fireproofing composition, included painting one-half of each of several wooden crates (conventional orange crates) with the composition, then dousing them with gasoline and setting fire. After substantially 30 minutes of fire, the unpainted halves of the crates were consumed while the painted halves were unburned and intact, although somewhat blistered.

Sodium Silicate

R.M. Hooks (U.S. Patent 2,804,398; August 27, 1957; assigned to Southwestern Petroleum Company, Inc.) describes a coating for impregnated wood, which, when subjected to pressure of gas evolution from within, will yield without rupture and separate from the wood surface, thereby reducing the temperature at the surface of the wood and preventing destructive distillation and charring of the wood.

The coating is comprised primarily of a sodium silicate solution with the addition of some mineral filler such as slate dust, fibrous talc, or ground limestone and asbestos fibers. Preferably a sodium silicate solution of 42.2° Baumé (approximately 38% by weight) and having an alkali to silica ratio of 1:3.22 is utilized. A satisfactory formulation is as follows:

	Percent
42.2° Baumé sodium silicate solution	60
Mineral filler (slate dust or fibrous talc)	35
7R asbestos fibers	5

The foregoing composition is of brushing or spraying consistency. If need be, it may be thinned slightly with water, but dries to form a hard coating due to the evaporation of water from the silicate solution. The silicate coating softens at approximately 1050°F. and continues to get softer up to approximately 1575°F., at which point it is softest. As the coating becomes soft, or just before it softens completely, the heat which causes such softening produces the formation of an excessive amount of gas from the preservative oils due to volatilization. With this volatilization and evolution of gas, the coating expands and swells in much the same manner as hot glass that is being blown by a glass blower.

As the silicate coating under the influence of heat expands away from the timber, in some cases as much as two inches or more, it remains intact and the temperature at the surface of the wood drops greatly, becoming sufficiently low to prevent destructive distillation and charring of the wood. The reduced temperature adjacent the surface of the impregnated wood slows down the volatilization of the oils considerably. Nevertheless, the heat absorbed within the timber continues to provide sufficient gas evolution to force the gas through pinhole openings which develop in the surface of the expanding coating. These pinholes permit diffusion of the gases into the atmosphere surrounding the coated timber.

The volatilization characteristics of the impregnating composition continue so long as any fire is present at the outer surface of the coating, due to the low flash point of the said composition. This continued evolution of gas from adjacent the timber surface excludes oxygen from the void or space created by expansion of the coating from the timber. Thus, there is no possibility of the timber itself burning. Even if an opening an inch or two in diameter should occur in the coating, by mechanical or accidental causes, gases from the preservative oils are sufficient to block the hold and prevent any combustion from occurring beneath the level of the expanded coating. A preferred coating, when protected by a vinyl aluminum weatherproof coating, has gone a period of 2,000 hours in the weatherometer and when tested by fire still shows good expansion properties. The formulation is as follows:

	Percent
42.2° Baumé sodium silicate solution	42.0
Slate dust or other suitable mineral pigment	22.0
7R asbestos fibers	6.0
Emulsified asphalt, 53% solids as asphalt	30.0

This formulation must be mixed in a special manner; the sodium silicate, slate dust, and asbestos must be mixed first, and then the emulsified asphalt is added. The vinyl aluminum weatherproof coating used to protect the preferred sodium silicate coating has the following formulation:

	Percent
High molecular weight vinyl chloride-vinyl-acetate copolymer resin	10.85
Medium molecular weight vinyl chloride-vinyl acetate copolymer resin	7.23
Tricresyl phosphate	7.23
Toluene	11.81
Isophorone	18.89
Methylisobutylketone	36.77
Aluminum powder	7.23

The aluminum powder used in this coating has a maximum retention of 15% on a 325 mesh screen. It has a bulking value of 0.047 gallon per pound and a weight of 21.24 pounds per solid gallon. The vinyl chloride-vinyl acetate copolymer resins have sufficient chlorine in them to prevent the coating from supporting combustion except when it is in the presence of fire.

Water Resistant Silicate Coating

A process described by G.F. Rolland and J.J. Seipel (U.S. Patent 3,442,671; May 6, 1969; assigned to Benjamin Foster Company) involves a protective coating composition which provides substantial incombustibility and water resistance so as to meet the requirements of Coast Guard Regulations pertaining to the insulation of finishing of heating, refrigeration and air-conditioning ducts and equipment on ships. The coating composition comprises:

(a) a binder component
(b) a suspending agent
(c) a chelating agent
(d) mica
(e) water

where the binder component comprises soluble alkali metal silicates wherein the alkali metal oxide to SiO_2 weight ratio is 1:1.8 to 3.75; the suspending agent comprises a colloidal magnesium aluminum silicate; and the chelating agent comprises a mixture of alkaline soluble salts of alkylenediaminepolyalkanoic acids and hydroxyalkyl substituted alkylenediaminepolyalkanoic acids. The following examples illustrate the process.

A comparison of the water resistant properties of dry applications produced from the formulations of this process and other silicate coating is given in the table below. The water resistance was determined in accordance with Federal Test Method Standard No. 141, Method 6162, as amended, utilizing a coating applied with a film applicator of 0.05 inch clearance and dried for 10 days at 73 ± 5°F. and 50 ± 2% relative humidity. The water resistance, as determined by the number of brush strokes withstood by the coatings, is recorded in the table.

Ingredients, percent by wt.	Examples					
	1	2	3	4	5	6
Potassium silicate	36	36	32	12		
Sodium silicate			4	12	36	36
Mica [1]	10		10	18	10	
Calcium carbonate		1				1
Titanium dioxide		4		5		4
Silica		2				2
Asbestos [2]		3		1		3
Suspending agents [3]	0.75	0.75	0.75	1	0.75	0.75
Chelating agents [4]	0.25	0.25	0.25	0.4	0.25	0.25
Water	53	53	53	50.6	53	53
Ratio $M_2O:SiO_2$	1:2.1	1:2.1	1:2.1	1:2.5	1:3.2	1:3.2
Water resistance (strokes)	>1,000	152	>1,000	823	718	177

[1] "Micro mica."
[2] Canadian Chrysotile Asbestos Grade 7D.
[3] The suspending agents consisted of a 5% aqueous solution of magnesium aluminum silicate having the following dry weight analysis: silicon dioxide, 61%; magnesium oxide, 13%; aluminum oxide, 9.2% balance alkali and alkaline earth metal oxides and water of combination; specific gravity at 25° C. 2.5; viscosity at 25° C. (Brookfield) 5% solution, 240 cps.
[4] The chelating agents were added as a 60% aqueous solution. Dry weight analysis 65% ethylenediaminetetraacetic acid; 30% hydroxyethyl-ethylenediaminetriacetic acid; 5% triethanolamine.

The compositions of Examples 1 to 6 inclusive were prepared by dissolving the silicate component in a portion of the water requirement. To this solution was then added the 5% aqueous solution of the suspending agents. The mica and/or the other mineral filler components, where employed, were then added with stirring, at room temperature, to yield uniform compositions to which were then added the respective amounts of chelating agents and the balance of the water to complete the formulations.

From the results of the water resistance tests it is readily apparent that the utilization of mica in conjunction with sodium and/or potassium silicate yielded formulations producing excellent water resistance properties which were not obtainable without the use of mica. It is also apparent from the results of Examples 1 through 6 inclusive, that optimum results were obtained utilizing potassium silicate as the sole silicate binder component.

Lithium-Sodium Silicate Solution

A process described by R.H. Patton, J.B. Cox and W.K. Stich, Jr. (U.S. Patent 3,180,746; April 27, 1965; assigned to Esso Research and Engineering Company) involves a single application liquid coating composition which comprises an aqueous solution of a mixture of lithium and sodium silicate, the lithium and sodium silicates being employed in amounts sufficient to provide a ratio of lithium oxide to sodium oxide to silicon dioxide, respectively, within the range from 0.75 to 1.00 : 0.05 to 0.25 : 2.5 to 5.0. The solution of the process is also capable of lending strength to wood as well as protection against rot, fire, water absorption, and may also be effective against termites, wood borers and other destructive insects.

It has been found that the incorporation of a dispersion of a metallic dust in the lithium-sodium silicate solution of the process, the lithium-sodium silicate solution being used as a liquid vehicle, with a small but effective amount of an alkaline metal dichromate provides a useful and important protective coating for metallic surfaces such as storage tanks for crude petroleum. The coating composition may also be employed to protect ships' bottoms and the interior of ships and tanks, providing protection from corrosion without contamination of hydrocarbons, aqueous solutions, and the like contained in the tanks. One aspect of the process is the making and use of the lithium-sodium silicate solution without the addition of any additive materials. The composition was made up in the following manner.

Example 1: Water, silica gel, lithium hydroxide, and sodium hydroxide are charged into a stainless steel pressure vessel equipped with an agitator. During the initial mixing of these ingredients, evolution of heat raises the temperature of the contents to 50° to 55°C. The reactor is then sealed and the temperature of the reactants, under continual agitation, is raised to 150°C. over a period of time, which may vary from thirty to ninety minutes. After the vessel reaches about 150°C. and about 70 pounds per square inch of pressure, the contents are slowly cooled at a rate of about 0.61°C. to about 1.83°C. per minute. This will result in a temperature drop from 150°C. to about 95°C. within about thirty to ninety minutes. This completes the initial part of the reaction which forms the solid sodium lithium silicate.

The second part of the operation (below 95°C.) requires a longer period of time in order to effectively dissolve the solid silicate at lower temperatures. The reduction of the temperature of the reactor contents from 95°C. to 40°C. at a rate of about 0.37°C. to about 0.146°C. per minute should be accomplished within 150 minutes to 240 minutes. The last temperature drop from 40°C. to 25°C. may be conducted over a period of 60 to 120 minutes at a cooling rate of about 0.25°C. to about 0.125°C. per minute.

Depending upon the amount of water used, the solution may be used in different solids ratios, i.e., the total solids of the solution may vary from about 10% to 45% by weight.

Example 2: A solution was prepared in a similar manner as described in Example 1, except that it was made from the following ingredients:

Sodium hydroxide	20 g.	Silica gel	300 g.
Lithium hydroxide monohydrate	62 g.	Water	1,146 g.

This solution can be described as $(0.75\ Li_2O + 0.25\ Na_2O) : 5\ SiO_2$ with a total solids content of about 25%. Cloth fabric and paper were saturated with this aqueous solution and allowed to air dry. After drying they were rigid, fireproof, and withstood 72 hours' exposure to 100% relative humidity at 40°C.

Example 3: Liquid compound and preparation —

Lithium hydroxide monohydrate	62.0 g.	Silica gel	270.0 g.
Sodium hydroxide	20.0 g.	Water	952.0 g.

These chemicals were placed into a stainless steel pressure vessel under continual agitation. The temperature was raised to 150°C., then the temperature was gradually reduced to 25°C. over an eight hour period in a manner similar to Example 1.

Example 4: Application of liquid compound to wood —

Compound (of Example 3)	500.0 g.	Water	100.0 g.

This mixture was placed into a container and three types of wood, (pine, oak, fir) were immersed for 30 minutes under a constant pressure of 25 mm. Hg. The wooden panels were removed and dried thoroughly, then subjected to an open flame. The wood was charred but at no time did it burn. A microscopic examination revealed that approximately 60% of the voids in the wood were occupied by silica.

Example 5: Lithium-sodium silicate vehicle —

Lithium hydroxide monohydrate	62.0 g.	Silica gel	270.0 g.
Sodium hydroxide	20.0 g.	Water	952.0 g.

These ingredients were placed into a stainless steel pressure vessel under continual agitation. The temperature was raised to 150°C., then the temperature was gradually reduced to 25°C. over a period of 8 hours in a manner similar to Example 1.

Polyvinyl acetate component —

Vinyl acetate	43.75 g.	"Tergitol" NPX	0.75 g.
2-ethylhexyl acrylate	6.25 g.	"Cellosize" WP-09	1.00 g.
Water	78.00 g.	Potassium persulfate	0.18 g.

The "Cellosize", a hydroxyethyl cellulose protective colloid; "Tergitol", an alkyl phenyl ether of polyethylene glycol which acts as a surface active agent emulsifier; and potassium persulfate, used as a catalyst, are blended into the water in a polymerization reactor equipped with an agitator and a condenser. The vinyl acetate and 2-ethylhexyl acrylate are combined and 10% by weight is added to the aqueous system. The aqueous system is warmed to 70°C. with agitation. The remaining vinyl acetate and 2-ethylhexyl acrylate mixture is added with stirring over a three hour period maintaining the temperature of the reaction at 70° to 80°C. After the addition is complete, agitation is continued for one-half hour. The contents are then cooled to room temperature.

Coating composition —

Lithium-sodium silicate vehicle	500 g.	Iron oxide	50 g.
Polyvinyl acetate emulsion	500 g.	Inert talc filler	200 g.

The iron oxide and talc are dispersed in the vinyl acetate emulsion and the silicate solution is then added. The coating mixture was applied over ferrous metal surfaces which had been wire brushed. Upon drying by water evaporation, the coating exhibited excellent adhesion, hardness and moisture resistance. It had an initial gloss of 30 as determined by a 60° gloss meter, withstood a 1/8 inch mandrel bend for flexibility and 8,000 scrub cycles in a 5% soap solution (ASTM D-16). The coating was not affected after 96 hours' exposure to 100% relative humidity at 40°C.

Nitrile Rubber Containing Silicate and Refractory

A process described by D.V. Surbuch and V.G. Boger (U.S. Patent 2,821,514, January 28, 1958; assigned to the B.F. Goodrich Company) involves a composition composed of from 12% to 52% by weight of at least one silicate of an alkali metal, from 10% to 34% by weight of a rubbery polymer comprising a butadiene hydrocarbon and the balance acrylonitrile, from 21% to 74% by weight of an insulating refractory and water in an amount at least sufficient to provide a tacky workable dispersion is readily applied to a base material and on drying will form a coating having excellent fire resistant and insulating properties to protect for extended periods of time base materials such as steel and wood. The composition on drying is adherent and impact resistant and exhibits little or no crazing. In very thin coatings it will protect base materials for some time at temperatures of about 4000° F. Further improvements are realized by incorporating in the composition a minor amount of a low temperature fusion point inorganic gassifying agent, which tends to increase the insulating values of the coating.

The method of preparing the composition is not too critical, it being only necessary that the refractory materials be added to the silicate prior to the addition of the latex to prevent coagulation of the latex. Alternatively all the materials can be mixed together except the latex which is finally added and dispersed throughout the system. The resulting composition can be termed either a dispersion or suspension since it appears to have properties of both types of systems. The base material to be coated need not be especially cleaned to receive the composition coating although grease and oil should be removed. The composition can be readily applied to degreased steel for example, which still contains mill scale or its oxide coating, yet no loss in adhesion is noticed over a steel surface which has been degreased and sand blasted.

A base material having a surface coating of the composition of the process may be permitted to air dry at room temperatures for a few hours (depending on the water content of the original composition) or be subjected to temperatures of the order of 212°F. until thoroughly dry. After drying, the coating obtained is hard, adherent, essentially check-free, and impact and fire resistant. It probably contains a minor amount of water in the form of water of crystallization of the silicate and possibly some entrapped free water mechanically held between the particles of the composition or in the polymer. However, the composition appears to be dry and to all intents and purposes, the bulk of the water has been removed. Successive coats of the composition can be built up by repeated spraying or dipping and then drying to obtain the desired thickness or can be achieved in one application by controlling the water content to the desired amount and, accordingly, the viscosity of the resulting composition.

The essentially dry composition coating is alkaline and can be painted or coated with a layer of any water or oil-base paint unaffected by an alkaline surface. The coating is unaffected by most commonly used organic solvents yet can be removed easily by soaking in warm water if desired. Furthermore, cylinders and panels coated with the composition herein described can be subjected to temperature of -60°F. for 24 hours, then to 160°F. for 24 hours, then back to -60°F. for weeks without affecting the adhesion of the composition or producing cracks as well as without destroying the original impact and fire resistant properties of the composition. The following examples illustrate the process.

Example 1: Water glass was poured into a container and a solution of a 50% : 50% by weight mixture of sodium and potassium carbonates in water added. Asbestos (7.8% retained on 18 mesh, 44.1% retained on 35 mesh, 16.6% retained on 70 mesh, and balance through 70 mesh, USS size) and fire clay (2.32% retained on 200 mesh, 3.23% retained on 325 mesh and balance through 325 mesh, USS size) which had been dry-mixed previously was then added to the carbonate-water glass solution with stirring. Butadiene-1,3 acrylonitrile latex was next added to the mixture and stirred. Finally, water was added in an amount in addition to that already present to give good troweling consistency and the composition thoroughly stirred to produce a homogeneous appearing dispersion or suspension. The resulting suspension had a tacky-like feeling to the touch, somewhat like wet clay or cement. The amounts of ingredients present in the composition were as follows:

Ingredients	Parts by Weight (grams) (Percent Same)	Dry Basis, H_2O Omitted, By Weight	
		Parts	Percent
Sodium disilicate solution (water glass, 39% total solids)	30.15	11.76	21.9
Hycar OR-15 (about 55% butadiene-1,3/45% acrylonitrile copolymer) latex (39% total solids)	16.13	6.29	11.7
Fire clay, particulate	30.15	30.15	56.3
Asbestos, refiberized	5.03	5.03	9.4
Potassium carbonate	.20	.20	.374
Sodium carbonate	.20	.20	.374
Water	18.14		
Totals	100.00	53.63	100.048

This composition was readily spread by troweling to form a coating 1/32 inch thick on a piece of sheet fender steel 1/32 x 4" x 12" which has been merely degreased and then allowed to air-dry. After drying, the coating was smooth, without any evidence of cold flow, checks or crazing, and was tightly bound to the steel as evidenced by the fact that it could not be removed by attempts to insert a knife between the coating and the surface of the sheet steel. Its impact resistance or resistance to mechanical shock was very high in that it did not separate from the metal base after repeated blows from a hammer on the coating (plate suspended on ends by two metal pads) just short of deformation of the metal. Its fire resistance, or resistance to thermal shock, was also very high, for it did not separate from the metal and prevented the metal from becoming hot enough to discolor for 3 seconds when subjected to temperatures of 3700+°F. in the reducing portion of an oxy-acetylene heating torch flame. On the other hand, an unprotected panel subjected to the same heat treatment melted through in less than 1 to 1 1/2 seconds. With high alloy steel (4,130), the coating prevented any heat discoloration for 3 seconds at oxygen flame temperatures up to 4000°F. while an uncoated panel of this steel was severely warped and discolored. The coating tended to fuse at such temperatures but did not separate on impact.

On the other hand, a coating containing only a dried water glass-refiberized asbestos composition had excellent adhesion but no impact resistance and blistered and chipped on the flame test. Even addition of fire clay to this two-component composition did not prevent chipping nor improve impact resistance. Although the high temperatures employed tend to cause slight fusion and discoloration of the insulation coating of the process, it does not blister or crack and will still perform its function of protecting the base material. At lower temperatures, for example, at 1900°F. for five minutes, the composition coating was only moderately discolored and could not be separated from the metal by impact. It exhibited no cracks or checks. Thicker coatings will, of course, provide protection for longer periods of time.

Example 2: The composition of this example was prepared in the same manner as that of Example 1 except that the amounts of the components were varied as shown below:

Ingredients	Parts by Weight (Grams)	Percent by Weight	Dry Basis, H_2O Omitted, By Weight	
			Parts	Percent
Sodium disilicate solution, water glass (39% T.S.)	40.33	36.60	15.73	27.3
Hycar OR-15 latex (39% T.S.—about 55% butadiene-1,3/45% acrylonitrile copolymer)	16.13	14.65	6.29	10.9
Fire Clay, pulverized	30.15	27.4	30.15	52.4
Asbestos, fibrous	5.03	4.57	5.03	8.7
Potassium carbonate	.20	.182	.20	.35
Sodium carbonate	.20	.182	.20	.35
Water	18.14	16.5	------	------
Totals	110.18	100.084	57.60	100.00

The above composition was then applied as a coating on steel plates and tested in the manner set forth in Example 1. It gave results equal to those exhibited by the composition of Example 1 even though the amount of sodium silicate solution was increased appreciably.

BORATES AND INORGANIC FILLERS

Alkali Metal Glycol Monoborates

D.L. Hunter, K. Kitasaki and G.W. Willcockson (U.S. Patent 3,131,071; April 28, 1964; assigned to United States Borax & Chemical Corporation) describe the use of the alkali metal salts of glycol monoborates as fire retardants in paints, lacquers and varnishes. The alkali metal salts of glycol monoborates useful in organic coating compositions can be prepared for example, by reacting an alkali metal hydroxide with a di-glycol boric anhydride, the di-glycol boric anhydride having the general formula:

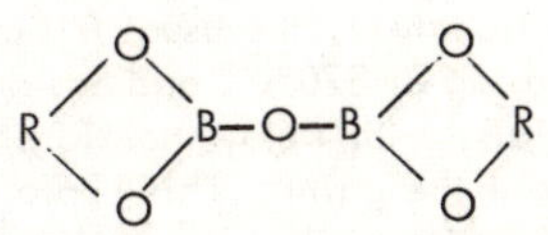

where R is an alkylene radical of 2 to 3 carbon atoms in length and having a total of from 2 to 10 carbon atoms. The following materials are typical examples of the 1,2- and 1,3-glycol boric anhydrides applicable to the process:

- di-(Propanediol-1,2) boric anhydride
- di-(Butanediol-1,2) boric anhydride
- di-(Butanediol-1,3) boric anhydride
- di-(2,2,4-trimethylpentanediol-1,3) boric anhydride
- di-(2,3,4-trimethylpentanediol-2,4) boric anhydride
- di-(Pentanediol-2,3) boric anhydride
- di-(Catechol) boric anhydride
- di-(2-methylpentanediol-2,4) boric anhydride

The example on the following page illustrates the process.

Example: Into a 2 l., three-necked, round-bottomed flask equipped with a paddle-type stirrer, a water condenser, and a pot thermometer, were added 270 g. of di-(2-methylpentadiol-2,4) boric anhydride and 500 mls. of methanol; 80 g. of sodium hydroxide was then added to the liquid mixture. When the initial exothermic reaction subsided, heat was applied and the reaction mixture was refluxed for about one-half hour. The methanol and water of reaction were stripped from the flask and the residue, sodium 2-methylpentanediol-2,4 monoborate, was then removed and vacuum dried for about 2 to 4 hours. Chemical analysis of the resultant sodium 2-methylpentanediol-2,4 monoborate revealed the following data: Na=13.76% (theory 13.86%); B=6.44% (theory 6.52%).

Essentially the same procedure was followed in experiments using the other alkali metal hydroxides. Using lithium hydroxide, the final product of the reaction was lithium 2-methylpentanediol-2,4 monoborate whose chemical analysis revealed the following data: Li=4.61% (theory 4.68%); B=7.12% (theory 7.22%).

It is generally accepted that metal salts of organic compounds are substantially insoluble in organic solvents. These alkali metal glycol monoborates, however, show remarkable solubilities in many organic solvents. For example, when sodium 2-methylpentanediol-2,4 monoborate is added in amounts of from about 0.5 to about 10% to oil base paints, lacquers and varnishes, which contain common commercial solvents, the sodium 2-methylpentanediol-2,4 monoborate dissolves in the solvents and imparts the property of flame resistance to the various paints, lacquers and varnishes. Sodium 2-methylpentanediol-2,4 monoborate when dissolved in an organic solvent such as hexane in amounts from about 0.2% to about 10% can be used to impregnate wood and fiberboard for increased fire retardance.

Ceramic Frit Comprised of Boric Acid, Lead Oxide and Sodium Fluoride

A composition described by G.O. Orth, Jr. and J. Spinelli (U.S. Patent 2,840,478; June 24, 1958; assigned to American Potash & Chemical Corporation) includes, in addition to the standard and ordinary paint containing ingredients, an additional fire retardant ingredient in the form of a low melting inorganic material which is highly effective in providing increased fire resistance. This mineral material is effective because, when subjected to heat, it provides a ceramic film on the painted surface which protects such painted surface. The frit fuses with the pigment particles present to provide a ceramic barrier to gases of decomposition given off by the carrier, e.g., a wood surface; the gases of decomposition, which are inflammable in most cases, are blocked and, in the case of wood or a like surface, inverted direction-wise into the wood where they are condensed or dispelled away from the burning surface.

The mineral material utilized in its simplest form is a finely divided mineral frit made up of carefully controlled amounts of boric oxide, lead oxide and sodium fluoride, heating these to an elevated temperature at which the mass is molten, then cooling and grinding the glass-like product to a desired size. If desired, the boric oxide can be provided by a boric acid, the lead oxide-boric acid-sodium fluoride mix being heated to drive off the water from the boric acid and form boric acid in situ. In addition to boric oxide and lead oxide, the frit can desirably contain a small amount of antimony oxide.

In general, the method of testing paints of this character has been to prepare panels which have been coated with one or two coats of the paint compositions, dried for a controlled period of time, after which they are burned under controlled conditions, the extent of protection of surface directly traceable to protective coating being observed. Suitable methods of testing are set forth in U.S. Department of Commerce Specification CS-42-49, and in Federal Specification TT-P-26a of January 13, 1954. The paints of the process pass these tests with the same general character of excellence as those paints comprising antimony oxide and other flame resistant materials and frequently excel those paints.

The frit should contain from 20 to 50 parts of B_2O_3 and from 25 to 60 parts of PbO and from 1 to 20 parts of NaF, all by weight. The preferred range is from 30 to 37 parts of B_2O_3, 48 to 60 parts of PbO, and 10 to 15 parts of NaF. The preferred composition containing 37.8 parts of B_2O_3, 48 parts of PbO, and 14.2 parts of NaF. The frits are added to the paint in such an amount that from 5% to 15% of the paint is frit on a wet basis; the preferred value being about 7%. On a dry basis, these values will change proportionately, the usual frit content being about 15% on this basis. The preferred composition is a clear, colorless glass, fusing at 370°C. and has negligible water-solubility. Its density is 3.58, which is in the range of common pigment materials, and it may therefore be incorporated with the pigment prior to incorporation of the latter into the liquid vehicle of the paint. This is preferably accomplished by mixing the pigment and the frit together in water, agitating these and permitting them to settle. To provide a standard water base paint for reference, the following composition was made up in a large quantity.

Pigment composition:	Parts
Titanium dioxide	15
Dicalite	10
Clay	25
Lithopone	40
Water ground mica	10
Total	100

Paint composition wet basis:	Parts
Resin solids — polyvinyl acetate (55%) dispersion	16
Pigment mixture	30
Water	54
Total	100

Stable dispersion at brushing consistency was obtained by one pass through a Morehouse Model A-200 laboratory mill.

Example 1: To illustrate use of the preferred frit composition given above, mixtures were made up in the following proportions by weight:

Ingredients	Paint 1	Paint 2	Paint 3
Polyvinyl acetate dispersion (Bakelite, 58.2% solids)	50.6	48.1	44.8
Water	75.8	72.1	67.1
Titanium dioxide	8.1	7.8	7.3
Silicon dioxide	5.2	5.0	4.6
Clay	13.5	12.8	12.0
Lithopone	22.0	20.1	19.6
Mica	5.2	5.0	4.6
Frit	20.0	30.0	40.0

These paints were applied at a level of 20 g. per square foot to both calendered and uncalendered Fir-Tex 18 pound density insulating board and compared in the Class F fire resistance test. A control was not used since the board would obviously be so severely charred without the frit that it would not provide a fair comparison. Char areas were as follows:

Paint No.	Calendered Board, sq. in.	Uncalendered Board, sq. in.
1	11	16
2	10	14
3	10	14

Example 2: To show the utility of the preferred frit in coating using an acrylic base as a vehicle, a paint was made up containing the following materials in the proportions indicated by weight, the frit being of the preferred composition given above.

Pigment Mixture:	Percent
Titanium dioxide	30
Lithopone	30
Zinc oxide	20
Calcium carbonate	20

Paint:	Parts
Pigment mixture	50
Acrylic base	70
Water	70

Paints were prepared containing 0, 5, 10 and 15 percent of the frit on a wet basis. The results were as follows:

Paints Prepared	Percent Frit (wet basis) added	Char area
1	5	11.7
2	10	11.2
3	15	11.2
4 (no addition)	----------	15.5

Alumina Particles Coated with Hydrolyzed Polyacrylonitrile

D.M. Harris (U.S. Patent 2,984,635; May 16, 1961, assigned to Monsanto Chemical Company) has found that fireproofing and heat insulating compositions may be produced from aqueous dispersions containing finely divided particles of alumina which are coated with a polyelectrolyte.

The laminar form of alumina which is employed in the process is composed of leaf or flake shaped particles having dimensions in the range of from 0.005 to 0.10 micron average thickness, while the average length and width of the flake has been bound to vary within the range of from 1 to 10 microns. This laminar form of alumina is prepared by a vapor phase oxidation method utilizing solutions of aluminum salts. Useful dispersing agents are the water-soluble polymeric polyelectrolytes, especially those having a weight average molecular weight of at least 10,000, and which contain substantially linear continuous carbon chains derived by the polymerization of an aliphatic unsaturated carbon to carbon double bond. Suitable water-soluble linear polymers are the synthetic linear polymers containing water-solubilizing radicals, such as carboxyl, carboxy acid salts, hydroxyl, amide and quaternary ammonium salt radicals. Since the polymers will only be used in small proportions and in the use are dissolved or suspended in water, the required extent of water-solubility is small.

A preferred polyelectrolyte for use in the coating of particles of laminar alumina is hydrolyzed polyacrylonitrile. The hydrolyzed polyacrylonitrile may be employed in the form of a salt or as polyacrylic acid obtained as the ultimate product of hydrolysis, and while preferably utilized in a water-soluble form, may also be employed as a water dispersible type. The hydrolyzed polyacrylonitrile may be obtained by the acid or alkali catalyzed hydrolysis of polyacrylonitrile as described in U.S. Patents 2,675,529 and 2,727,835. The polyacrylonitrile may be either a high or low molecular weight type, and the hydrolyzed product may have a wide range of viscosity.

The particles of laminar alumina are coated by suspending them in a solution or slurry of the hydrolyzed polyacrylonitrile so that the individual particles are completely coated with the solution. In order to aid in wetting of the particles with the solution, suitable wetting agents may be present in the solution or dispersion. The concentration of the polyelectrolyte such as hydrolyzed polyacrylonitrile in the solution should be such as to deposit from 3% to 25%, or preferably from 5% to 15% by weight of the hydrolyzed polyacrylonitrile relative to the weight of the modified alumina. After coating with the water solutions, the particles are removed, allowed to drain and then dried by any suitable means. If a dry milling method is desired, the dry polyelectrolyte and the laminar alumina may be ground together for eventual slurrying in water.

The alumina particles which have been coated and dried as above may readily be resuspended in an aqueous suspending medium. The suspension may be made directly into water or into aqueous solutions or suspensions containing other pigments or suspending agents. The suspension so prepared may be applied directly as a coating composition upon normally combustible surfaces or structures. Application may be made by brushing, spraying or dipping in the usual manner. The products are particularly useful for the protection of surfaces against sparks or heated metal particles which are generated during welding or cutting operations, and which are often a source of fire when these operations are carried out adjacent to combustible materials. The coating compositions provided are readily redispersible in water, and accordingly may be quickly removed, once their purpose has been served. The following examples illustrate the process.

Example 1: An aqueous dispersion was prepared utilizing 100 g. of water containing 10 g. of laminar alumina. The particles had an average thickness of about 0.02 micron and the greatest dimension of the irregularly shaped particles was about 15 microns. The dispersion of alumina and water also contained 1 g. of hydrolyzed polyacrylonitrile. The three components were mixed in a high speed blender, after which films were dispersed upon wooden panels to give a dry film thickness of about 0.003 inch. It was found that the coated panels were resistant to ignition by a flame, so that the coated panels were greatly improved in fire retardancy. It was also found that sodium silicate, e.g., as a 1 weight percent addition, is a desirable additive to aqueous dispersions of the combination of alumina and hydrolyzed polyacrylonitrile to further improve the abrasion resistance of the coating.

Example 2: The coated laminar alumina particles in suspension as described in Example 1 were utilized as a surface coating on iron panels which were to be welded. It was found that the alumina coating reduced the adhesion of spatter particles from the welding operation so that the iron particles were readily cleaned by a simple wire brushing operation without the necessity of grinding or filing in order to remove spatter particles of the welding metal.

Reflective Coatings Containing Fibrous Potassium Titanate

A process described by F.E. Edlin (U.S. Patent 3,311,585; March 28, 1967; assigned to E.I. du Pont de Nemours and Company) is directed to aqueous compositions comprising water-insoluble fibrous potassium titanate, colloidal silica and a latex-type elastomeric material, which upon drying form an adherent heat reflectant coating.

Flames emit radiant energy for the most part in the visible and near infrared wave bands generally shorter than six microns. A surface having the highest effective reflectance will possess the greatest capability to reject this radiant energy. Since no perfect reflectors are known, some amount of the radiant energy will always be absorbed by a surface. This energy which is absorbed, can be re-emitted as radiation of a much longer wavelength generally corresponding to the temperature of the surface. Therefore, a surface which is capable of efficiently emitting the radiation of the longer wavelength corresponding to its temperature is also very desirable to aid in forestalling combustion temperatures.

A composition has been found, which, when dried to an adherent coating, reflects from 93% to 95% of all radiation having a wavelength between 0.35 and 2.1 microns. This wavelength includes the visible wave band which is between 0.4 and 0.7 microns and also the near infrared wave band usually considered to extend from 0.7 to 2 microns. Further, these coatings emit radiation from bodies at near ambient temperature up to 400°F. with high efficiency. At 140°F. the coatings emit radiation equivalent to 96% to 98% of the radiation which would be emitted by a perfect black body.

To put these excellent optical properties into perspective, the reflectance over the broad wave band of 0.35 to 2.1 microns of the compositions of this process, when dried to a coating, is equal to or better than the reflectance of silver in telescope optics; and these coatings emit radiation at 140°F. more completely than the finest carbon black. The compositions comprise from 25 to 35 parts by weight dry basis of fibrous potassium titanate, sufficient colloidal silica aquasol to give from 25 to 35 parts by weight dry basis of colloidal silica and sufficient latex-type elastomer emulsion to give from 1 to 3 parts by weight of a latex-type elastomeric polymer, all combined in an aqueous suspension.

One of the essential ingredients of the compositions of this process is a water-insoluble fibrous potassium titanate such as described in U.S. Patent 2,841,470. The ratio of TiO_2 to K_2O in the potassium titanate can vary from 4.25 to 1 to 6.5 to 1, but ordinarily a ratio of from 4.8 to 1 to 5.5 to 1 is most satisfactory for use in the process. These compositions may be applied to almost any type surface in which a fire resistant thermal insulating or light reflectant coating would be desirable. Metal structure such as steel columns, beams, and panels may be coated in order to effectively insulate them with a thermal radiation reflector. Good bonding occurs between films formed from the compositions of this process and such metal structures. Cellulosic materials such as wood, plywood, Masonite, Celotex, Homosote, Upson board and other such materials are equally satisfactory substrates for the compositions of this process. The following illustrative examples are given. In the examples, radiant panel flame spread tests are made in accordance with ASTM E162-60T.

Example 1: 30 g. of PKT which is a pigmented fibrous potassium titanate having a ratio of TiO_2 to K_2O of 5.3 to 1, a diameter of approximately 1/4 micron and a length of 10 to 20 microns is dispersed in 100 g. of water in a paddle agitator. The resulting slurry has a pH of 11. This slurry is mixed with 100 g. of an aqueous colloidal silica sol containing approximately 30% SiO_2 and having an ultimate particle size of 10 to 15 mμ, a viscosity of 5 cps., a pH of 9.8 and having an SiO_2 to alkali oxide ratio of 95 to 1. To this mixture is added 4 g. of a 50% emulsion of a neoprene polymer latex composed of about 44% chloroprene and about 56% dichlorobutadiene. The pH of the final mixture is adjusted with 1N HCl to 10.0.

Example 2: 30 g. of fibrous potassium titanate as described in Example 1 are dispersed with vigorous agitation in 200 g. of a colloidal silica aquasol containing about 15% SiO_2, having an ultimate particle size of 7 to 9 mμ, a viscosity of 4 cps., a pH of 8.5, and an SiO_2 to alkali oxide mol ratio of about 150 to 1. To this mixture is added with agitation 4 g. of a 50% emulsion of a neoprene polymer latex composed of about 44% by weight chloroprene and about 56% dichlorobutadiene. The pH is adjusted to 10.9 with 1N HCl.

Example 3: 50 g. of PKT as described in Example 1 are extracted with sufficient HCl to give a TiO_2 to K_2O ratio of 6.3 to 1 and dried. 30 g. of this material is combined with agitation with 200 g. of aqueous colloidal silica sol as described in Example 2, and 7 g. of the 50% neoprene emulsion described in Example 2.

Example 4: 30 g. of fibrous potassium titanate as described in Example 1, but having a TiO_2 to K_2O ratio of 4.5 to 1, are dispersed with agitation in 200 g. of a colloidal silica aquasol as described in Example 2, and 6 g. of the 50% neoprene emulsion described in Example 2.

Example 5: The formulation as described in Example 2 is sprayed to a thickness when dried of 2 mils on a pressed bonded cellulosic composition board commercially available as Celotex. The coated composition has a flame rating of 8 when tested in accordance with the procedure of ASTM E162-60T.

Example 6: The procedure of Example 5 is repeated with the formulation of Example 2 to give a coating of a thickness of 4 mils when dry. The coated composition has a flame rating of 4 when tested in accordance with the procedure of ASTM E162-60T.

Example 7: The procedure of Example 5 is repeated with the formulations of Examples 1,3 and 4 to give a coating thickness of 4 mils when dry. The coatings exhibit similar outstanding flame ratings to that of Example 6.

MISCELLANEOUS

Epoxy Novolac-Polyhalogenated Phenol Reaction Products

A.M. Partansky (U.S. Patent 3,239,476; March 8, 1966; assigned to The Dow Chemical Company) describes a coating composition which comprises an epoxy resin-based oleoresinous composition (referred to as an oleresinous polyester) consisting of the reaction product of the adduct, of an epoxy novolac and a polyhalogenated phenol, and one or a plurality of relatively long carbon chain unsaturated fatty acids. The acid must contain at least 1 olefinic bond per mol and from 8 to 22 carbon atoms. The proportions of halogenated phenol adduct and fatty acid selected should provide enough halogenated phenol adduct to render the oleoresinous polyester self-extinguishing. The percentage by weight of the adduct and the fatty acid employed is usually between 70% and 40% of the adduct and between 30% and 60% of the adduct and between 35% and 50% of the fatty acid. Any bromo- or chlorphenol may be employed as the halophenol in the preparation of the resinous adduct, so long as it provides at least about 10% bromine or about 18% chlorine by weight of the final coating vehicles, when the coating film includes no additament, e.g., a pigment, or at least half of these amounts when the coating film also contains auxiliary fire retardant agents such as Sb_2O_3. When both brominated and chlorinated phenols are employed, a rising intermediate value directly proportional to the percent of chlorine to bromine, may be used. The following procedure is illustrative of a mode of carrying out the synthesis.

An epoxy novolac-halogenated monophenol adduct is prepared by charging a suitable reaction vessel, e.g., one provided with a means for stirring and heating and for controlling and recording temperature, with an epoxy novolac resin, as above described, and heated to between about 90°C. and 110°C. Thereafter, while accompanied by stirring and while being maintained substantially uniformly at this temperature, a predetermined amount of halophenol is added, usually at a rate just fast enough for it to be dissolved in the epoxy novolac. At this time, there is mixed with the contents of the reaction vessel an etherification catalyst, usually of a tertiary amine, such as triethylamine (preferably previously dissolved in a small amount of a solvent such as xylene) in an amount sufficient to provide between about 0.01% and 0.05%, based on the weight of the reaction mixture. Because this reaction is exothermic, the temperature rises adiabatically and, if unchecked, may reach 200°C. or higher resulting in self-polymerization. To prevent this, cooling is applied to check the temperature at about 150°C. It is then maintained at between about 150°C. and 160°C. for about 20 to 30 minutes. Heating is then discontinued and the mixture is cooled. The adduct so made contains substantially no solvents, by-products, or other extraneous materials and is ready for esterification with fatty acids without any further treatment. The following examples are illustrative of the process.

Example 1: 713 g. (1.25 mols) of an epoxy novolac, a commercial product sold under the designation D.E.N. 438 were reacted, following the procedure as described above, with 504 g. (2.0 mols) of 2,4-dibromophenol in the presence of 2 cc of 10% triethylamine solution, as catalyst. The peak temperature reached due to exothermic heat was 168°C. A resinous product began to form after about 45 minutes. The product was removed from the reaction vessel and examined. It had the following properties:

Molecular wt.	975	40% butyl Carbitol solution showed:	
Avg. epoxy groups per molecule	1.48	Viscosity	50 cps.
Avg. hydroxyl groups per molecule	1.60	Color, Gardner	2
Esterification equivalent (based on OH calculation)	203	Bromine content	26.3%

Example 2: Following a similar procedure to that of Example 1, 500 g. (0.877 mol) epoxy novolac, D.E.N. 438, were reacted with 374 g. (1.404 mols) of pentachlorophenol. The resulting resinous adduct had the following properties:

Molecular wt.	1,100	40% butyl Carbitol solution showed:	
Avg. epoxy groups per molecule	1.6	Viscosity	75 cps
Avg. hydroxy groups per molecule	1.9	Color, Gardner	10
Esterification equivalent (based on OH calculation)	216		
Chlorine content	28.9%		

Example 3: Following a similar procedure to those in the preceding examples, 1,300 g. (2.25 mols) of D.E.N. 438 were reacted with 700 g. (2.78 mols) of 2,4-dibromophenol in the presence of 0.3 g. of triethylamine to give a resin with the following characteristics:

Molecular wt.	950	Esterification equivalent (based on OH calculation)	170
Avg. epoxy groups per molecule	2.0	Bromine content	22.2%
Avg. hydroxyl groups per molecule	1.6	Durran melting point	62°C.

Example 4: 385 g. of the adduct of an epoxy novolac and 2,4-dibromophenol, prepared as above described in Example 3, were mixed with 315 g. of castor oil fatty acids, (i.e., the hydrolyzed triglycerides of dehydrated castor oil) in a 1 l., 3-neck, round-bottom flask, provided with a stirrer, thermocouple, and a reflux condenser and equipped with a water-separating device. The resulting reaction mixture was heated with a Glas-Col electric heater. About 10 ml. of xylene containing a few drops of an antifoam agent were also added to the mixture. The temperature of the mixture was then raised to 240°C. over a period of about 1.5 hours and maintained at that temperature for an additional 3.5 hours. It was then cooled and diluted by adding 300 g. of petroleum naphtha, accompanied by stirring. During the reaction period, small amounts of the reaction mixture were withdrawn at intervals and the viscosity, color, and acid number determined. The length of time preceding each sampling, measuring from the time the temperature reached 200°C., and the test results obtained are set out below:

Reaction period at between 200° and 240° C., in hours	Acid No.[1]	Viscosity in cps.[2]	Color according to the Gardner Scale[2]
1.0	20.4	450	6-7
2.0	8.7	1,250	8
3.0	6	2,100	8-9
3.75	3.8	5,000	8-9
5.0	3.0	7,000	9

[1] Initial calculated acid number of the mixture was 87.0.
[2] Viscosity and color were determined in 70% solution of naphtha.

Additional polyesters were prepared by esterification of resinous adducts described in Examples 1, 2 and 3 using unsaturated acid mixtures. These data are summarized in the table which follows:

Example No.	Resin used is from Example—	Fatty acid component employed to make polyester		Esterification cooking schedule, hrs./° C.	Final acid No.	Kind and percent of halogen
		Its oil origin	Weight percent			
5	1	Linseed	36.5	2.5 hrs./235° C	4.9	16.7% Br.
6	1	D.C.O.[1]	36.9	2.0 hrs./238° C	5.9	16.6% Br.
7[2]	1	D.C.O	45	4.3 hrs./240° C	3.2	14.0% Br.
8	2	D.C.O	45	4.1 hrs./240° C	7.2	15.9% Cl.
9	3	Soya	45	3.2 hrs./240° C	4.1	12.2% Br.
10	3	D.C.O	45	3.5 hrs./240° C	3.8	12.2% Br.
11	D.E.R. 664[3].	D.C.O	45	4.7 hrs./245° C	3.9	None.

[1] D.C.O. means dehydrated castor oil.
[2] 5% 4,4'-isopropylidenediphenol was added to the reaction mixture in this example.
[3] D.E.R. 664 is a commercial solid epoxy resin used in oleoresinous coatings; it has an epoxide equivalent. weight of between 875 and 975.

For testing oleoresinous polyester as surface coatings they were mixed with the conventional paint driers in the following proportions: 0.2% lead, 0.1% zinc, and 0.02% cobalt, all in the form of their naphthenate salts. For convenience of application, the viscosity was further reduced to about 400 cps. by the addition of more solvent. Drying characteristics and properties of the coatings made are noted in the following table.

Example # from previous table	Wt. percent of halogen	Touch Dry rm. temp. hrs:min.	Sward Rock Hardness[1]	1/8" Mandrell flexibility test[2]	2 hrs. in 5% aq. NaOH	24 hrs. in in 10% acetic acid	24 hrs. in naphtha	Fire resistance
100% #6	16.6% Br	1:20	38	Pass	No effect	No effect	No effect	SE[3]
100% #8	15.9% Cl	1:25	25	Pass	No effect	Sl. hazy	No effect	SE mixed with 2% Sb_2O_3
66% #8 and 34% #11	10.5% Cl	1:20	30	Pass	No effect	No effect	No effect	SE mixed with 5% Sb_2O_3
100% #9	12.2% Br	3:10	12	Pass	Softened[5]	Sl. hazy	No effect	SE
66% #9 and 34% #11	8.1% Br	2:10	19	Pass	No effect	Sl. Hazy	No effect	SE mixed with 2% Sb_2O_3
100% #10	12.2% Br	1:25	29	Pass	No effect	No effect	No effect	SE
66% #10 and 34% #11	8.1% Br	1:20	34	Pass	No effect	No effect	No effect	SE mixed with 2% Sb_2O_3

[1]This test is described in Physical and Chemical Examination of Paints, Varnishes, Lacquers and Colors, 11th ed. (1950) by H.A. Gardner.
[2]This test is a modification of ASTM D1640-59, "Drying Time for Varnishes" Method A on "Dust-Free" drying time.
[3]SE means self-extinguishing according to ASTM Test No. D635-56T.
[4]Slightly hazy indicates a very mild attack on the resin.
[5]Softened indicates a moderate attack on the resin.

Referring to the data in the above table, the "touch dry" time and Sward Rocker hardness were determined on 3 mil thick film cast on plate glass. For flexibility, the films were cast on a 10 mil thick tin plate. The chemical resistance was determined on the films so formed by dipping the closed end of 1" diameter test tube into the diluted varnish for a distance of about 3", draining the surplus of the liquid, and inverting the tube for 48 hours air drying at room temperature and then immersing the film coated end of the tube into a test liquid for specified lengths of time.

The fire resistance tests were made essentially in accordance with the procedure of the ASTM Test D635-56T. Test samples were prepared by dipping strips of 181 type, Volan A chromate treated glass cloth into the varnish solution, evaporating the solvent, making four-ply laminate and heat-curing in a steam press.

Barium Sulfate and Chlorinated Wax

R.W. Craig (U.S. Patent 2,627,475; February 3, 1953, assigned to Diamond Alkali Company) describes the use of barium sulfate in an oil based paint containing a chlorinated paraffin. The following method of testing the paint is set forth.

Red cedar shingles, 5 inches wide and 16 inches long, are selected as test panels. An area of 3 x 16 inches is painted by brush with the fire retardant paint to be examined, leaving an area of 2 x 16 inches uncoated. Where one coat is applied to the shingle, the panel is permitted to air-dry for 30 days. Where two coats are applied, a space of 5 days intervenes between the two coats. At the end of the drying period, the panels are tested as follows: The test shingle is held in a vertical position by means of clamps with the thin end of the shingle at the bottom. A lighted Bunsen burner with a two inch blue cone flame is placed under the uncoated edge of the test shingle. The shingle is lit with the tip of the Bunsen burner flame just touching it for a period of 30 seconds, after which the flame is removed. The shingle is permitted to burn until it extinguishes itself and the weight of the undestroyed area comprising the treated area is compared with the weight of that portion of the shingle prior to burning. Where less than 5% of the weight is lost in the course of the burning of the untreated area, the test is considered to be excellent. The paints of the process pass this test with the same general character of excellence as those paints comprising antimony oxide and other flame resistant materials.

Example 1: A composition comprising the following ingredients is suitably milled until a homogeneous mixture is obtained:

	Lbs.	Gallons
Anatase Titanium Dioxide	100	2.85
35% Leaded Zinc Oxide	429	8.85
Magnesium Silicate	100	4.20
Barytes	510	14.25
Chlorinated Paraffin (Average C_{24}) Containing about 70% Chemically Combined Chlorine	238	17.20
Alkali Refined Linseed Oil	188	24.20
Kettle Bodied Linseed Oil	50	6.25
Mineral Spirits	135	20.80
24% Lead Naphthenate	3.8	.40
6% Manganese Naphthenate	1.9	.20

Shingles painted with this composition and tested in accordance with the test set forth above show excellent results in that less than 10% of the weight of the shingle is lost in the test period.

Example 2: The following ingredients are suitably combined in a paint mill until the same are absolutely homogeneous:

	Lbs.	Gallons
Titanium Dioxide	270	7.7
Barium Sulfate	482	13.4
Precipitated Calcium Carbonate	197	8.9
Asbestine	122	5.1
Alkyd Resin Comprising 70% of Vehicle Solids	229	29.02
Chlorinated Paraffin Having 70% of Chemically Combined Chlorine	68	5.06
Mineral Spirits	191	29.45
Driers	10.3	1.25

It will be recognized that in this formulation, the total vehicle solids comprise 30% of the chlorinated material and the barium sulfate comprises 40% by weight of the total pigment present. This material, when tested in accordance with the test procedures set forth above, is superior with respect to fire retardancy to the paint of Example 1 and compares favorably with antimony-containing paints.

Chlorinated Rubber and Mineral Fillers

A.M. Shenk (U.S. Patent 2,938,937; May 31, 1960; assigned to Ideal Chemical Products, Inc.) describes flame resistant, high-heat insulating composition which is cold setting air drying and comprises a mixture of a filler, a vehicle including a binder and mineral materials, and a solvent. The compositions has a paste-like consistency so that it may be troweled or sprayed by the use of heavy duty spray equipment. The amounts of filler and vehicle including the binder and mineral materials are preferably in the following ranges:

Filler	15 to 35%
Vehicle:	
Binder	18 to 35%
Minerals	30 to 65%

Examples of satisfactory compositions have been found to be as shown on the following page.

General Coating Formulations

Binder	Mineral Materials	Filler Materials
24.25	54.0	21.75
18.0	65.0	17.0
35.0	50.0	15.0
35.0	30.0	35.0

The filler which is used in the composition is preferably a mixture of expanded micaceous material such as vermiculite and asbestos in the following ranges:

Expanded vermiculite	10 to 55%
Asbestos	90 to 45%

The binder which is utilized in the vehicle comprises a mixture of chlorinated rubber, chlorinated resin and a plasticizer in the following ranges:

Chlorinated rubber	27 1/2 to 55%
Chlorinated resin	10 to 50%
Plasticizer	22 1/2 to 35%

The chlorinated resin may be chlorinated terphenyl or chlorinated paraffin. The chlorinated terphenyl is a commercially available material containing approximately 42% to 60% chlorine. The chlorinated paraffin is a commercially available material containing approximately 70% chlorine. The plasticizer may comprise chlorinated biphenyl, butyl benzyl phthalate, tricresyl phosphate, triphenyl phosphate, cresyl diphenyl phosphate or tritolyl phosphate. The chlorinated biphenyl is a commercially available material containing 48% to 60% chlorine. The mineral materials comprise inert, inorganic materials and preferably a mixture of antimony oxide, calcium carbonate, titanium dioxide and asbestine in the following ranges:

Antimony oxide	5 to 15%
Calcium carbonate	15 to 50%
Titanium dioxide	40 to 80%
Asbestine	10 to 40%

Asbestine is a milled mixture of talc and asbestos. The percentage composition of the binder and mineral materials in the vehicle based on the total weight of the vehicle is 20% to 75% binder and 25% to 80% mineral materials. The composition is applied to various objects as a coating by troweling or spraying. After the paste has been applied, the coating will become set upon evaporation of the volatile solvent. This occurs at room temperature; but if desired, a small amount of heat, on the order of approximately 180°F., may be applied to accelerate the setting action. The coating may be applied in various thicknesses ranging from 1/32 of an inch to 4 inches or more, depending on the insulating properties required. The coating is preferably applied in successive layers to provide the most efficient bond. The following example illustrates the process.

Example:

Vehicle A	72.7%
Vermiculite	4.3%
Long fiber asbestos	23.0%
	100.0%

In preparing the above composition, the vermiculite is slowly added to the vehicle and thoroughly dispersed therein by slow stirring. The asbestos is then slowly added and the mixture stirred until a substantially complete dispersion is obtained. The consistency of the mix is fairly heavy, but it may be applied easily by troweling or by spraying with special equipment of the type used in spraying cement mixtures. Vehicle A, used in this example, is composed of the following ingredients, using the proportions indicated.

Chlorinated rubber	7.3%
Chlorinated paraffin A	3.4%
Tricresyl phosphate	3.1%
Adhesive A	3.7%
Xylene	23.0%
Antimony trioxide	10.1%
Whiting	3.4%
Titanium dioxide	10.5%
Asbestine	35.5%
	100.0%

This vehicle may be prepared in the following manner: using approximately 2/3 the total amount of xylene (aromatic solvent), the chlorinated rubber and chlorinated paraffin A is put into solution by stirring. The tricresyl phosphate and adhesive A are placed in a heavy duty mixer, the motor started and the antimony trioxide, whiting, titanium dioxide and asbestine added. These materials are mixed to a thick paste, using a sufficient amount of the solution of chlorinated rubber and chlorinated paraffin. The thick paste is then passed through tight milling rolls, preferably three heavy rolls, into a container where the remainder of the rubber-paraffin solution is added. The balance of the xylene is then added with stirring, and the mixture filtered through a coarse screen.

The chlorinated rubber used in the above described vehicle is a commercially available product containing 67% chlorine and having a viscosity range of 15 to 23 centipoises in a 20% solution in toluene. The chlorinated paraffin A used in the foregoing example is a commercially available wax-like resin containing 69% to 72% chlorine. Tricresyl phosphate is an excellent plasticizer for use in a composition of this type. However, there are other fire resistant plasticizers which may be used, such as cresyl diphenyl phosphate, triphenyl phosphate, etc. The above designated adhesive A is formed from the following ingredients, using the proportions indicated:

Nonochlorterphenyl	22%
Chlorinated vegetable oil	56%
Xylene	22%
	100%

The chlorinated vegetable oil used in adhesive A is oiticica oil, containing approximately 40% chlorine. In addition to xylene in the vehicle composition, other volatile aromatic solvents such as toluene, benzene, etc. may be used. In applying the above coating composition to a metallic surface, the surface to be coated is first primed with a coat of vehicle A. On surfaces other than metal, such as wood, cement, brick, etc., no primary coat is necessary. The coating composition is then sprayed onto the surface and built up to the desired thickness. The material sets in about thirty minutes (more or less, depending upon the thickness of coating desired) to a solid mass having excellent adherence to the surface coated, as well as resilient properties with resultant high resistance to cracking or chipping. The material is white in color and has a surface somewhat rough in texture. A smooth surface may be obtained, however, by applying a finish coat of vehicle A, if desired.

When applied to structural materials such as steel, the structural materials are not only insulated and prevented from melting, but in addition, the loss of the structural properties of the material is prevented. This has been shown by various tests. For example, a 2" x 2" x 1/4" angle iron, 10 feet long, was prepared with two 1/2" metal plates approximately 18" square, welded on each vertical end. The entire angle, but not the plates, was reinforced by wrapping with diamond shaped expanded metal spot welded. The entire angle and underside of plates were troweled and sprayed with a 1/4" to 1/2" thickness of the composition. First application was permitted to cure approximately three weeks. Second application was permitted to cure approximately two and one half weeks. The angle was placed in a vertical position in a deep pit. Lead ingot weights of 150 lbs. were placed topside of the steel plate. The angle was secured level and plumb. The pit was filled with water up to within ten to twelve inches of the top of the pit. Gasoline and other chemicals were poured on surface of water and set afire.

At the conclusion of the fire tests and immediately following cessation of flame, 800 lbs. of water pressure was turned directly on the coating of the angle. Some small portions of outer layer of coating tore off, exposing bottom layer of coating, which was still in excellent condition. At the expiration of seventeen minutes burning time, during which temperatures of 3000° to 4000°F. were reached, the composition had done a very satisfactory job in preventing the weighted steel angle from bending or buckling or losing any of its structural strength. Similar tests conducted on uncoated angles resulted in buckling and bending of the steel to the ground in one minute and thirty seconds.

Bentonite Thickened Compositions

E.A. Behr, H.M. Tobey and R.E. Hill (U.S. Patent 3,220,858; November 30, 1965; assigned to Chapman Chemical Company) describe a grease-like wood fire retardant composition comprising an oil-soluble, halogenated organic compound containing bromine or chlorine, which contains phosphorus or which is mixed with organic phosphorus compounds, and a thickening agent of a compound of bentonite and an organic-nitrogen base in an amount sufficient to impart a grease-like consistency to the composition. The fire retardant composition preferably includes organic compounds containing bromine or chlorine or mixtures of the two halogen compounds.

The fire retardant ingredients in the gel or grease compositions are relatively water-insoluble and may be used to effectively impart fire resistance to exterior structures exposed to various climatic conditions. The gel or grease compositions are also resistant to weathering conditions and the action of water does not dissolve or destroy their gelling properties. The thickening agent employed in the process for obtaining a grease-like consistency in the final grease composition is preferably a compound of bentonite and an organic-nitrogen base as described in U.S. Patent 2,904,467. In addition, the class of organic ammonium montmorillonites, including benzyl stearyl dimethyl ammonium bentonite may be employed.

Example: An overhead panel test was designed to evaluate fire control agents in field burning tests and to simulate the portion of a ballasted deck trestle exposed to the hazard of weed fires. The panels were weighed before and after exposure to fire to determine weight loss data during the burning of tumbleweed. Also, the time required for the panels to become self-extinguishing is recorded in the accompanying tables and, in addition, the depth of charring is so recorded. In performing the panel tests, the panels were built three feet wide and fourteen feet long from Douglas fir and southern pine lumber of 2 inch thickness which had been treated one to four years previously with a 50/50 mixture of coal tar creosote-petroleum. Three fire retardant grease-like compositions were prepared and used in these tests. Each grease-like composition was sprayed by means of Alemite equipment on one of the two wide surfaces (50 square feet) of two southern pine panels and two Douglas fir panels for a total of 12 panels. The compositions of these greases are set forth below.

	Gel-A, percent	Gel-B, percent	Gel-C, percent
Monophenyl-dicresyl phosphate containing 8.6% phosphorus	78.75	50.25	40.75
Technical pentachlorophenol	6.00	6.00	6.50
2,4,6-tribromophenol	-	27.50	-
Aroclor 1262	-	-	34.25
D-71A	-	-	10.00
Stearyl ammonium bentonite	10.00	11.00	-
Benzyl stearly dimethyl ammonium bentonite	-	-	7.25
Methanol	5.25	5.25	1.25

Gel A contained only about one part phosphorus to 0.6 part by weight of chlorine while Gel B contained one part phosphorus to 4.6 parts bromine by weight in addition to the 0.6 part of chlorine, whereas Gel C contained one part phosphorus to 7.3 parts chlorine including the pentachlorophenol and the term D-71A is a trade name designating a commercially available composition that consists essentially of a low boiling hydrocarbon petroleum solvent. Application data for these compositions is given below.

Fire Retardant Gel	Original Weight of Gel Applied to 50 Sq. Ft. of Area in Pounds[1]	Calculated Pounds of Elements Applied Per Square Foot of Area			Total Phosphorous, Chlorine and Bromine Applied Per Square Foot	Species of Treated Wood
		Phosphorous	Chlorine	Bromine		
Gel A	20	0.024	0.016		0.040	Douglas fir, Southern pine.
Gel B	23	0.018	0.018	0.080	0.116	Do.
Gel C	20	0.014	0.102		0.116	Do.
None	Control					Do.

[1] The weight of gel applied to each panel corresponded to a 1/16 inch layer over the 50 sq. ft. of surface area of the panels.

After two and one half months of exposure (summer), the panels were weighed and installed in an open end section replica representative of closed deck bridge construction. The permanent portion of the structure was protected from the fire with asbestos. Tumbleweeds were piled two to three feet high under the panel with the gel coated side down, and the weeds were ignited. The results of these overhead panel burning tests along with control panels are shown in the table below. The panels treated with the compositions of the process were self-extinguishing, and the flames died out after a fairly short period of time, especially the panels treated with Gel B and Gel C. The fire retardant effect was especially evident in the test employing southern yellow pine panels when the results employing these treated panels are compared with the control panels. Even though the amount of available phosphorus per square foot of area was less in Gel B and Gel C, the latter two were more effective in their fire retarding effects. Gel A contained only one part phosphorus to 0.6 part by weight of chlorine whereas Gel B contained one part phosphorus to 4.6 parts bromine by weight in addition to 0.9 part of chlorine. Gel C contained one part phosphorus to 7.3 parts chlorine including the pentachlorophenol.

Gel	Time After Ignition for Weeds to Burn Out		Time After Ignition for Panels to be Self-Extinguishing		Weight Less During Fire in Pounds[1]	Depth of Char in Inches
	Minutes	Seconds	Minutes	Seconds		
Gel A	1	30	16	00	24	Slight.
	2	00	Removed ater 20 min.[2]		27	Slight.
Gel B	1	30	4	00	13	Slight.
	1	30	5	00	26	Slight.
Gel C	1	45	13	00	25	Slight.
	2	00	7	13	22	Slight.
None	2	00	23	00	22	1/16 inch.
	1	30	Extinguished after 1 hr. 27 min.[3]		50	1/16 to 1/2.

[1] The weight loss of the gel coated panels included the residual gel which was burned away during the fire. This caused the weight losses to be high when compared to the untreated controls.

[2] The panel was almost self-extinguished when removed after 20 minutes.

[3] The panel was still burning after one hour and 27 minutes and was removed and extinguished.

These compositions were also very effective in a full scale bridge replica test and a fire cabinet test.

ADHESIVES

LAMINATES

Brominated Soybean Oil

G. J. Peterson (U.S. Patent 3,305, 431; February 21, 1967; assigned to Swift & Company) describes a process for coating and impregnating fibrous absorbent cellulosic materials, particularly in web form, with a smooth mixture containing a major amount of an adhesive base and lesser amounts of halogenated, aliphatic, organic compositions and water-insoluble salts of antimony, arsenic, bismuth and tin. The composition is applied on the surface of the fibrous material, such as paper or wood, in the form of a continuous film, and the coated cellulosic base is then bonded either to another cellulosic material or to a noncellulosic substrate. Laminated materials represent a favored area of use of these compositions, particularly laminated webs or sheets such as are employed in the preparation of insulation materials in the building trades and also where sheets of wood are bonded together to form a veneer. A preferred flameproof adhesive containing:

	Percent by Weight
Brominated soybean oil (30% bromine)	20
Everflex-B polyvinyl acetate emulsion[1] (55% solids)	65
Antimony oxide	15

[1] Vinyl acetate-maleic anhydride copolymer-Everflex-B described in Bulletin E-13, March 1955, of the Dewey and Almy Chemical Co., Cambridge, Mass.

when spread on 30 lb. kraft paper and the paper laminated to aluminum foil provided good flameproofing properties. The flameproof adhesive was employed at a level which provided about 13% brominated oil based on the weight of the paper. The following examples illustrate the process.

Example 1: The following ingredients were mixed to form a smooth dispersion —

	Parts by Weight
Brominated soybean oil (35% bromine)	22.5
Everflex-B polyvinyl acetate emulsion[1] (55% solids)	66.5
Antimony oxide	11.0

[1] See footnote in previous table

This formula was applied on the surface of 0.00035" aluminum foil and the foil was immediately bonded to a web of 30 lb. unbleached sulfite kraft paper. After smoothing the paper, the lamination was placed in the oven and dried for 2 minutes at about 200°F. Twenty-four hours later, the lamination was tested for flameproofness according to ASTM Method D777-46. The laminated paperfoil showed excellent flameproofing properties. The laminate contained about 13% by weight of the brominated oil based on the weight of the paper.

Example 2: The formula shown on the following page was prepared substantially in accordance with the method set forth in Example 1 above. When this material was deposited on the surface of 30 lb. kraft paper, this paper laminated with another untreated web of 30 lb. kraft paper, and the laminated material dried, the laminate showed excellent resistance to the development of combustion when subjected to the standard ASTM flameproofing test.

	Parts by Weight
Water	42.5
Dextrin	18.0
Borax	2.0
Sodium metasilicate	2.5
Antimony oxide	15.0
Brominated soybean oil (35% bromine)	20.0

Example 3: A hot melt adhesive was prepared by dispersing antimony oxide in brominated soybean oil and this dispersion was mixed with melted microcrystalline wax and W. wood rosin and also ethyl cellulose. The formula of the hot melt, flameproof adhesive is as follows:

	Parts by Weight
W. wood rosin	46
Microcrystalline wax (M.P. 190° to 195°F.)	10
Ethyl cellulose	9
Antimony oxide	15
Brominated soybean oil (35% bromine)	20

This composition was applied to the surface of a web of 30 lb. kraft paper, a thin film of the adhesive was formed, and glass fiber scrim was placed on the adhesive surface, followed by a web of aluminum foil. The laminate was pressed and then subjected to drying at about 200°F. in the oven. This laminate exhibited very satisfactory flameproofness.

Polyvinyl Acetate — Fire Barrier

A process described by R.L. Muri and R.E. Collette (U.S. Patent 3,202,567; August 24, 1965; assigned to Ludlow Corporation) provides improved flame retardant barrier constructions employing emulsion type laminants. The flame retardant barriers commonly employed for flame resistant jacketing, facing materials in duct insulation and similar uses, embodying plies of combustible sheet material or plies of such material laminated to foil, are of two general types.

In one type the laminant comprises hot melt type adhesives which are nonburning and incorporate substantial amounts of snuffing action components. Because of the relatively large amounts of laminant employed in such hot melt type it is possible to incorporate sufficient amounts of the potentially reactive fire-retarding agents to mitigate or overcome the burning of various untreated substrates. The second type of construction comprises various paper, film and foil combinations laminated with an emulsion type adhesive. Such adhesives are nonburning but, as contrasted to the hot melt type, possess little if any potential fire-retarding action. Because of the relatively small amounts of the emulsion type adhesive employed in fire barrier plied constructions and the correspondingly minor fire retardant action the substrates must have poor burning characteristics as presented for example by foil, vinyl film and krafts pretreated to impart flame resistance.

The requirements of certain uses of flame barrier construction impose specifications which are difficult to satisfy. Thus the Underwriters Laboratory 90A Ratings require a flame spread of 25 or less, smoke development of 50 or less and the less demanding 90B rating requires a flame spread of 50 or less and smoke development of 100 or less based on the following standards:

Cement board	0	Flame	0	Smoke
Red Oak	100	Flame	100	Smoke

The test procedure to determine these values is set forth in a publication, Underwriters Laboratory Inc., Subject 723, published August 1950. A satisfactory flame resistant foil to paper barrier should also display a moisture vapor transmission resistance of 0.5 perms or less at 73°F. and 50% relative humidity and possess all these values after accelerated ageing for a period of one month at 120°F. and 100% relative humidity. Furthermore the adhesive qualities of the emulsion type laminant must be such as to insure an intimate and permanent bond to foil and paper substrates. Finally, and not the least in importance, the emulsion type adhesive should have no corrosive action on the metal foil of the laminant, or deleterious effect on the other portions of the laminated construction.

Previously, flame retardant barriers comprised of aluminum foil and flame resistant kraft laminated with emulsion type adhesives have been produced which has passed Underwriters Laboratory requirements as to flame spread and smoke development. In such products the adhesive is compounded with neoprene and other nonburning polymers and contain substantial amounts of sodium antimonate and metallic oxides, such as oxides of antimony, arsenic, bismuth, lead, iron and zinc. While these prior laminants do present satisfactory flame spread and smoke development characteristics they do have an outstanding disadvantage which is not apparent under the described testing procedure, namely their corrosive action on aluminum foil over a period of time.

The reasons for this are not entirely clear but may be due in some considerable part to the fact that the sodium antimonate, being a basic salt, contributed to a corrosive alkaline environment for the contacting foil, or in some cases may be due to a galvanic action between the metal oxide and the foil at high humidities. An improved laminate has been found which is based on the concept of using a noncorrosive, nonchlorinated polymer as the base adhesive in combination with selected halogenated plasticizers such as chlorinated and brominated flame blanketing plasticizers and adhesive modifiers to attain the desired low flame, low smoke and corrosive resistance in the adhesive system.

General a priori considerations would lead to the expectation that latices of acrylic resins and polyvinyl acetate copolymers would constitute most satisfactory adhesive components for emulsion type laminants. These resins provide excellent bonding to various substrates, excellent water resistance and aging characteristics. It was found however that they do not prove satisfactory for use in fire barrier constructions. Specifically it was ascertained that a fire barrier construction comprised of a 65 lb. intumescent coated kraft, glass scrim and foil laminated with an emulsion comprised of 37.6% polyacrylate, 7.7% polyvinylidine chloride, 27.5% chlorinated paraffin (70%), 13.6% antimony oxide and 13.6% titanium dioxide when tested by the Underwriters Laboratory procedure gave a smoke rating of 75 and a flame spread of 40, which were considerably in excess of the permissible high standard of rating 90A.

It was evident from such tests that despite their intrinsic satisfactory adhesive characteristics, these compounds, because of their chemical structures are relatively flammable and require a large and inordinate amount of nonburning material to effectively stop their burning. In continuing study of the problem of developing satisfactory emulsion type laminates it was found that, although not possessing the inherent adhesiveness and water resistant properties of the copolymers, the polyvinyl homopolymers and homologues such as polyvinyl butyral, polyvinyl acetals and polyvinyl alkyl ethers, have better flame resistance and lower smoke values and that effective adhesiveness, and water resistance could be conferred on the laminant by the incorporation of functional adjuvants in the emulsion.

According to the process, improved nonburning emulsion type laminants may be formulated by blending selected hydrogen halide-evolving plasticizers, tackifiers and fillers in a dispersion of polyvinyl acetate to produce a plasticized polyvinyl acetate emulsion which may readily be applied as a laminant to foil and paper substrates to produce a barrier construction characterized by low flame spread, low smoke development. The flame retardant plasticizing component of the adhesive preferably comprises one or more halogenated phosphate esters. Particularly effective members of this group are tris-dibromo propyl phosphate, tris-beta-chloroethyl phosphate, tris-dichloro propyl phosphate and tris-chlorobromo propyl phosphate. The following examples are illustrations of emulsion type adhesives produced according to the process.

Example 1:	Polyvinyl acetate homopolymer emulsion (55% solid)	59.4 parts by dry weight
	Triethyl phosphate	18.2 parts by dry weight
	Tris-chlorobromopropyl phosphate	4.3 parts by dry weight
	Sulfonamide-formaldehyde resin	11.5 parts by dry weight
	Titanium dioxide	6.6 parts by dry weight
Example 2:	Polyvinyl acetate emulsion (55%)	60.5 parts by dry weight
	Triethyl phosphate	14.6 parts by dry weight
	Tris-dibromo propyl phosphate	7.3 parts by dry weight
	Sulfonamide-formaldehyde resin	11.1 parts by dry weight
	Titanium dioxide	6.5 parts by dry weight
Example 3:	Polyvinyl acetate emulsion (55%)	75.7 parts by dry weight
	Tris-beta chloroethyl phosphate	18.3 parts by dry weight
	Sulfonamide-formaldehyde	6.0 parts by dry weight
Example 4:	Polyvinyl acetate emulsion (55%)	67.6 parts by dry weight
	Triethyl phosphate	14.9 parts by dry weight
	Tris-chlorobromo propyl phosphate	4.9 parts by dry weight
	Sulfonamide-formaldehyde resin	4.9 parts by dry weight
	Titanium dioxide	7.7 parts by dry weight
Example 5:	Polyvinyl acetate emulsion (55%)	66.0 parts by dry weight
	Tris-beta chloroethyl phosphate	16.0 parts by dry weight
	Tris-dibromo propyl phosphate	3.8 parts by dry weight
	Sulfonamide-formaldehyde resin	7.7 parts by dry weight
	Clay	6.5 parts by dry weight
Example 6:	Polyvinyl acetate emulsion (55%)	73.0 parts by dry weight
	Tris-beta chloroethyl phosphate	19.3 parts by dry weight
	Sulfonamide-formaldehyde resin	5.7 parts by dry weight
	Clay	2.0 parts by dry weight

Example 7:	Polyvinyl acetate emulsion (55%)	71.1 parts by dry weight
	Tris-beta chloroethyl phosphate	15.5 parts by dry weight
	Sulfonamide formaldehyde resin	8.3 parts by dry weight
	Titanium dioxide	5.1 parts by dry weight

Such adhesives may readily be made up by blending the plasticizers, adjuvants and fillers together with water into the 55% polyvinyl acetate dispersion to produce a plasticized polyvinyl emulsion of suitable working viscosity as for example between 1000 and 5000 cp. In manufacturing a typical fire barrier construction, aluminum foil of various weights is coated with a selected emulsion adhesive on a reverse roll laminator, reinforced with fiberglass scrim and laminated to kraft and then dried either at ambient temperatures or, preferably, over steam heated drums providing a temperature of 200° to 250°F. The emulsion is applied to the substrates at the rate of 7 to 13 lbs. (dry weight) per 3,000 square foot. In such typical construction the foil may be 1/3 mil or 2/3 mil aluminum foil, weighing respectively 15 lbs. and 29 lbs. per 3,000 square foot. The kraft may be a flame resistant treated kraft or an intumescent coated kraft. Fire resistant barrier constructions made of 2/3 mil foil, glass scrim and flame resistant treated kraft laminated with the adhesives of Examples 1 to 7 were tested by the Underwriters Laboratory procedure and the following results were observed:

	Flame Spread	Smoke Developed
Example 1	23	21
Example 2	26	24
Example 3	26	18
Example 4	39	49
Example 5	27	26
Example 6	29	22
Example 7	31	17

Since the Underwriters Laboratory ratings round off to the nearest 5, it will be seen that barrier constructions laminated with the adhesive of Examples 1, 2, 3 and 5 fully comply with the established difficultly satisfied commercial specifications of U.L. Rating 90A and that the products of Examples 4, 6 and 7 comply with Rating 90B. The results of the tests of the several laminants also illustrate the specificity of action or effect of the several components on the pyrogenic characteristics of the barrier construction and particularly the marked functional benefits conferred by the association of the chlorinated and brominated secondary plasticizers with a base polymer having the characteristics of polyvinyl acetate. The laminates described are durable, strong, flexible and are characterized by moisture resistance and moisture vapor penetration resistance and when incorporating aluminum foil develop no corrosion in use.

TAPES

Pressure-Sensitive Adhesive

A process described by E.J.M. Rosenthal and J.T. Bullock (U.S. Patent 3,212,925; October 19, 1965; assigned to Rotunda Limited, England) provides a pressure-sensitive tape which comprises a relatively noninflammable backing carrying a relatively noninflammable adhesive comprising one or more elastomers, fillers, softeners and resins, together with (a) antimony trioxide and (b) polyvinyl chloride or one or more copolymers of vinyl chloride. The backing is preferably a material based on a polymer or copolymer of vinyl chloride. It may be plasticized and may contain a flame retardant additive or additives, but other relatively noninflammable backing materials may be used. The following example illustrates the process.

Example: The composition of the adhesive in parts by weight is as follows –

Smoked sheets (or other form of natural rubber)	100
Antimony trioxide (rubber-compounding grade)	50
Zinc oxide	8
Lanolin	4
Polyvinyl chloride (paste-making polymer)	42.5
BX Terpene resin No. 10	30
BX Terpene resin No. 115	85
Butylene polymer of average molecular weight approximately 1200	30
Antioxidant	3
Petroleum solvent	420

The term "Paste-Making Polymer" refers to the type of polymer sold as being suitable for the preparation of the so-called organosols and/or plastisols, a specific example being that sold by Imperial Chemical Industries Limited as Corvic P.65/50. Examples of suitable butylene polymers are those sold as Polybutene No. 32 and Indopol H 300 by the American Oil Co. Both of the BX Terpene Resins listed are believed to be Beta-Pinene polymers, the No. 10 being a viscous liquid with a Ball and Ring melting point of 10°C. and the No. 115 having a Ball and Ring melting point of 115°C. The latter can be replaced by Piccolyte S115 which has the same melting point. A suitable petroleum solvent is that sold by Shell Mex and B.P. Limited as S.B.P.-5.

The method of making the adhesive mass is as follows: Compound the zinc oxide, antimony trioxide and antioxidant into a portion of the smoked sheet on conventional rubber compounding machinery, adding part of the lanolin if desired. Lightly mill the remainder of the smoked sheet and dissolve together with the previously mentioned portion in a portion of the solvent. Disperse the polyvinyl chloride paste-making polymer into all of the two liquid resins, i.e. BX Terpene Resin No. 10 and Polybutene No. 32, in the manner usually employed when making polyvinyl chloride spreading pastes. Dissolve the BX Terpene Resin No. 115 in a portion of the solvent. When the rubber solution is smooth, stir in the lanolin (or the remainder of the lanolin) then the polyvinyl chloride paste, and then the solution of Terpene Resin No. 115. Finally thin to spreading consistency with more petroleum solvent.

The adhesive mass is spread on to a backing material by a normal method, for example, a layer of 1 mil to 3 1/2 mils is spread on sheets of plasticized polyvinyl chloride, previously anchor coated, of thickness 3 mils to 17 mils, the thicker sheets normally carrying the thicker layer of adhesive. The composition of the backing material in parts by weight is as follows:

Polyvinyl chloride	100
Tritolyl phosphate	50
Basic lead carbonate	6
Calender lubricant	0.5
With or without antimony trioxide	5

The following test results on four varieties of adhesive tape, two in accordance with the process and two other types, indicate the improvement obtained by the process. In these tests two types of backing material were used, of the two alternative compositions specifically referred to above, Composition X containing antimony trioxide and Composition A not containing antimony trioxide. Also two different adhesive compositions were used, the composition set out in detail above, referred to as Y, and a typical, relatively inflammable adhesive of the following composition in parts by weight, referred to as B.

Smoked sheets (or other form of natural rubber	100
Zinc oxide	32
Lanolin	13
BX Terpene resin No. 115	95
BX Terpene resin No. 10	9.5
Butylene polymer of average molecular weight approximately 1200	28
Petroleum solvent	465

The four varieties of tape were as follows:

(1) Backing A, adhesive B
(2) Backing X, adhesive B
(3) Backing A, adhesive Y
(4) Backing X, adhesive Y

In all cases the approximate thickness of the backing was 0.006" and the approximate thickness of the adhesive layer after evaporation of the solvent was 0.002". Samples from a number of coils of each variety of tape were flame tested, five 1 ft. lengths from each coil being tested. The results obtained were as follows:

Variety	1	2	3	4
No. of secs. to extinction:				
Coil 1	(1)	9.3	6.6	6.7
Coil 2		11.1	8.4	6.0
Coil 3		13.5	8.2	6.5
Coil 4		11.2	9.7	5.8
Coil 5		13.5	8.0	------
Coil 6		10.1	6.3	------
Coil 7		-----	7.2	------
Average No. of secs. to extinction	(2)	11.4	7.8	6.2
Corrected extinction time in secs	13	7.4	3.8	2.2
Composition	AB	XB	AY	XY

[1] Very variable.
[2] About 17.

The figures given on the previous page for the number of seconds to extinction are each the average of the five 1 ft. samples from each coil. The test on each sample was carried out as follows: The sample was suspended vertically by sticking one end to a metal bar placed with its ends resting on top of a metal screen in the form of a metal box, open at the top and at the bottom, 30" high, 12" wide and 12" deep with a window 15" x 6" cut centrally in one face for observation purposes and with the lower edges of each side of the box cut away to a height of 1", except for legs left at each corner, to provide air inlets when the box stands on a solid surface.

A fuse was attached to the lower end of each sample to enable it to be ignited. This fuse consisted of a 1" length of 1/2" wide cellophane self-adhesive tape of the variety sold as Sellotape, folded in half with the adhesive surfaces in contact to form a half inch square with each half overlapping the end of the sample by a distance of 1/8" and adhering to it. The test was carried out by igniting the fuse by means of a match or a very small Bunsen flame, removing the flame immediately the fuse ignites and taking the time in seconds from the first ignition of the fuse until the complete extinction of the flame, whether or not ignited material fell from the tape. The corrected extinction times in the table shown are derived from the average extinction times by subtracting 4 secs., the figure of 4 secs. representing an estimate of the burning time for the fuse alone. It will be seen that the corrected extinction time for variety 3 (AY) is less than 1/3 of that of variety 1 (AB) and variety 4 (XY) is less than 1/3 of that for variety 2 (XB).

Fire Retardant Tape

W. Juda (U.S. Patent 3,336,153; August 15, 1967; assigned to Prototech Incorporated) describes a fire retardant tape comprising, in combination, a backing material capable of charring but infusible and dimensionally stable below its charring temperature; an intumescent coating bonded upon one surface of the backing material, the coating having an incipient intumescing temperature below the charring temperature; and an adhesive disposed upon the other surface of the backing material, the adhesive being temperature stable at least up to the incipient intumescing temperature range.

Typical preferred fire retardant coatings may begin to intumesce at about 200°F., for example, although coatings with higher intumescent temperatures may be used with higher temperature stable adhesives. Referring to Figure 7.1a, a fire retardant tape 1 is shown comprising a backing material 2, which may, for example, be paper, cellulose fabric or a plastic film, preferably of 0.5 to 50 mils in thickness, serving as a supporting means. The backing material 2 is capable of charring but is infusible and dimensionally stable below its charring temperature in order to maintain its bond to a surface, such as the wall 5, in the presence of heat. An intumescent fire retardant material is bonded or applied in a substantially thin layer or coating 3, preferably from 1 to 30 mils in thickness, to one major surface of the backing material 2. A temperature stable adhesive 4 is applied upon the other major surface to enable bonding of the tape 1 to a combustible to be protected.

The tape may be applied to the combustible itself, such as the wall 5, or it may be applied to a housing within which the combustible is disposed, as in the case of the walls of a safe within which papers or the like are stored. Upon the application of flame temperature to the intumescent coating 3, it changes to a thick carbonaceous foamed layer, shown at 6 in Figure 7.1b, having good insulating characteristics. Paper has been found to be especially suitable for use as backing material 2 both because of its stability up to its charring temperature and its relatively low cost.

A useful paper for this purpose, as an illustration, is a wet strength crepe paper, such as #311 Aqualized Duracel crepe paper, having a ream weight of 27 pounds (weight of 480 sheets, 24" x 36"), marketed by Brown Paper Company. It is dip-saturated with a curable butadiene-acrylonitrile copolymer latex (as Hycar OR-25 marketed by B.F. Goodrich Chemical Company). The paper is then coated on one of its major surfaces with about two ounces per square yard of a rubbery based pressure sensitive adhesive, preferably one which is normally tacky.

Before use the adhesive may be covered, for convenience, with a paper layer. A product particularly satisfactory for the process is a normally tacky and pressure sensitive adhesive coated creped paper having a thickness of approximately 9 mils, a machine direction tensile strength of about 19 pounds per inch of width, an elongation at break of 10%. Such a tape is available commercially from Minnesota Mining and Manufacturing Company as Scotch Brand Pressure Sensitive Tape #209 (4MSK).

The intumescent material is applied as a liquid, but is often somewhat brittle once it sets or cures. When flexible coatings and flexible tapes are desired, it has been found that the intumescent material may be rendered pliable by the addition of a small amount of compatible plasticizer. The addition of between 1 and 10 percent by weight of the liquid of a commercially available plasticizer has been found to render the intumescent material adequately pliable. For example, when Citroflex (that is, acetyltriethyl citrate ester plasticizer) or triethylene glycol or like compatible plasticizers is added to Alibi 107A, a commercially available intumescent fire retardant paint of the type described, for example, in United States Patent 2,984,640 (Examples 6 and/or 7), the material may be applied to a flexible backing 2 and the resultant product is sufficiently flexible to be rolled without cracking the intumescent layer. The plasticizer may be mixed with the intumescent material just prior to application to the backing material 2 or long prior to such application. The plasticizer, in the amounts above indicated, effects no significant change in the fire retardant characteristics of the intumescent material.

FIGURE 7.1: FIRE RETARDANT TAPE APPLIED TO WALL SURFACE

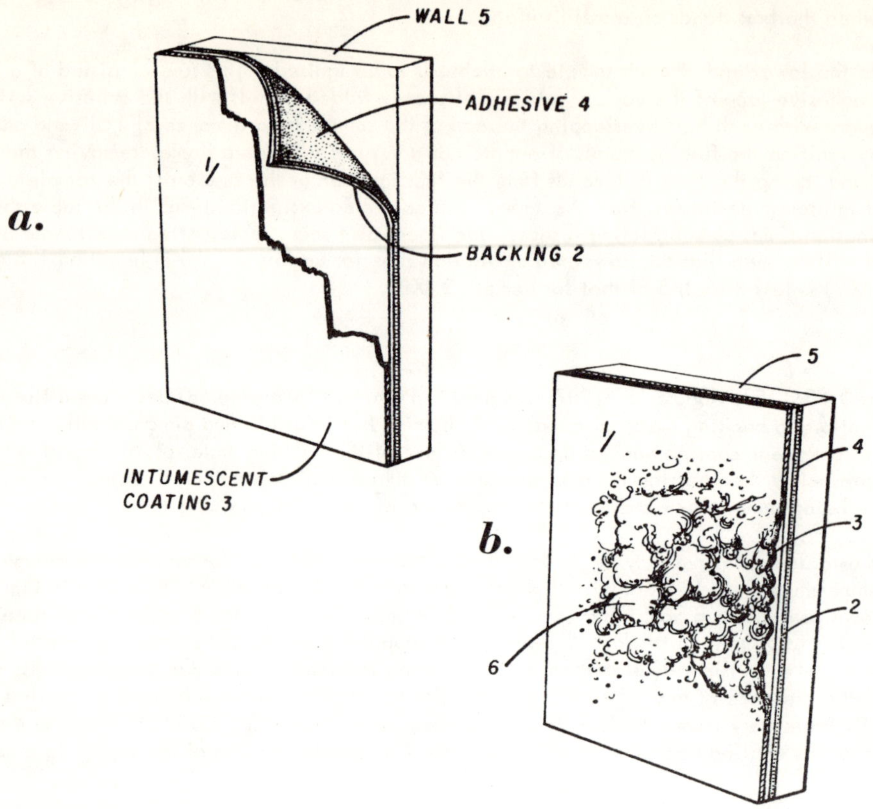

Source: W. Juda; U.S. Patent 3,336,153; August 15, 1967

The adhesive 4 must be temperature stable within the temperature range of interest, as has been discussed, and that is at least up to the incipient intumescing temperature and preferably through the intumescing temperature range. It is necessary that the adhesive be temperature stable up to the intumescing temperature in order to maintain the bond between the tape and a surface to be protected, as the wall 5. It is desirable that the adhesive 4 be temperature stable up to nearly the charring temperature of the backing 2, since continued exposure to the flame subjects the heat insulating foam to a gradual increase in temperature. Initially, the heat energy of the flame is used in causing the formation of the heat insulating thick foam. The tape affords continued protection until the backing 2 chars, provided that the adhesive 4 is temperature stable up to the charring temperature.

The adhesive 4 is preferably one that is pressure sensitive, that is, one that is normally tacky and requires merely that a pressure be exerted to effect adherence thereof to a surface. Typical tapes with pressure sensitive adhesives as well as examples of pressure sensitive adhesives may be found in Encyclopedia of Chemical Technology, vol. 1, pp. 199-201, published by The Interscience Encyclopedia, Inc., in 1947.

An example of the efficacy of the process is afforded by tests on birch panels about 12 inches by 6 inches in dimension, coated by fire retardant tape of the above described character, as shown in Figure 7.1a. The method of testing was the standard Cabinet Method specified by the Department of Defense ASTM D1360-58. A weight loss of approximately 3 g. in this test was obtained, which is far better than the permissible 15 g. Other tests of a different character with the tape applied to cardboard strips that, without the tape would burst into flame after about 5 seconds of exposure to a Bunsen burner flame, permitted exposure to the flame for 4 to 6 minutes before the cardboard even commenced to char on the back surface.

MISCELLANEOUS

Polyisocyanate-Phosphorus Derivatives

A process described by J.R. Caldwell (U.S. Patent 3,041,207; June 26, 1962; assigned to Eastman Kodak Company) involves treating combustible, nonvolatile organic materials with a mixture of a trivalent or pentavalent phosphorus derivative and a polyisocyanate. The phosphorus derivative is cemented to the surface by the polyisocyanate and the coating has substantial resistance to removal by water, organic solvents, and abrasion. Suitable classes of compounds are as follows:

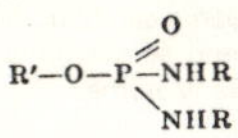

Ann. 407, 290 (1915)

$(RNH)_3PO$ Ann. 326, 129 (1903)

R is hydrogen, alkyl containing 1 to 4 carbon atoms, or aryl containing 6 to 8 carbon atoms. R′ is alkyl containing 1 to 4 carbon atoms, or aryl which may be substituted with halogen, amide, or sulfonamide. Certain isocyanate compounds such as those which contain two or more isocyanate groups or latent isocyanate groups can be used. In general, any diisocyanate or triisocyanate may be used. Suitable examples include hexamethylenediisocyanate, 2,4-tolylenediisocyanate, p,p′-diisocyanatodiphenylmethane, 1-chlorophenylene-2,4-diisocyanate, and naphthalene triisocyanate.

In the system, the amount of polyisocyanate required to produce a satisfactory adhesion to the treated material is from 20 to 100% based on the weight of the phosphorous derivative. In order to obtain adequate retardant action, from 1 to 20% of the phosphorous derivative is applied based on the weight of the textile, paper, or the like. However, the preferred range is from 5 to 10% of the phosphorous derivative. The phosphorous compounds and polymers may or may not be soluble in the common solvents. If they are soluble, they can be dissolved in a common solvent with the polyisocyanate for application. If they are insoluble, they can be ground or pulverized and mixed as slurries with the polyisocyanate. The following example illustrates the process.

Example: One molecular equivalent of triethylene glycol and two molecular equivalents of 2,4-tolylenediisocyanate were heated in dioxane solution at 80°C. for one hour. This product (A) can be used to anchor or cement a wide variety of phosphorous derivatives to textiles, paper, wood, etc. One part of product (A) and two parts of

$$C_4H_9O{-}PO(NHC_2H_5)_2$$

were mixed in tertiary butyl alcohol and the mixture was applied to cellulose acetate, cotton, and viscose fabrics. It was also applied to paper. The presence of 6 to 8% of the mixture, on a dry weight basis, retarded burning substantially. A linear polyester was made by heating $C_6H_5POCl_2$ with hydroquinone as described in U.S. Patent 2,435,252. One part of product (A) and two parts of the polyester were mixed in dioxane and applied to various textile fabrics, to give gain in weight of 10 to 12%. The fabrics were then heated at 100°C. for forty minutes to cure the coating. The fabrics showed improved resistance to burning after being laundered or dry cleaned five or six times.

COMPANY INDEX

The company names listed below are given exactly as they appear in the patents, despite name changes, mergers and acquisitions which have, at times, resulted in the revision of a company name.

INVENTOR INDEX

U.S. PATENT NUMBER INDEX

NOTICE

Nothing contained in this Review shall be construed to constitute a permission or recommendation to practice any invention covered by any patent without a license from the patent owners. Further, neither the author nor the publisher assumes any liability with respect to the use of, or for damages resulting from the use of, any information, apparatus, method or process described in this Review.

NUCLEOTIDES AND NUCLEOSIDES 1970

by S. Gutcho

Basic biological research is one of the areas of greatest scientific progress today. The introduction of new equipment and new technologies in the middle of this century increased the pace of biochemical research with resultant major forward strides into the search for the basic compounds of life.

The discovery of inosinic acid in 1847 can be cited as the initial event not only with respect to the history of the nucleotides but also for the history of the nucleic acids. The nucleotides are the fundamental units of nucleic acids. The nucleotides found in nucleic acids are phosphate monoesters of nucleosides, themselves compounds important in medical and physiological research. The nucleosides are obtained during partial hydrolysis of nucleic acids.

Some nucleotides are also important coenzymes. They include a class of compounds not found in nucleic acid. Nucleotides of this type are modified vitamins and function as coenzymes, as for example, coenzyme A.

The nucleic acids and the nucleotides were dominant during this time of major discoveries. However, in the midst of all this research, the now all-so-simple nucleotide, inosinic acid, whose discovery over a century ago marked the beginning for many of today's exciting discoveries in molecular biology, has itself come into its own.

A highly purified inosinic acid could now be prepared in tremendous quantities and advantage taken at last of its long known property, flavor enhancement. The popularity of these nucleotides as flavoring agents has resulted in their present day production as pure compounds on a commercial scale. IMP (inosene monophosphate) along with GMP (guanylic acid) have therefore become important industrial commodities which developed along with biochemical research on the nucleic acids and nucleotides.

A perusal of the accompanying Table of Contents will indicate the wealth of data offered by this book. 159 processes are described. The numbers in ( ) which follow each compound indicate the number of preparation processes covered for these compounds. This volume therefore becomes an invaluable guide to these important compounds.

200 pages

$35

SOIL RESISTANT TEXTILES 1970

by Dr. M. W. Ranney

Textile Processing Review No. 5

This latest in the series of Textile Processing Reviews summarizes recent developments in textile improvement in the area of soil resistance. As an adjunct to wash and wear properties, textiles have become not only less able to repel soil, but in many cases actually tend to attract and hold the soil. Naturally this has been one of the major consumer complaints against wash and wear textiles. In order to enhance consumer acceptance of these treated fabrics and thereby increase consumer demand for them, agents capable of preventing soil attraction and processes for applying these agents had to be found.

Ideal soil release finishes must be capable of releasing stains readily and preventing redeposition of soil during laundering. Treatments should render man-made fibers and durable press reactants less attractive to oily stains and should be more easily wetted.

In addition, the increasing use of synthetic fibers and blends of wool and synthetics in the carpet industry has resulted in the great need for effective and durable soil retardants. The irregularities in the fiber surfaces of carpeting materials readily tend to collect dirt and dirt particles resulting in a color change of the fiber. Therefore, it is obvious that the need for soil repellants is of extreme importance in making carpeting using synthetics acceptable to the consumer. Thus another facet in the search for effective soil retardants appears to be in the carpet industry.

This report summarizes the developments in soil retardant and soil release finishes in both the carpet industry and in textile manufacture. It includes the newest technology associated with the use of acrylates and fluorochemical treatments. Fluorochemicals have received the greatest emphasis in the past few years as demonstrated by the number of processes included under this heading, indicated in the Table of Contents. The abbreviated Table of Contents shown also gives a clue to the amount of valuable processing material provided by this book. The numbers in ( ) following each treating agent indicate the number of processes covered for that particular compound.

216 pages

$35

DENTIFRICES 1970

by T. Jefopoulos

Dentifrice compositions are used today as therapeutic and cosmetic aids. Toothpastes and powders must therefore incorporate as active ingredients both cleaning and polishing materials as well as various antibacterial and anticaries agents for the prevention of tooth decay.

The causes of tooth decay are well known. The search for effective agents to retard and/or prevent decay is ceaseless, since tooth decay represents to most people not only a source of pain and discomfort but also a large outlay of both time and money.

In addition, modern advertising methods have made the need for healthy teeth and gums apparent to a vast consumer population. The contribution of good dental care to overall health has become common knowledge. Furthermore, the desire for a sparkling smile has come to be regarded as a birthright.

The most common dental therapeutic agents in use today are the fluorides, ammonium compounds, penicillin and antienzyme agents. The well publicized decay preventing value of the fluoride toothpastes has made them the most widely used anticaries material today. Dental formulations also contain ingredients which contribute to the polishing and cleaning functions of many dentifrices. In addition to functioning as a cleaning, polishing and therapeutic agent, any dental formulation must offer the assurance of a non-toxic, non-sensitizing, non-irritating and stable composition.

This book offers a guide to the large store of information available in the U.S. Patent literature in this important field with regard to both therapeutic and cosmetic agents useful for modern dentifrices. It also covers information concerning dental preparations available for professional prophylactic use as well.

The following abbreviated Table of Contents will give an indication to the valuable and timely information it contains. The numbers in ( ) after each entry indicate the number of processes or formulations included in the topics covered.

191 pages

$35

ELECTROLUMINESCENT MATERIALS 1970

by M. Sittig

Electronics Materials Review No. 6

Electroluminescence occurs when electric fields cause the excitation of electrons in semiconductor solids. The phenomenon of electroluminescence offers a variety of proven and potential applications such as illuminated display panels and color TV tubes. Since the general proposition of obtaining light from solid bodies without heating them to incandescence is an attractive one, much research and development is being put into the evolution of this new breed of luminescent materials.

This latest in the Electronics Materials Review series is concerned with manufacturing processes for producing electroluminescent materials. The great quantity of manufacturing processes provided by this review is indicated by the Table of Contents shown below. The numbers in ( ) indicate the number of processes given for producing each of the indicated electroluminescent materials. This book, based as it is on the latest U.S. patent literature, will serve as an invaluable guide to a productive new industry.

306 pages

$35

FRESH MEAT PROCESSING 1970

by Dr. E. Karmas
Food Processing Review No. 12

This Food Processing Review, deals with 106 detailed processes covering essential developments in the fresh meat processing industry since 1960. The book provides a well-organized tour through the field; the processes included are well researched and presented as an easy-to-use guide to what is being done in this vital field today.

The material has been divided into two parts; processes for enhancing palatability, and preservation processes. The numbers in ( ) after each heading indicate the number of processes for each entry.

A. Palatability: Tenderness (33), Flavor and Tenderness (8), Flavoring (12), Color (13), Integral Texture (6). B. Preservation: Moisture Retention (9), Antimicrobial Treatment (10), Ionizing Radiation (7), Other Methods of Preservation (8). 236 pages. **$35**

SOLUBLE TEA PRODUCTION PROCESSES 1970

by Dr. N. Pintauro
Food Processing Review No. 11

This book describes production processes for producing soluble tea and offers a wealth of detailed practical information based primarily on the U.S. patent literature. Describes 73 specific processes in this field with substantial background information. The Table of Contents is listed below. The numbers in ( ) indicate the number of processes in that category.

- Withering and Rolling (4)
- Fermentation, Firing and Sorting (8)
- Extraction (13)
- Recovery of Aroma (10)
- Tannin-Caffeine Precipitate (Cream) (15)
- Filtration and Concentration (8)
- Dehydration Process (6)
- Agglomeration and Aromatization (9)

Illustrations. 183 pages. **$35**

ANIMAL FEEDS 1970

by M. Gutcho
Food Processing Review No. 10

This is a significant work, based on the patent literature, that shows you how to prepare a wide variety of modern feed products from numerous sources. It discusses the use of many chemicals, additives, and supplement used for animal feeds.

The 278 processes covered in this book are listed in the Table of Contents below. The numbers in ( ) indicate the number of processes described under that heading.

Introduction; Forage and Fodder (32), Fats and Oils (21), Molasses and Flavoring (15), Estrogens as Growth Stimulators (17), Antibiotics as Anabolic Stimulators (33), Antioxidants in Feeds (6), Minerals and Vitamins (13), Growth-Promoting Chemical Additives (24), Poultry Feeds (40), Ruminant Feeds (23), Feed for Swine (10), Pet and Other Feeds (19), Feed Products from Industrial Waste and By-Products (25). Indexes. 350 pages. **$35**

CONFECTIONARY PRODUCTS MANUFACTURING PROCESSES 1969

by M. Gutterson
Food Processing Review No. 6

This book is of technological significance in that it details over 200 processes for producing confections, based on the U.S. patent literature since 1960.

Based solely on new technology, this book offers substantial manufacturing information relating to this field. The wide scope of detailed data can be seen by the chapter headings indicated below:

- Candy
- Chocolate Products
- Whipped Products
- Icings
- Gels
- Coatings and Glazes
- Gums and Stabilizers
- Egg Products
- Marshmallows and Meringues
- Puddings
- Frozen Confections
- Chewing Gum
- Other Confections
- Indexes

Illustrations. 321 pages. **$35**

EDIBLE OILS AND FATS 1969

by Dr. N. E. Bednarcyk
Food Processing Review No. 5

This book describes in detail 225 recent process developments.

Shortenings; Fluid, Plastic, Miscellaneous: Margarine and Spreads; Margarine Oils, Highly Nutritional Oil Blends, Antispattering Agents, Fluid and Whipped Margarines, Flavor, Color, and Texture Modifications, Low Calorie Spreads: Salad Oils, Mayonnaise and Emulsified Dressings; Crystallization Inhibitors, Emulsified Dressings, Flavored Salad Oils, Low Calorie Dressings: Frying and Cooking Oils; Equipment, Breakdown Inhibitors, Antispattering Additives, Other Additives: Hard Butters; Preparation by Fractional Crystallization, Preparation by Ester Exchange, Miscellaneous: Oil Processing; Antioxidants and Stabilizers; Emulsifiers and Emulsions; Mixed Ester Emulsifiers, Dried Emulsion, Miscellaneous: Peanut Butter and Spreads; Chocolate Products; Indexes. Illustrations. 404 pages. **$35**

SNACKS AND FRIED PRODUCTS 1969

by Dr. A. Lachmann
Food Processing Review No. 4

The sales of snack foods in the U.S. may reach the two billion dollar mark in 1969. Many companies are actively working on new snack foods or on improved processes. The patent literature on french fried potatoes, potato chips, corn chips and other crisps is continually growing and it is the purpose of this book to present this literature in easy readable form.

French fried potatoes and their methods of production are described in the second chapter. The next chapter deals with potato chips, still the most popular product of the snack food industry. The U.S. market for potato chips is estimated to be approximately 600 million dollars in 1969. In Chapter Four the processes for corn chips are covered; in Chapter Five, apple crisps. Chapter Six describes processes for expanded chips and some specialty items; and the last chapter deals with batter mixes. Many Illustrations. 181 pages. **$35**

BAKED GOODS PRODUCTION PROCESSES 1969

by M. Gutterson
Food Processing Review No. 9

This book describes 201 recent processes for the production of baked goods. Based on the patent literature, it offers an up-to-date comprehensive publication of manufacturing processes.

There is a substantial amount of information in this book relating to the use of various chemicals and related additives.

Contents: Bread, Yeast Leavened Products, Chemically Leavened Products, Leavening Agents, Air Leavened Products, Non-Leavened Products, Refrigerated Doughs, Emulsifiers and Dough Improvers, Miscellaneous, Indexes. Illustrations. 353 pages. **$35**

SOLUBLE COFFEE MANUFACTURING PROCESSES 1969

by Dr. N. Pintauro
Food Processing Review No. 8

This book describes significant manufacturing processes for producing soluble coffee, and offers a wealth of detailed practical information based primarily on the U.S. patent literature. Describes 114 specific processes in this field with substantial background information.

Introduction: Roasting, Extraction, Filtration and Concentration, Recovery of Aromatic Volatiles, Spray Drying and Other Dehydration Processes, Freeze Drying Processes, Aromatization of Soluble Coffee Powder, Agglomeration Techniques for Soluble Coffee, Decaffeinated Soluble Coffee, Packaging of Soluble Coffee. Illustrations, Indexes, 254 pages. **$35**

ALCOHOLIC MALT BEVERAGES 1969

by M. Gutcho
Food Processing Review No. 7

The traditional brewing process is a batch operation, costly and time consuming. There would be economic advantages to improved continuous processes which would require less capital investment for plant and equipment, give savings in labor, better use of raw materials, shorter processing time, and a more uniform product.

Detailed descriptive process information is found in this review, based on 157 U.S. Patents in the brewing field, issued since 1960. The 157 processes are organized in 7 chapters which tend to follow the steps in the brewing process.

Contents: Malting, Wort, Hops, Fermentation, Freeze Concentration and Reconstitution of Beer, Chillproofing, Preservation against Microbiological Spoilage, Foam, Indexes. Illustrations. 333 pages. **$35**

PROTEIN FOOD SUPPLEMENTS 1969

by R. Noyes
Food Processing Review No. 3

The 126 Processes in this book are organized in 8 chapters by raw material source including the important newer processes for producing protein by fermentation of hydrocarbons. Another chapter on textured foods describes in detail a number of processes for producing these products that simulate meat. Indexes by company, inventors and patent number help in providing easily obtainable information.

This book is based upon the patent literature and serves a double purpose in that it supplies detailed technical information and can be used as a guide to the U.S. Patent literature on processes to obtain protein materials.

Contents: Hydrocarbon Fermentation, Fish-Based Protein, Soybeans, Cottonseed, Other Oilseeds and Legumes, Wheat and Gluten, Milk-Based Protein, Textured Foods, Miscellaneous, Indexes. Many illustrations. 412 pages. **$35**

DEHYDRATION PROCESSES FOR CONVENIENCE FOODS 1969

by R. Noyes
Food Processing Review No. 2

Describes 236 up-to-date dehydration processes for producing specific foods. Most detailed body of information ever published.

The detailed, descriptive process information in this book is based on 236 U.S. patents in the food dehydration field—issued between January 1960 and May 1968. This book serves a double purpose in that it supplies detailed technical information, and can be used as a guide to the U.S. patent literature on dehydration of foods. By indicating only information that is significant, and eliminating much of the legal jargon in the patents; this book then becomes an advanced commercially oriented review of food dehydration processes.

Dry Milk Products, Cheese and Yoghurt, Eggs, Fruit and Vegetable Juices, Fruits, Potatoes, Vegetables, Coffee, Tea, Miscellaneous. Many illustrations. 367 pages. **$35**

FREEZE DRYING OF FOODS AND BIOLOGICALS 1968

by R. Noyes
Food Processing Review No. 1

The detailed, descriptive process information in this book is based on 105 U.S. patents in the freeze drying field—issued between January 1960 and May 1968. Serves a double purpose in that it supplies detailed technical information, and can be used as a guide to the U.S. patent literature on freeze drying.

Higher costs of freeze drying are due to high capital costs for equipment, as well as higher operating costs due to high energy consumption and more limited output. This book contains many cost-cutting ideas.

Since the products obtained by freeze drying are considerably superior in most cases than those produced by other processes, considerable research and development work has been carried out attempting to lower costs. This book will give you the latest developments and recent advances. Numerous illustrations. 313 pages. **$35**

PHTHALOCYANINE TECHNOLOGY 1970
by Y. L. Meltzer
Chemical Processing Review No. 42

Advances in phthalocyanine technology have been truly explosive during the past few years. New phthalocyanine products, processes and applications have poured forth from industrial, governmental and academic laboratories at a rapid pace. These advances in technology have made themselves felt in the market place and in government programs, and have contributed to corporate sales and profits. At the same time, however, competition has become more intense in the phthalocyanine field making it imperative to keep up with the latest technological advances.

Examines recent developments in phthalocyanine technology as reflected in U.S. patents and other literature. The first 23 chapters discuss up-to-date manufacturing processes for phthalocyanine pigments and dyes. Chapters 24 through 31 discuss unusual new applications for phthalocyanines. 390 pages. **$35**

RADIATION CHEMICAL PROCESSING 1969
by R. Whiting
Chemical Processing Review No. 41

A number of radiation induced chemical processes are already operating commercially. The radiation processing of chemicals has been growing at an annual rate of about 25% per year. Currently, $100 to 150 million worth of irradiated products are produced in the United States per year, however, it has been forecast that by 1980, the value of products receiving radiation treatment will be close to $1,000 million per year.

This book surveys the radiation processing field and is based on the U.S. patent literature since 1960. Over 250 separate processes are described in detail in the chemical, polymer, rubber, petroleum, textile and other fields.

Contents: Polyolefins, Other Polmers. Elastomers, Hydrocarbons, Organic Chemicals, Inorganic and Organo-Metallic Compounds, Other Processes, Indexes. 377 pages. **$35**

CITRIC ACID PRODUCTION PROCESSES 1969
by R. Noyes
Chemical Processing Review No. 37

Detailed descriptions of production processes for citric acid, based on the patent literature. Particular emphasis is placed on deep fermentation techniques. The use of citric acid as an acidulant in the food industry will increase, even though other acids are making in-roads in citric acid consumption. It is expected that the use of citric acid in non-food industries will increase in the future. The Table of Contents, describing specific processes in considerable detail, is indicated below:

Processing, Iron Impurities, Other Microorganisms, Recovery and Purification, Other Processes. Indexes. 157 pages. **$24**

AMINES, NITRILES AND ISOCYANATES PROCESSES AND PRODUCTS 1969
by M. Sittig
Chemical Process Review No. 31

This book describes processes for producing amines, nitriles, and isocyanates, the backbones of many important commercial polymeric materials. The diamines are the fundamental building blocks for the polyamides (nylons). The nitriles (particularly acrylonitrile) are the basic raw materials for acrylic fibers and plastics. The isocyanates (particularly TDI) are the building blocks for polyurethanes.

Acrylonitrile will continue to expand due to its increasing availability at lower cost resulting in new derivatives. New low cost routes to isocyanates, based on oxidation technology will result in lower costs.

Introduction: Manufacture of Amines, Manufacture of Mono-Nitriles, Acrylonitrile Derivatives, Isocyanate Manufacture, Future Trends. 62 illustrations. 201 pages. **$35**

PRACTICAL DETERGENT MANUFACTURE 1968
by M. Sittig
Chemical Process Review No. 27

Reasonably well accepted theories indicate that linear carbon chains are destroyed completely by bacteria whereas branched chains or benzene rings cannot be digested by bacterial attack. Hence, the pressure is on to produce, at minimum, aromatics with wholly linear side chains and, at most, wholly linear detergent molecules.

The problem of replacing nonbiodegradable materials with new materials is not just one of a simple conversion of manufacturing operations but involves performance and economics.

Introduction, Manufacture of Branched-Chain Olefins, Manufacture of Linear Alpha Olefins, Manufacture of Linear Paraffins, Competitive Routes to Straight-Chain Alcohols, Competitive Routes to Alkylaromatics, Routes to other Detergent Raw Materials, Sulfation and Sulfonation Processes, Detergent Formulation, Future Trends. 76 illustrations. 212 pages. **$35**

POLYACETAL RESINS, ALDEHYDES, AND KETONES 1968
by M. Sittig
Chemical Process Review No. 26

The aldehydes are used almost entirely as intermediates for the production of condensation products of various molecular sizes. The phenol-formaldehyde or phenolic resins helped to establish the foundations for the modern plastics industry. Urea and melamine combine with formaldehyde to give other useful polymers. Now, polyformaldehyde or polyoxymethylene, the homopolymer of formaldehyde is assuming increased importance, both in its own right and as the starting backbone for a whole new family of copolymer resins which may someday rival the polyolefins in volume and breadth of application.

Contents; Introduction, Production of Aldehydes, Production of Ketones, Reactions of Aldehydes, Reactions of Ketones, Aldehyde Polymers, Future Trends 81 illustrations. 237 pages. **$35**

PHOTOCHEMICAL PROCESSES 1969
by B. Albertson
Chemical Processing Review No. 36

The purpose of this Review is to provide an up-to-date description of industrial photochemical technology as recorded in the U.S. Patent literature since 1960. Describes in detail 210 photochemical production processes.

Photochemistry, long known as a selective powerful tool, has excited the imagination of chemical engineers as a method of manufacturing various chemicals. From the economic point of view, photochemistry has the following advantages: 1. Many materials cannot be obtained in any other way. 2. Specificity of effects cannot be obtained by other methods. 3. Less need for high pressures and temperatures can lower costs.

Introduction, Photohalogenation, Photonitrosation, Organic Photochemical Reactions, Inorganic Photochemical Reactions, Photopolymerization, Indexes. Illustrations. 185 pages. **$35**

ALKALI METAL PHOSPHATES 1969
by Dr. M. W. Ranney
Chemical Processing Review No. 34

This book describes recent processes for production of alkali metal phosphates. The 97 processes are based on U.S. patents issued since 1960 and offer a comprehensive treatment of up-to-date technical information. The numerous equipment drawings, flow sheets and processing data add to a completely rounded picture of latest developments.

Contents:
- Orthophosphates
- Metaphosphates
- Pyrophosphates
- Tripolyphosphates
- Phosphites/Hypophosphites

Numerous illustrations. 344 pages. **$35**

CHLORINE AND CAUSTIC SODA MANUFACTURE RECENT DEVELOPMENTS 1969
by Dr. R. Powell
Chemical Processing Review No. 33

Contents:
- General Considerations
- Preparation of Brine for Electrolysis
- Brine Electrolysis in Diaphragm Cells
- Brine Electrolysis in Mercury Cells
- Combination Mercury Cell and Amalgam Fuel Cell System
- Recovery of Mercury from Brine
- Production of Caustic Soda in the Decomposer Unit
- Coordinated Operation of Diaphragm and Mercury Cells
- Platinum-Coated Titanium Anodes
- Electrolysis of Sea Water
- Electrolysis of Hydrogen Chloride
- Catalytic Oxidation of Chlorides
- Cracking of Hydrogen Chloride
- Nitrosyl-Chloride Route
- Cooling, Drying and Purification of Chlorine
- Purification of Caustic Soda

Numerous illustrations. 265 pages. **$35**

MONOSODIUM GLUTAMATE AND GLUTAMIC ACID 1968
by Dr. R. Powell
Chemical Process Review No. 25

This technical Review with numerous flow diagrams and equipment designs, describes in detail processes for manufacture of MSG.

U.S. demand in 1970 is expected to be 80 million pounds. World production of MSG was 240 million pounds in 1967.

Numerous illustrations. 256 pages. **$35**

HYDRAZINE MANUFACTURING PROCESSES 1968
by Dr. R. Powell
Chemical Process Review No. 28

General Considerations, Raschig Process, Bayer Process, Processes Based on Urea, Electrical Glow Discharge Processes, Chemo-Nuclear Processes, Direct Production of Anhydrous Hydrazine, Other Processes, Recovery, Concentration, Stabilization. Numerous illustrations. 252 pages. **$35**

ALCOHOLS, POLYOLS, AND PHENOLS MANUFACTURE AND DERIVATIVES 1968
by M. Sittig
Chemical Process Review No. 23

Contains process facts you need: For each process, considerable information is presented under these and other headings: Feed Materials, Pressure, Reaction Medium, Reactor Design, Temperature, Reaction Time, Catalyst, Product Recovery.

In combination with either organic acids or inorganic reactants, the alcohols, polyols, and phenols produce esters which have a substantial impact on the consumer market in the form of: flame retardants, synthetic resins, plasticizers, hydraulic fluids, solvents, synthetic lubricants, surface-active agents.

Contents: Introduction, Manufacture of Alcohols, Manufacture of Glycols, Manufacture of Polyols, Manufacture of Phenol, Reactions of Alcohols, Reactions of Glycols and Polyols, Phenol Derivatives. 71 illustrations. 201 pages. **$35**

ACETYLENE PROCESSES AND PRODUCTS 1968
by M. Sittig
Chemical Process Review No. 22

Describes specific processes for acetylene production, and acetylene derived products.

Contents: Introduction, Manufacture of Acetylene, Separation of Acetylene from Gas Mixtures, Handling of Acetylene, Reactions of Acetylene, Future Trends. 90 illustrations. 214 pages. **$35**

AROMATICS MANUFACTURE AND DERIVATIVES 1968
by M. Sittig
Chemical Process Review No. 17

Contents: Introduction, Production of Aromatics, Separation of Aromatics, Purification of Aromatics, Reactions giving Hydrocarbon Products, Other Reactions, Phenol Production, Styrene Manufacture and Derivatives, Future Trends. 73 illustrations 232 pages. **$35**

CARBON BLACK TECHNOLOGY RECENT DEVELOPMENTS 1968
by Dr. R. Powell
Chemical Process Review No. 21

This technical Review, replete with numerous flow diagrams and equipment designs, describes the latest developments in carbon black manufacture. The information is organized by company, and type of process.

Carbon black growth parallels that of the rubber industry, since 95% of production is used in that industry. For the United States an average growth rate of 5% per year seems likely; and for the rest of the world, demand should increase by 10 to 15% per year for the next few years.

Introduction; Feedstocks, Channel Blacks, Furnace Blacks, Thermal Blacks, Acetylene Blacks, High Structure Carbon Blacks, Low Structure Carbon Blacks, Unconventional Processes, Carbon Black Pelletizing, Other Finishing Treatments. Numerous illustrations. 242 pages. **$35**

HYDROGEN PEROXIDE MANUFACTURE 1968
by Dr. R. Powell
Chemical Process Review No. 20

There are both chemical and electrolytic processes for manufacturing hydrogen peroxide. This Review is concerned exclusively with the chemical processes, since all new plants are being built based on chemical methods of production. The electrolytic process will eventually be displaced.

Major emphasis is placed on the anthraquinone processes. Newly developed chemical processes are also described. A portion of the book is also devoted to purification, stabilization, and concentration techniques.

Shows processes of such firms as FMC Corp., Air Liquide, CIL, Laporte, Ugine, Montecatini, PPG, duPont, Shell, Kali-chemie, etc.

Contents: General Considerations, The Anthraquinone Process, Other Processes, Purification, Concentration, Stabilization. Numerous illustrations. 221 pages. **$35**

PARAFFINS AND CYCLO-PARAFFINS MANUFACTURE AND DERIVATIVES 1968
by M. Sittig
Chemical Process Review No. 19

Exciting new product developments are planned as linear paraffins are available in quantity from improved separation processes. The higher linear paraffins have assumed new importance as economic raw materials for the production of biodegradable detergents.

Cyclohexane, as the most important cycloparaffin, is the second most important benzene derivative after styrene in terms of volume production. Based on the ever-growing needs of the synthetic fiber industry, cyclohexane production and prices are attracting interest in other cyclohexane-based processes.

Contents: Introduction, Production of Paraffins and Cycloparaffins, Separation of Paraffins and Cycloparaffins, Purification of Paraffins, Reactions of Paraffins, Reaction of Cycloparaffins. 92 illustrations. 234 pages. **$35**

ELECTRO-ORGANIC CHEMICAL PROCESSING 1968
by Dr. C. Mantell
Chemical Process Review No. 14

This volume has been written from the viewpoint of the chemical engineer, with emphasis on plant processes, operating data, and plant design. The commercial successes in this field have been attained by the chemical engineering approach.

Contents: 1. History, Potentials of the Elements, 2. Reaction Potentials and Overvoltage, 3. Controlled Potentials, 4. An Electrolytic Approach to Simple Problems. 5. Electrolytic Amalgams as Reductants, 6. Selectivity of Electro-organic Processing, 7. Experimental Investigation, 8. Engineering Materials for Scaleup, 9. Oxidation and Regeneration of Oxidants, 10. Dialdehyde Starch and Cell Design, 11. Oxidation of Sugars, 12. Hydroisomerization, 13. Tetra Alkyl Leads by Electrolysis-Commercial Plant, 14. Electrolytic Treatment of Municipal Water Wastes, 15. Electroplating of Organosols, 16. Future. 186 pages. **$35**

OLEFINS-MANUFACTURE AND DERIVATIVES 1968
by M. Sittig
Chemical Process Review No. 12

This Review concentrates on the olefins which are available or potentially available in large volume. These are the materials of most immediate interest to industry. They are the most rewarding materials from a research standpoint in that development of an interesting end-use will not be hindered by the unavailability of an exotic raw material.

New techniques for producing olefins (such as oxidative dehydrogenation) are the subject of intense research and development, and of increasing commercial application. Dimerization, disproportionation, and polymerization, are all used to tailor available olefin feed materials to commercial requirements for olefins of various chain lengths.

Contents: Introduction, Manufatcure of Olefins, Separation of Olefins, Purification of Olefins, Reaction of Olefins. Future Trends. 213 pages. **$35**

CATALYSTS AND CATALYTIC PROCESSES 1967
by M. Sittig
Chemical Process Review No. 7

In this book, catalysts are discussed in terms of the processes which use the catalysts. In each case major emphasis is placed on the catalyst itself—on its composition, preparation and handling in the process. The balance of the process operating conditions are discussed as well, so that the catalyst will be highlighted against a background of discussion of the integrated overall process, including vital catalyst activation and recovery operations, for example.

More and more specific catalysts are being developed to operate on particular molecular species in a feed mixture and to convert them to specific types of products.

Contents: Hydrocarbon Conversion Processes, Hydrocarbon Polymerization Processes, Hydrocarbon Oxidation processes Future Trends. 109 illustrations 303 pages. **$35**

TITANIUM DIOXIDE AND TITANIUM TETRACHLORIDE 1968
by Dr. R. Powell
Chemical Process Review No. 18

A technological review of two important titanium chemicals.

Titanium dioxide is the most important white pigment. The history of titanium dioxide manufacture is based on the sulfuric acid process from ilmenite ore. However since 1954, the chlorination process from rutile ore has gradually increased in importance. Today, all new plants are based on the chlorination process, offering better pigment properties, and potentially lower costs.

This book is a technological review of titanium dioxide production processes, based on the chlorination of rutile ore. It describes the various processing steps (a) chlorination of the ore, (b) separation of sublimated solids, (c) purification of crude titanium tetrachloride, and (d) conversion of titanium tetrachloride, to titanium dioxide. Numerous illustrations. 306 pages. **$35**

DIOLEFINS—MANUFACTURE AND DERIVATIVES 1968
by M. Sittig
Chemical Process Review No. 16

Diolefins will assume increasing importance due to new processes, new raw materials, and new products.

A major avenue of opportunity which has opened for the diolefins in the last few years is polymerization to a slight degree, or oligomerization, to "big-ring" componds having 8 or 12 carbons atoms in the ring as a result of dimerization or trimerization of butadiene, for example. These materials, themselves olefinic in nature, offer a fantastic range of synthetic possibilities. Hence both the manufacture and utilization of this new family of organic intermediates are covered in some detail in this Review.

Contents: Introduction, Manufacture of Diolefins, Separation of Diolefins, Purification of Diolens, Manufacture of "Big Ring" Compounds, Reaction of Diolefins, Future Trends. 61 illustrations. 197 pages. **$35**

SUGAR ESTERS 1968
by Research Corporation

Already approved for use in foods in a number of countries, use of sugar esters in food in the United States awaits FDA clearance. Sugar esters are an important new raw material for the food industry.

This book contains the papers presented at a Symposium held in California during 1967. The Contents are shown below:

Early History, Nebraska-Snell Process, Present Status of Processes, Sugar Outlook, Use of Sucrose Monopalmitate, Applications in Foods, Animal Feeding, Sucrose Esters in Cosmetics, Sucroglycerides in Foods, The Enhancement of Herbicide Sprays, Herbicide Adjuvants, Effectiveness of Fat-Sugar Derived Surfactants, Laundering Performance, Sucrose Ester Detergents, Evaluation of Sucrose Monotallowates, Water Quality, Water Pollution Problems. 134 pages. **$15**

INDUSTRIAL GASES MANUFACTURE AND APPLICATIONS 1967
by M. Sittig
Chemical Process Review No. 4

This book discusses conventional cryogenic air separation and purification techniques in considerable detail. Such plants have now reached a high state of refinements as regards low cost, efficient, safe operation. Many of the patents on air separation processes have expired and this opens up opportunities for smaller firms to enter the business.

In order to overcome problems associated with choice of materials of construction, and in refrigeration requirements, engineers and scientists are working on radical new processes, operating at temperatures nearer to room temperature. This book discusses newer techniques such as adsorption using molecular sieves, and permeation using various membrane materials.

This is a down-to-earth practical book summarizing the present state of the art. 313 pages. 103 illusrtations. **$35**

COMBINE HYDROCARBONS AND OXYGEN FOR PROFIT 1968
by M. Sittig
Chemical Process Review No. 11

Contents: Introduction; Cyclic Oxides from Olefins and Diolefins; Acids, Anhydrides and Esters; Olefins and Diolefins by Oxdehydrogenation; Aldehydes and Ketones from Olefins; Alcohols from Paraffins and Cycloparaffins; Peroxy Compounds; Future Trends. 214 pages. **$35**

COMBINE HYDROCARBONS AND NITROGEN FOR PROFIT 1967
by M. Sittig
Chemical Process Review No. 8

Products and Derivatives from: Ammonia, Diethylamine, Nitric Acid, Alkanolamines, Pyridines, Urea, Nicotinic Acid, Calcium Cyanamide, Caprolactam, Ethylene Imine. 64 illustrations. 200 pages. **$35**

COMBINE HYDROCARBONS AND HALOGENS FOR PROFIT 1968
by M. Sittig
Chemical Process Review No. 10

Production of Halogenated Methane Derivatives, Halogenated Ethanes and Ethylenes, Halogen Derivatives from Propylene, Halogen Derivatives of Butadiene, Halogen Derivatives of Higher Paraffins, Halogenated Aromatics and Future Trends. 62 illustrations, 181 pages. **$35**

SYNTHESIS OF ACRYLIC ESTERS BY TRANSESTERIFICATION 1967
by J. K. Haken

Transesterification is of more general application than often suggested, and can be utilized for commercial synthesis of many homologous series of esters.

Contains available literature, including patents. Patents have been examined, extracted and tabulated. The literature has been surveyed through Jan. 1967 Chem. Abstracts. 95 pages. **$18**

ANTIOBESITY DRUG MANUFACTURE 1970
by Dr. B. Idson
Chemical Processing Review No. 43

This book discusses 162 processes in the field of anorectic drug production technology. The numbers in ( ) in the Table of Contents show the number of preparations discussed for each heading.

N-Alkyl Amines (2), Substituted Aminopropanes—Arylaminopropanes (17), Substituted Aminopropanes—1-Heterocyclic-2-Aminopropanes (2), Substituted Aminopropanes—Bicyclic Aminopropane (1), Substituted Aminopropanes—1-Heterocyclic Oxoaminopropanes (1), Substituted Aminoalkenes (2), Aralkyl Amines (22), Aralkyl Hydrocarbon (1), Ring Substituted Heterocyclic Compounds (20), Nitrogen Substituted Heterocyclic Compounds (43), Bicyclic Compounds (2), Azacyclic Compounds (5), Miscellaneous Functional Substitutions (24), Alkaloids (1), Antibiotic (1), Resins (3), Steroids (6), Amphetamine Salts (2), Compositions (8), Appendix. 193 pages. **$35**

TETRACYCLINE MANUFACTURING PROCESSES 1969
(2 Volumes)

This work contains 196 U.S. patents for the production of tetracyclines and derivatives, and is published in two sections. An exhaustive attempt was made to include all U.S. tetracycline patents issued up to the Summer of 1969. Only those patents relating to fermentation or chemical reactions were selected. Those U.S. patents that are merely compositions of tetracyclines with other materials, addition compounds, or salts, are not included in this work.

Chlortetracycline, Oxytetracycline—5-Hydroxytetracycline, Tetracycline, Demethyl Tetracyclines—DMTC and DMCTC, 2N-Derivatives, Position 4 Derivatives, 6-Methylene Derivatives, 6-Deoxy Derivatives, Anhydrotetracyclines, 7-and/or 9-Derivatives, 11a-Halo Derivatives, 5a, 11a-Dehydrotetracyclines, 12a-Derivatives, Epimers, Mechanism Study Intermediates. Indexes. 931 pages. **(2 Volumes) $45**

BASIC PATENTS FOR MAJOR DRUGS 1969
(3 Volumes)

This book contains basic manufacturing patents for the 301 major drugs most frequently prescribed in the U.S. The actual patents themselves are reproduced. Contains 341 U.S. Patents.

A valuable feature is that you can easily obtain patent expiration dates and determine what drugs will be "out-of-patent" in the near future. A therapeutic index is also included. 1705 pages. 3 bindings. $75

VITAMIN B_{12} MANUFACTURE 1969
by R. Noyes
Chemical Process Review No. 40

Vitamin B_{12} active substances are important therapeutic products for treatment of pernicious anemia. Also used for treatment of various other human ailments, and as a veterinary growth factor. This book offers various methods of producing vitamin B_{12} active substances. 327 pages. **$35**

NEW FERTILIZER MATERIALS 1968
by C. I. E. C.

This book is invaluable to fertilizer and chemical manufacturers who are looking to produce or market potentially useful new fertilizer materials.

Ureaform, Crotonylidenediurea (CDU), Isobutylidene Diurea, Superphosphoric Acid, Triple Superphosphate of 55% P_2O_5 Content, Ammonium Phosphates, Ammonium Polyphosphate, Nitrophosphates, Nitrate of Potash, Potassium Metaphosphate (Potassium Polyphosphate), Monopotassium Phosphate and Monoammonium Phosphate, Magnesium as a Plant Nutrient, Magnesium Phosphates, Magnesium Ammonium Phosphate and Related Compounds, Sulfur as a Fertilizer, Oxamide, Urea Nitrate, Urea Phosphate, The Hydrides of Phosphorus, Red Phosphorus, Fertilizer Application Equipment, Modern Use of Fertilizer. Numerous illustrations. 430 pages. **$35**

FERTILIZER DEVELOPMENTS AND TRENDS 1968
by A. V. Slack

This book analyzes the opportunities for improving fertilizer technology based on current world-wide trends in research and development.

Chemistry of fertilizers and current manufacturing methods are discussed only to the extent necessary for adequately relating new departures to standard practice. The major emphasis is in departures that have been introduced in the past five or six years, and new developments that will assume major importance in the future.

Contents: Research and Development, Ammonia, Ammonium Nitrate, Urea, Ammonium Sulfate, Slow Release Nitrogen, Other Nitrogen Fertilizers, Phosphoric Acid, Ammonium Phosphate, Nitric Phosphates, Superphosphate, Thermal and Miscellaneous Phosphate Processes, Potassium Fertilizers, Fluid Fertilizers, Bulk Blending, Minor Nutrients. 98 illustrations. 406 pages. **$35**

PESTICIDE PRODUCTION PROCESSES 1967
by M. Sittig
Chemical Process Review No. 5

Production processes for these pesticides are described: Olefin-based Insecticides; Olefin-based Fungicides; Olefin-based Herbicides; Diolefin-based Insecticides; Diolefin-based Fungicides; Diolefin-based Herbicides; Aromatic-based Insecticides; Aromatic-based Fungicides; Aromatic-based Herbicides 48 flow diagrams. 200 pages. **$35**

UREA PROCESS TECHNOLOGY 1968
by Dr. R. Powell
Chemical Process Review No. 13

The information in this book is up-to-date, and is based on information made available since 1960. The information in this volume includes a huge variety of numercial data.

General Considerations, Processes, Low Biuret Processes, Product Finishing and Controlled Release Fertilizer. Illustrations. 324 pages. **$35**

HAIR PREPARATIONS 1969
by A. Williams

Up-to-date technology of hair care products.

The rapidly rising demand for hair care preparations has resulted in a steady stream of new products for both the consumer and the professional beauty shop operator over the past few years. Active research and new product formulation have been carried out by a number of firms. This book discusses a significant portion of these later developments.

This book provides a detailed technological summary of recent developments based on 138 U.S. patents, since 1960, covering all aspects of hair preparations for the head, beard, eyelashes, and eyebrows.

Introduction, Dyeing, Bleaching, Waving, Setting, Shampoos-Rinses, Grooming-Tonics, Shaving Assistants, Other, Indexes. 208 pages. **$35**

SUSTAINED RELEASE PHARMACEUTICALS 1969
by A. Williams

This book is based on the U.S. patent literature, and presents substantial technical information for production of these products.

TABLETS-PILLS
- Cellulosic Coatings
- Lipid Coating
- Gels-Gums
- Specific Medicaments
- Tablet Design
- Miscellaneous

CAPSULES
- General Coatings
- Particle Size Control
- Capsule Design & Manufacture

INJECTABLES
- Antibiotics
- Miscellaneous

MISCELLANEOUS RELEASE PREPARATIONS
- Suppositories
- Powders
- General

Indexes. 273 pages. **$35**

VITAMIN E MANUFACTURE 1969
by T. Rubel
Chemical Processing Review No. 39

This review relates the known methods for the preparation of tocopherols from natural products or by synthetic means, and conversion of non-alpha tocopherols.

Introduction; Tocopherols From Deodorizer Sludge, Conversions to Alpha Tocopherol, Synthesis of Tocopherols, Miscellaneous Related Processes, Indexes. 114 pages. **$24**

FIBRINOLYTIC ENZYME MANUFACTURE 1969
by T. Rubel
Chemical Processing Review No. 38

This Review presents some methods for the production and purification of fibrinolytic agents, their precursors and activators.

Plasminogen and Fibrinolysin, Urokinase, Streptokinase and Streptodornase, Other Fibrinolytic Enzymes. Indexes. 139 pages. **$24**

AMMONIUM PHOSPHATES 1969
by Dr. M. W. Ranney
Chemical Processing Review No. 35

This book describes recent processes for production of ammonium phosphates.

Introduction; Ammonium Orthophosphates, Diammonium Orthophosphates, Ammonium Polyphosphates, Metal Ammonium Phosphates, Ammonium Phosphate—Ammonium Nitrate Mixtures. Many illustrations. 278 pages. **$35**

CONTROLLED RELEASE FERTILIZERS 1968
by Dr. R. Powell
Chemical Process Review No. 15

This book offers you complete technical data on numerous processes and products in this field. The two major approaches are (a) compounds of low solubility, and (b) coated granules.

Introduction; Compounds of Low Solubility, Coated Granules, Prevention of Nitrogen Losses, Rapid-Release Fertilizer. 279 pages. **$35**

NITRIC ACID TECHNOLOGY RECENT DEVELOPMENTS 1969
by Dr. R. Powell
Chemical Process Review No. 30

Ammonia Oxidation Process, Wisconsin Thermal Process, Nitrogen Fixation by Shock Waves, Nitrogen Fixation in a Nuclear Reactor, Absorption of Nitrogen Oxides in Water, Concentration of Dilute Nitric Acid Solutions, Direct Production of Concentrated Nitric Acid, Purification of Nitric Acid, Stabilizers for Nitric Acid. Numerous illustrations. 245 pages. **$35**

PHOSPHORIC ACID BY THE WET PROCESS 1967
by R. Noyes
Chemical Process Review No. 9

This book is devoted to production of phosphoric acid by the wet process. Production technology is changing rapidly, with new single tank reactors, the advent of superphosphoric acid, new commercial processes for acidulation with hydrochloric and nitric acids, new semihydrate and anhydrite processes, etc. 282 pages. **$35**

AMMONIA AND SYNTHESIS GAS 1967
by R. Noyes
Chemical Process Monograph No. 26

Outlines tehnology and economics pertaining to ammonia production, with many flow diagrams and tables. Various methods of hydrogen and synthesis gas preparation, shift conversion techniques, various carbon dioxide removal systems, final purification, compression, and ammonia synthesis. Bibliography. Many illustrations. 175 pages. **$20**

POTASH AND POTASSIUM FERTILIZERS 1966
by R. Noyes
Chemical Process Monograph No. 15

Introduction: Markets, Future Projections; Geology and Geography; Underground Deposits of Soluble Minerals; Brines, Other Sources of Potash; Flotation; Crystallization; Other Refining Processes; Potassium Sulphate; Potassium Nitrate; Potassium Phosphates; Final Product. 210 pages. **$15**

WATERPROOFING TEXTILES 1970

by Dr. M. W. Ranney

Textile Processing Review No. 4

This Textile Processing Review summarizes the technology of water resistant treatments for textiles and fabrics as described in the U.S. patent literature since the early 1950's. 246 waterproofing processes are included—64 relate to use of fluorochemicals.

The numbers in ( ) after each entry in the Table of Contents, where the treatment processes are organized by the agent used, indicates the number of production processes for each agent.

Production Processes for Waterproofing Textiles using the following Agents: Metal Salts and Wax-Containing Formulations (44), Silicones and Alkyl Polysiloxanes (53), Organofunctional Silicones and Fluorosilanes (20), Acrylics (8), Nitrogen Containing Compounds (30), Fluorochemical Compounds (64), Elastomer, Vinyl, Polyolefin Vapor Permeable Fabrics (27), Miscellaneous Treatments (9). 353 pages. **$35**

FLAME RETARDANT TEXTILES 1970

by Dr. M. W. Ranney

Textile Processing Review No. 3

Describes 177 commercial processes to produce flame retardant textiles and fabrics.

Most activity is based on chemical modification of cellulose through hydroxyl groups. Use of phosphoric acid, urea-phosphates, and other phosphorylating agents all confer flame retardant properties to cellulose. A significant portion of this book is devoted to the latest in application of phosphorous containing materials.

Numbers in ( ) indicate the number of processes described.

Ammonium Salts, Borates (12); Antimony, Titanium Metal Oxides (25); Amine-Phosphorus Products (21); Aziridines, APO, APS (21); Methylol-Phosphorus Polymers, THPC (27); Phosphonitrilic Chlorides (9); Triallyl Phosphates and Phosphonates (26); Silicones Isocyanates, Miscellaneous (10); Nylon, Acrylics (18). 373 pages. **$35**

CREASEPROOFING TEXTILES 1970

by Dr. M. W. Ranney

Textile Processing Review No. 2

Summarizes with detailed process information relating to textile creaseproofing agents used to obtain wash and wear, or permanent press fabrics. Over 300,000 words, describes 343 processes in this field. Shows you chemical agents used, and processes by which they are applied.

Dimethylolethlene Urea and Related Compounds, Aldehyde-Urea Condensates, Uron Resins, Aminoplasts—Catalyst Performance, Melamine Derivatives, Triazones, Carbamates, Other Nitrogen-Containing Compounds, Phosphorus-Amino Compounds—Aziridines, Aminoplast-Thermoplastic Resin Compositions, General Processing Techniques and Formulations, Aldehydes, Acetals, Epoxies, Epihalohydrins, Sulfones, Sulfonium Salts, Cross-Linking Agents, Miscellaneous, Polymeric Coatings—Rubber, Vinyl, Silicones, Radiation Curing, Wool, Nylon, and Others. Indexes. 460 pages. **$35**

PLASTIC FILMS FROM PETROLEUM RAW MATERIALS 1967

by M. Sittig

Chemical Process Review No. 6

This Review is devoted to the production of synthetic polymer films. Discusses polyolefins, vinyls, and other important polymer film-forming materials, in an easy to read format.

In this, as in many fields of chemical product technology, there is a great scarcity of practical information dealing with the materials, equipment, techniques and operating conditions used in actual manufacturing processes. This Review will, it is hoped, begin to overcome that need and will begin, at least, to introduce the reader to the practicalities of polymer manufacture, film production, and film treatment processes.

Contents: Raw Materials, Polymer Manufacture, Film Production Processes, Film Compositions, Film Treating Processes, Future Trends. 118 illustrations. 276 pages. **$35**

POLYOLEFIN PROCESSES 1967

by M. Sittig

Chemical Process Review No. 2

This book is devoted to manufacturing techniques and processes for polyolefin resins—from the olefinic feed material to the purified raw material. Thus, it includes catalyst recovery, solvent recovery, and the attendant steps between olefin and polyolefin. This review is based primarily on the patent literature highlighting the combination of intense research effort and major commercial potential.

Includes: Current Status of Various Polyolefins, Processes Using Non-Metallic Catalysts, Process Using Metal Oxide Catalysts, Processes Using Metal Alkyl-Reducible Metal Halide Catalysts, Polymer After-Treatment, Polymer Fabrication, Future Trends.

For each process, data is presented under these headings: Feed Materials, Temperature, Pressure, Reaction Time, Reaction Medium, Catalyst, Reactor Design, Product Recovery. 96 illustrations. 234 pages. **$35**

PLASTICIZER EVALUATION AND PERFORMANCE 1967

by I. Mellan

This book will help you evaluate: 1. A new plasticizer with a known resin; 2. A known plasticizer with an unknown resin; 3. A new plasticizer with a new resin.

The 110 tables of performance data, and the standard methods of testing described in this book, provide data from which one can estimate what a particular plasticizer will do in a specific resin.

The plasticizers were chosen for inclusion, primarily by their greatest demand in the industry.

Chapters: 1. Introduction; 2. Testing; 3. Comparative Performance Data; 4. Performance Data of Individual Plasticizers; 5. Brand Names and Their Manufacturers; 6. Abbreviations for Coding Plasticizers; 7. Chemical Names of Plasticizers and Their Brand Names; 8. Brand Names of Plasticizers and Their Chemical Names; 9. Bibliography. 178 pages. **$20**

DYEING OF SYNTHETIC FIBERS 1969

by C. Whiting

Textile Processing Review No. 1

The dyeing of synthetic fibers continues to be a challenge. Successful methods however, have been developed. Chapters 2 through 7 of this Review are divided into two sections. The first presents various new dyes and dyeing processes which are applicable to synthetic fibers. The second section concerns many auxiliary products available to aid in production of an acceptably dyed product. These include dye improving agents, leveling and retarding agents, agents used in after treatment to achieve optimum fastness properties.

Introduction: Dyeing Polyolefin and Polypropylene Fibers, Dyeing Polyamide Fibers, Dyeing Polyester Fibers, Dyeing Acrylic Fibers, Dyeing Hydrophobic Fibers, Dyeing Glass Fibers, Dyeing Miscellaneous Fibers, Dyes Applicable to More than One Type of Fiber, Dyeing Fiber Blends. 257 pages. **$35**

TEXTURED AND NOVELTY YARN PROCESSES 1967

by M. Harrison

In the world today, hundreds of millions of pounds of synthetic fibers are being textured by a wide variety of processes. Yet, for those of you who wish to explore this important subject, there is very little published material. Texturing is a process where cold and slick continuous filament yarns of synthetic fibers are converted into yarns with greatly improved bulk, stretch, adsorbency, and hand.

This book has a novel approach in organizing information. By using this book, you will be able to research, develop, adapt, and solve problems in this field.

This book has been written so that it is useful to both those well versed in the field, as well as those who have had little experience with these processes, and who want to become more familiar with them. 316 pages and many illustrations. **$35**

SYNTHETIC LEATHER FROM PETROLEUM 1969

by M. Sittig

Chemical Process Review No. 29

The newer synthetic leathers offer permeability to water vapor and air, as does natural leather. In addition, the newer products offer good tear strength and softness. It is expected that these newer synthetic leathers will make substantial inroads in shoe and other markets.

These modern synthetic leathers consist, in general, of nonwoven mats of such fibers as polyesters impregnated with such binder resins as polyurethanes. The fiber-binder combination is then rendered porous by one of a number of different processes.

Introduction: Manufacture of Fiber-Forming Polymers, Types of Structural Fibers Used, Mat Formation, Mat Treatment, Binder Polymer Formulation, Other Ingredients, Fiber-Binder Combination, Consolidation, Coating, Introduction of Porosity, Surface Modification, Product Properties, Future Trends. 84 illustrations. 214 pages. **$35**

COMPATIBILITY AND SOLUBILITY 1968

by I. Mellan

Normally, it requires laborious testing to determine compatibility of polymers, resins, elastomers, plasticizers, and solvents. Predictions made without testing or literature searching, are usually unreliable. We all know of the many commercial products that were failures, when compatibility and solubility considerations were ignored.

This book helps you evaluate proper materials by the use of 224 tables. The tables were originally published by manufacturers of various materials, and reproduced in this book, to give you a reference work to this important subject. The tables are organized in three sections—polymers and resins, plasticizers and esters, and solvents. The tables in each section indicate the solubility and compatibility of the particular material with a wide range of other materials. 304 pages. **$20**

SYNTHETIC FIBERS FROM PETROLEUM 1967

by M. Sittig

Chemical Process Review No. 1

Production processes for the four major synthetic fibers: nylon, polyesters, acrylics, and polyolefins. Contains a wealth of processing data with particular emphasis on process conditions. A technical evaluation of the synthetic fiber industry. 116 illustrations. 275 pages. **$35**

STEREO-RUBBER AND OTHER ELASTOMER PROCESSES 1967

by M. Sittig

Chemical Process Review No. 3

Synthesis of new polymers with rubber-like elasticity is a vital, growing industry. Modification of molecular architecture of elastomers for specific properties is subject of worldwide research. Review summarizes state of the art in elastomer industry; new routes to monomers and polymerization techniques. 215 pages. 77 illustrations. **$35**

PRODUCING FILMS OF ELECTRONIC MATERIALS 1970
by M. Sittig
Electronics Materials Review No. 5

Films of varying thickness have a multitude of applications in the fabrication of modern electronic devices. This book describes numerous processes for film production as indicated in the Contents below. The numbers in ( ) indicate the number of processes described under that heading.

- Cathode Sputtering (10)
- Vacuum Evaporation (18)
- Gas Plating (2)
- Electroplating (6)
- Electroless Deposition (6)
- Film Deposition From Reacting Vapors (32)
- Explosive Evaporation (1)
- Squeezing on Optical Flats (1)
- Electron Deposition (2)
- Ion Beam Deposition (1)
- Reaction & Deposition From Solution (5)
- Oxide Films by Reaction at Solid Surfaces (7)
- Carrier Transport in Vapor Phase (6)

97 Illustrations. 295 pages. **$35**

DOPING AND SEMICONDUCTOR JUNCTION FORMATION 1970
by M. Sittig
Electronics Materials Review No. 4

Donor or acceptor impurities may be added to the liquid phase from which semiconductor crystals are grown or they may later be deposited on the crystal surface and diffused inward. In either case the addition of such impurities is called doping. Where there is formed a transition between a P-type and an N-type material as a consequence of one of these operations, a semiconductor junction is formed. This book describes 115 processes for doping and semiconductor junction formation.

Production of Alloyed Junctions, Diffusion Processes, Melt Grown Junctions, Doping During Melting, Simultaneous Dopant and Substrate Deposition, Spark Doping Processes, Doping by Particle Bombardment, Hydrothermally Grown Junctions, Doping Epitaxially Grown Layers, Future Trends. Indexes. 132 illustrations. 318 pages. **$35**

SEMICONDUCTOR CRYSTAL MANUFACTURE 1969
by M. Sittig
Electronics Materials Review No. 3

1. Introduction
2. Verneuil Method
3. Other Fusion Processes
4. Spark Discharge Process
5. Single Crystal Pulling From a Crucible
6. Single Crystal Production by Zone Melting
7. Other Melt Processes
8. Dendritic Crystal Production From Melts
9. Dendritic Crystal Production From Vapor
10. Single Crystals by Sintering
11. Melting Under Pressure
12. Crystals by Deposition From Vapor Phase
13. Crystals by Deposition From Vapor Phase Reaction
14. Hydrothermal Growth
15. Crystal Growth in Solution Melts
16. Shaped Crystals
17. Single Crystal Film
18. Future Trends

Indexes

106 illustrations. 303 pages. **$35**

TEXTILE GUIDE TO EUROPE 1970
by Noyes Data S. A.

This is the fourth in our series of directories dealing with key industries of Western Europe.

The detailed company information provided falls into two categories. In Part 1, entries are listed alphabetically under country and are described wherever possible, with the following data: full name and address; principal executives; product range; domestic and foreign subsidiaries and affiliates; plant location; latest sales figures and numbers of employees. In Part 2, the arrangement of companies is again by country but this time classification is according to the type of textile products manufactured.

Contains no advertising material. Layout is simple and easy to follow. Entries can be located rapidly and information extracted without reference to complicated lists of explanatory symbols and abbreviations. A French, German and Spanish vocabulary of key words has been included. 220 pages. **$20**

ELECTRONICS GUIDE TO EUROPE 1969
by Noyes Development S. A.

Electronics is currently one of the most dynamic sectors of European industry, with a growing number of participating companies manufacturing an ever expanding range of products.

It contains up-to-date company profiles of over 600 leading electronics manufacturers in 14 countries.

The comprehensive company information included in this guide will greatly assist in the planning of effective marketing operations. Wherever possible, entries are accompanied by the following key data arranged for easy comparison: Full name and address, Ownership, Principal executives, Product range, Domestic and overseas subsidiaries, Plant location, Latest sales figures, Number of employees.

This book will help you: Increase sales, Prepare market reports, Know company officials, Locate suppliers, Organize joint ventures, Make licensing arrangements. 150 pages. **$20**

EUROPEAN CHEMICAL MARKET RESEARCH SOURCES 1969
by Noyes Development S. A.

This is a book which will prove an invaluable aid to those undertaking research into the rapidly expanding chemical markets of Western Europe. Designed to serve as a guide to the numerous sources of statistical and market information currently available on this industry, it will enable the researcher to pinpoint, with the minimum of delay, those publications most likely to be of assistance as well as indicate a range of sources which might previously have remained unknown.

Each country is treated separately, sources being grouped under the following headings: Government statistics and reports, Other statistics and reports, Trade associations, Trade journals, Other newspapers, and magazines, Directories, Advertising statistics, Bank reviews. 102 pages. **$19**

MANUFACTURE OF SEMICONDUCTOR COMPOUNDS 1969
by M. Sittig
Electronics Materials Review No. 2

Manufacturing processes for semiconductor compounds:

Aluminum Arsenide, Aluminum Phosphide, Aluminum Telluride, Bismuth Telluride, Boron Nitride, Boron Phosphide, Cadmium Oxide, Cadmium Phosphide, Cadmium Selenide, Cadmium Sulfide, Cadmium Zinc Arsenide, Calcium Borate Glass, Ferrous Sulfide, Gadolinium Nitride, Gallium Antimonide, Gallium Arsenide, Gallium Arsenide Phosphide, Gallium Phosphide, Gallium Sulfide, Indium Antimonide, Indium Arsenide, Indium Arsenide Phosphide, Indium Phosphide, Lead Selenide, Magnesium Silicide, Germanide and Stannide, Silicon Carbide, Silicon-Germanium Alloys, Silver Indium Selenide, Thallium Telluride, Titanium Dioxide, Vanadium Phosphate Glass, Zinc Arsenide, Zinc Oxide, Zinc Selenide, Zinc Sulfide, Organic Semiconductors. Indexes. 106 Illustrations. 326 pages. **$35**

PURE CHEMICAL ELEMENTS FOR SEMICONDUCTORS 1969
by M. Sittig
Electronics Materials Review No. 1

This book gives detailed descriptive manufacturing techniques for producing pure chemical elements for use in semiconductors. In order to give up-to-date information, the book is based on the U.S. patent literature since 1960.

Due to rapid commercial growth, some of these materials, particularly silicon are in short supply. Demand is growing rapidly and offers an interesting and profitable market for those alert chemical firms looking for new ventures.

Introduction: Antimony, Arsenic, Bismuth, Boron, Gallium, Germanium, Indium, Phosphorus, Selenium, Silicon, Tellurium, Thallium, Zinc, Future Trends. 112 illustrations. 335 pages. **$35**

OTHER BOOKS RELATING TO EUROPEAN INDUSTRY

EUROPEAN PETROCHEMICAL FACTS 1968 $35

FERTILIZER INDUSTRIES OF EUROPE 1968 $35

COSMETICS INDUSTRY OF EUROPE 1968 $35

EUROPEAN MAN-MADE FIBER MARKET REPORT 1968 $35

EUROPEAN AGRICULTURAL CHEMICALS SURVEY 1969
by S. A. Mann

This is a guide to the European Agricultural Chemicals Industry that will help you in sales, market research, development, planning, acquisitions and mergers. The pesticides considered in this survey include all plant or crop protection chemicals such as insecticides, herbicides, fungicides, etc.

There are three types of information presented for each country: 1. Discussion of the industry, including market structure. 2. Statistical material on production, consumption, imports, and exports. 3. The names and addresses of the leading pesticide producers in each country. This book contains a total of about 1,000 names and addresses of Europe's leading pesticide producers.

This book contains considerable statistical material, invaluable for market research studies. Much of the statistical material is broken down in considerable detail. 129 pages. **$35**

FOOD GUIDE TO EUROPE 1969

This is a directory of the European Food Processing Industry that will help you in sales, market research, development, planning, acquisitions, and mergers. It is expected that the manufacture of processed foods in Europe will be a substantial growth industry.

Describes the 800 leading European food processing firms in 18 countries: Name and Address, Ownership, Plant Locations and Products, Local Subsidiaries and Affiliates, Foreign Subsidiaries and Affiliates, Principal Executives, Annual Sales, Number of Employees.

A valuable marketing guide. Only the 800 major food processing firms are included.

Will help you: Increase sales, Concentrate on the big buyers, Plan joint ventures, Make licensing arrangements, Know company officials, Prepare market research reports, Talk intelligently about the European food processing industry. 130 pages. **$20**

EUROPEAN FOOD PROCESSING INDUSTRY 1968
by S. A. Mann

A guide to European Food Processing Industry that will help in sales, market research, development, planning, acquisitions, and mergers in this large growth industry relating to manufacture of processed and convenience foods.

For 18 countries, includes discussion, statistical material, and names and addresses of about 2,000 major food processers. 197 pages. **$35**

CHEMICAL GUIDE TO EUROPE 1968
Fourth Edition

Describes the 1,000 leading European chemical manufacturers in 18 countries. Gives name and address, ownership, plant locations, products, local and foreign subsidiaries and affiliates, principal executives, annual sales and number of employees of the firms tht have most to offer by way of sales contacts, licensing arrangements and joint ventures. 207 pages. **$20**

INDUSTRIAL SOLVENTS HANDBOOK 1970
by I. Mellan

A handbook with complete, up-to-date, pertinent data regarding industrial solvents.

821 tables contain pertinent data concerning physical properties of solvents and degrees of solubility of materials in these solvents. Numerous graphs included giving a great deal of data concerning various parameters. Also includes phase diagrams for multi-component products.

The vast amount of information contained in this book is shown in the abbreviated Table of Contents in the next column. The numbers in ( ) after each entry indicate the number of tables.

Hydrocarbon Solvents (14); Halogenated Hydrocarbons (130); Nitroparaffins (5); Organic Sulfur Compounds (5); Monohydric Alcohols (122); Polyhydric Alcohols (150); Phenols (6); Aldehydes (10); Ethers (53); Glycol Ethers (79); Ketones (44); Acids (18); Amines (124); Esters (61). 478 pages. **$25**

WATER POLLUTION CONTROL AND SOLID WASTES DISPOSAL 1969
by M. Sittig
Chemical Processing Review No. 32

Describes specific water pollution processes, a substanial growth industry. Much of the technology necessary to effect water pollution control and solid wastes disposal has already been developed. In many cases these environmental problems could be resolved by intelligent process choice, and economic application of the proper control equipment.

This book in most cases covers processes and equipment which can be licensed or purchased.

Introduction: The Water Problem, The Solid Waste Problem, Types of Water Contaminants, Removing Specific Inorganic Water Contaminants, Removing Specific Organic Water Contaminants, Removing Specific Solid Contaminants from Water, Handling Liquid and Solid Radioactive Wastes, Sewage Disposal, Solid Waste Disposal Processes, Future Trends. 78 illustrations. 244 pages. **$35**

AIR POLLUTION CONTROL PROCESSES AND EQUIPMENT 1968
by M. Sittig
Chemical Process Review No. 24

It is to a great extent application of known apparatus, processes, and concepts that will enable us to control air pollution.

Most complete descriptions of processes and equipment are available in the patent literature. These then, in most cases, cover processes and equipment which can be licensed or purchased today to combat environmental control problems. Only through sound engineering principles in process design are we going to control air pollution at minimum cost. This book gives detailed process descriptions.

Contents: Sources of Air Pollution, Air Pollution Control Devices, Removing Specific Gases and Vapors from Air, Equipment for Removing Solids and Liquids from Air, Removing Specific Solids and Liquids from Air, Removing Automotive Exhaust Fumes from Air. Future Trends. 102 illustrations. 260 pages. **$35**

POLYMETHYLBENZENES 1969
by H. W. Earhart

This book presents detailed physical property data on the polymethylbenzenes (PMB's) and their derivatives, discusses the chemistry of the PMB's, and presents known as well as suggested end-uses for numerous PMB derivatives.

Presents a comprehensive compilation of technical information concerning: Benzene, Toluene, Xylene, Mesitylene, Pseudocumene, Hemimellitine, Durene, Isodurene, Prehnitene, Pentamethylbenzene, Hexamethylbenzene.

Complete physical property data, established chemical reactions giving relative kinetic rate data, yields, etc. and end-use applications are presented and discussed in detail. Other related information is also presented.

This book contains a substantial amount of data on this subject. There are 63 detailed tables. The bibliography is extensive and includes 549 literature references. **$20**

GUIDE TO CHEMICAL PLANT PLANNING 1969
by Dr. R. Lobstein

The assemblage of information needed for selecting materials and equipment, and for fitting them into the dimensional scheme, can present a major problem to the most experienced of engineers and architects. Endless perusal of catalogs, technical magazines, and handbooks often not specifically concerned with the problem at hand becomes necessary. One purpose of this book is to make much of the searching effort unnecessary. The required information has been collected and condensed into clear, quickly available essential data.

Much of the book is intended as an aid in the selection of mechanical equipment. Many tables and diagrams in current usage by equipment and machinery suppliers are included.

Contents: Introduction, Piping, Pumps, Heat, Power Transmission, Mechanical Equipment, Electrical Engineering, Operational Requisites. 452 charts. 524 pages. **$24**

ORGANIC CHEMICAL PROCESS ENCYCLOPEDIA
by M. Sittig
Second Edition 1969

Gives the key processing facts for instant reference for 711 industrial organic chemical processes—with 711 large flow diagrams.

This book offers a handy desktop reference to organic chemicals and their industrial processes. The primary purpose of this book is to present the significant processing conditions all in one place —saving you many hours of work trying to obtain specific numbers.

The format of the book is such that the key process facts can be picked out quickly—they are on the same place on every page. Process facts included are: equation; feed materials; coproducts; type of catalyst; phase, reactor type; solvent used; temperature; pressure; reaction time; heat required or heat evolved; product yield; product purity; materials of construction; major product uses; and reference. 712 pages—8½" x 11"—hard cover. **$35**

SOLAR CELLS 1969
by Dr. M. W. Ranney

This report summarizes recent process development and application work directly related to the use of photoresponsive materials in solar cells systems. While commercial development has centered mainly on silicon semiconductor type cells, advances in the use of other inorganic and organic semiconductors as well as photoemissive devices for solar cell application are included to give a complete picture of this rapidly developing area of technology. This technology will find increasing application in power generating stations in remote or underdeveloped regions in the world.

Semiconductors—Silicon, Cell Panuel Fabrication Techniques—Silicon, Semiconductors—Cadmium, Gallium, Photoemission, Organic Photochemical Cells, Miscellaneous.

Indexes by subject, company, and individual, help in quickly determining who is doing what in this field. Indexes, 85 Illustrations, 271 pages. **$35**

MICROENCAPSULATION TECHNOLOGY 1969
by Dr. M. W. Ranney

This book is based on U.S. patent technology in the microencapsulation field. Detailed descriptions and illustrations of processes and products are given. An abbreviated table of contents is outlined below:

Phase Separation Methods; Coacervation-Aqueous Phase Separation, Organic Phase Separation, Spray Drying, Miscellaneous: Interface Reactions—Polymerization; Dissolved Monomer Polymerization, Interfacial Polymerization, In-situ-Polymerization, Vapor Deposition: Physical Methods; Fluidized Bed, Electrostatic, Multi-Orifice Centrifugal, Vacuum Metallizing, Coating Fusible Material: Applications; Xerographic Toner, Light Sensitive-Photographic Materials, Heat Sensitive, Transfer Sheets—Dyes, Miscellaneous, Indexes. Illustrations. 275 pages. **$35**

FUEL CELLS RECENT DEVELOPMENTS 1969
by Dr. M. W. Ranney

This book describes in detail a wide range of recent developments in fuel cell technology, based on the latest U. S. Patents.

Fuels; Hydrocarbons, Nitrogen Compounds, Miscellaneous: Electrolytes; Solid Electrolytes, Electrolyte Additives, Liquid Electrolytes, High Temperature Electrolytes, Miscellaneous: Membranes-Separators; Ion-Exchange Membranes, Separators, Barriers: Electrodes; Catalysts, Membranes-Hydrophobic, Carbon Electrodes-Wetproofing, Control of Porosity, Miscellaneous Electrode Construction: Fuel Cell Construction; General, Control Systems, Water Removal-Control, Pressure Variation and Control: Biochemical and Thermocells: Biochemical Cells, Thermocells: Regeneration and Reactivation; Regeneration, Reactivation, Indexes. Many illustrations. 325 pages. **$35**

MEMBRANES TECHNOLOGY AND ECONOMICS 1967
by Dr. R. Rickles

The membrane process industry will develop into a major business within the next decade.

This book presents both the technical and economic factors that govern the present and future prospects in this fast-growing field. The present status of membranes, equipment and processes are discussed.

A complete summary of those companies which are involved in the industry is presented—together with current and projected economics, and extent of the industry.

This book will assist you in undertaking profitable projects in this field that appears to have a brilliant future.

Chapters: Membranes Theory, Electrodialysis, Ultrafiltration, Dialysis and Diffusion Control, Membrane Filtration, Gas Permeation, Medical Applications, Preparation of Synthetic Polymeric Membranes. 197 pages. **$24**

CORROSION RESISTANT MATERIALS HANDBOOK 1966
by I. Mellan

This book helps you cut corrosion costs by using the proper commercially tried and accepted corrosion resistant materials. 95 Tables selected by the author from manufacturers literature, (at no cost to, nor influence from, the manufacturer of the materials): Resins (54 Tables); Elastomers (13 Tables); Cements & Mortars (4 Tables); Glass (2 Tables); Wood (4 Tables); Metals & Alloys (18 Tables).

The key to the value of this book is the extensive double indexing. The 95 Tables themselves are arranged by type of corrosion resistant material. The extensive Index is organized by corrosive chemical—and refers you to the specific recommendations in the Tables.

If you have a problem with a corrosive chemical, you will be able to quickly determine the most suitable corrosion resistant material for your requirements. 240 pages. **$25**

INORGANIC CHEMICAL AND METALLURGICAL PROCESS ENCYCLOPEDIA 1968
by M. Sittig

Today, there is more "glamour" in many inorganic chemicals and metals than there is in organic chemicals. For example, gallium arsenide, now important in the laser and semi-conductor field was of no significance until only recently. The same can be said for many other inorganic and metallurgical chemicals that are now assuming importance due to rapidly changing developments in electronics, lasers, integrated circuits, cryogenics, superconductivity, materials development, nuclear energy, and the space program.

This book is organized in an unusual format. There is one inorganic chemical or metallurgical process on each page. At the top of the page an equipment drawing or flow diagram is shown, and underneath a description of the process is given. 883 pages —8½" x 11"—hard-cover. **$35**